SIMPLICIUS

On Aristotle Physics 8.6-10

SIMPLICIUS

On Aristotle Physics 8.6-10

Translated by
Richard McKirahan

B L O O M S B U R Y
LONDON • NEW DELHI • NEW YORK • SYDNEY

Bloomsbury Academic
An imprint of Bloomsbury Publishing Plc

50 Bedford Square
London
WC1B 3DP
UK

1385 Broadway
New York
NY 10018
USA

www.bloomsbury.com

First published in 2001 by Gerald Duckworth & Co. Ltd.
Paperback edition first published 2014

British Library Cataloguing-in-Publication Data
A catalogue record for this book is available from the British Library.

ISBN HB: 978-0-7156-3039-6
PB: 978-1-7809-3897-4
ePDF: 978-1-7809-3896-7

Acknowledgements
The present translations have been made possible by generous and imaginative funding from the following sources: the National Endowment for the Humanities, Division of Research Programs, an independent federal agency of the USA; the Leverhulme Trust; the British Academy; the Jowett Copyright Trustees; the Royal Society (UK); Centro Internazionale A. Beltrame di Storia dello Spazio e del Tempo (Padua); Mario Mignucci; Liverpool University; the Leventis Foundation; the Arts and Humanities Research Board of the British Academy; the Esmée Fairbairn Charitable Trust; the Henry Brown Trust; Mr and Mrs N. Egon; the Netherlands Organisation for Scientific Research (NWO/GW). The Editor wishes to thank István Bodnár, John Ellis, Edward Hussey, Ben Morison, Don Morrison and Marwan Rashed for their comments, and Han Baltussen for preparing the volume for press.

Typeset by Ray Davies

Contents

Introduction	1
Textual Emendations	11
Translation	13
Chapter 6	15
Chapter 7	33
Chapter 8	51
Chapter 9	96
Chapter 10	105
Notes	159
Bibliography	182
Appendix: Notes on the text of Aristotle's *Physics*	184
English-Greek Glossary	187
Greek-English Index	197
Subject Index	235
Index of Names	243
Index of Passages Cited	245

Introduction

After a millennium and a half, Simplicius' commentary on the *Physics* still stands up well against even its most recent rivals. The magnitude of the work is impressive in its own right, but sheer quantity does not make for a good commentary. However, in it Simplicius constantly brings to bear his thorough knowledge of Aristotle and the entire Greek philosophical tradition, as well as his acuity in dissecting arguments. He makes frequent use of earlier commentaries now lost, expecially those of Eudemus, Alexander and Simplicius' own teacher, Ammonius, not hesitating to quote them at length. His independence is instanced by his occasional disagreement with Alexander, whom he frequently cites with approval. He takes strong exception to Philoponus, his contemporary Christian rival in Alexandria, and in the present volume he engages in extensive criticism of one of Philoponus' lost works in passages notable for their invective. As a Neoplatonist, Simplicius attempts to reconcile the doctrines of Aristotle with those of Plato, and this interpretive programme is prominent in the present volume. In what follows, for each of the five chapters of the *Physics* covered here, I briefly summarize the contents of each and indicate some of the features of Simplicius' commentary that are of historical and philosophical interest.

In Chapter 6, Aristotle argues that there is an eternal and unmoved primary mover; it causes a single, eternal, continuous motion, and therefore what is primarily moved by it is eternal as well, and it is the motion of this that causes generation, perishing, and other kinds of change to occur in other things. He also analyses the nature of self-movement which animals possess and concludes that it is not continuous, whereas even when not undergoing this motion animals undergo other motions such as breathing and growth, which are due not to their own agency, but to the changing environment, whose motion is ultimately due to the unmoved primary mover.

Simplicius' contributions in his treatment of Chapter 6 consist largely in explicating Aristotle's arguments, supplying missing steps where necessary, and offering supplementary arguments for some of Aristotle's claims (1250,35-1251,4, 1252,10-11, 1253,3-12, 1255,21-33).

At 1252,18-23, he argues that the unmoved elements in self-movers fail to satisfy the account of what it is to be a principle of motion. At 1255,34-

1256,30, he raises and then solves a serious objection: that although Aristotle's argument for the existence of an eternal unmoved mover is based on the existence of continuous eternal motion, the points for which Aristotle argued in Chapter 1, that prior to any motion hypothesized to be first there is always another motion, and likewise there is always a motion posterior to any motion hypothesized to be last, establish only that motion is consecutive, not that it is continuous. At 1259,15-28, he offers several interpretations of Aristotle's problematic claim that the soul moves itself by leverage (259b20). At 1260,22-35, he quotes with approval Alexander's account of how a cause of circular motion, while being located in the body that is in motion, can be 'in the same place' throughout the motion. He also rejects Alexander's view that the souls of the planetary spheres are moved incidentally (1261,30-1262,13). He likens Aristotle's view that changes in the sublunary sphere are due to the variations in position of the sun, moon and other planets, with Plato's views in *Phaedrus* 246B on the motion of the soul in the heaven. Throughout, he interprets Aristotle's abstract arguments in terms of Aristotle's cosmology, a strategy which gives considerable clarity to several of the individual arguments as well as the overall direction of Aristotle's thought.

Chapter 7 establishes by means of several arguments that locomotion is the primary kind of change and the only kind that can be continuous, and hence that this is the kind of motion that the first mover causes. At 1265,26-8, Simplicius contrasts Aristotle's account of growth here (260a29-b5) with that in the *Categories*, saying that here Aristotle is speaking 'more like a natural scientist'. At 1267,15-19, he makes an argument for a point not made by Aristotle, that considerations of combination and separation, which Aristotle used in proving that generation and perishing are posterior to locomotion, can also be used to prove that growth and decrease are posterior to locomotion. At 1267,19-28 (also 1272,38-1273,12), he claims that in holding that locomotion is prior to other forms of change, Aristotle agrees with Plato (*Laws* 10, 893E-894D). Whereas at 260b17-19 Aristotle lists three ways in which a thing may be prior, at 1268,7-1269,5 Simplicius goes through other significations of priority found in *Categories* 12 and *Metaphysics* 5,11. At 1270,4-13, he fills in missing steps in the elliptical argument at 260b29-30. At 260b30-261a12, Aristotle argues that even though in individuals that are subject to generation locomotion is posterior to other changes, nevertheless the cause of generation of the individual will undergo locomotion prior to the individual's generation. Whereas Aristotle merely envisages an infinite chain of ancestors, where in each case the parent undergoes locomotion prior to generating offspring, Simplicius, bearing Aristotle's cosmology constantly in mind, refers the entire series of generations to an eternal, prior cause, the motion of the heavens (1270,25-6, 1270,35-7). At 1273,28-33, he carefully shows the relation between Aristotle's argument that only locomotion can be continuous (261a31-b22) and his argument in Chapter

8 that among locomotions only circular locomotion can be continuous. At 1274,16-25, he offers arguments for the proposition that a single continuous motion cannot arise from opposite or contrary motions, which Aristotle uses tacitly in his argument. At 1274,25-8, he describes lateral motion as a compound of upward and downward motion and compares it to a compound made out of contrary elements. At 1274,33-1275,4, he provides a badly needed explication (securely based on Chapter 8) of Aristotle's oracular claim that 'a thing that is not always undergoing a particular motion, but that existed previously, must previously have been at rest' (261b1-2). At 1277,31-3, he supports Aristotle's claim that it is 'absurd if something that was generated had to perish immediately, and could persist for no time interval' (261b23-4) through considerations drawn from Neoplatonic epistemology.

Chapter 8 establishes that circular locomotion is the only kind of motion that can be eternal, single and continuous. At 1278,10-13, Simplicius problematizes the claim (which Aristotle apparently considered obviously true) that if either of the components of a combined (rectilinear and circular) motion is not continuous, the combined motion is not continuous either, and he proposes a response to someone who denies this claim. At 1279,22-32, we have a useful treatment of the difference between arguments based on 'signs' (*sêmeia*) and demonstrations. At 1292,24-1293,5, in discussing Aristotle's statement that 'it is an accidental attribute of a line to be an infinite number of halves' (263b7-8), Simplicius points out that this is not an accidental property in Aristotle's normal use of that expression, and suggests that Aristotle here introduces a different kind of accident, namely, what belongs to something potentially. He quotes with approval Alexander's use of Aristotle's claim that the instant of change (e.g. the instant when Socrates died) is the last instant of the process of dying and does not belong to the interval in which the thing is in the state that results from the change (when Socrates is dead), to solve sceptical arguments against the possibility of dying or being born (1296,18-35), and to show that propositions of the form 'If Dion is alive, Dion will be alive', which some people claimed change truth value indeterminately, do not change truth value at all (1299,36-1300,30), and he offers a cogent modification of this latter claim of Alexander (1300,30-6). At 1301,19-29, he tells us that a 'dialectical' (*logikos*) argument is one based on terms that are not 'appropriate' but more common and more general, and capable of applying to other things too, and says that they are so called because they arise through accepted (*endoxos*) arguments. At 1303,12-24 (cf. also 1304,39-1305,6), he states and refutes an objection that one might make to one of Aristotle's arguments that backward-turning rectilinear motion stops at the endpoint (264a18-19), and at 1303,27-33 he quotes an additional argument of Alexander's for the same thesis.

In Chapter 9 Aristotle proves that circular motion is the primary kind of motion, since it alone is simple and complete, it alone can be eternal, one, continuous and uniform, and it is the measure of other motions. At 1317,18-28 and 1317,33-1318,7, in commenting on Aristotle's claim (265b12-13) that things in rectilinear motion undergo locomotion nonuniformly, Simplicius offers some valuable comments on non-uniform (i.e. accelerated) motion, both violent and natural. At 1318,10-15, he gives us his opinion of why Aristotle introduces the opinions of his predecessors after his own demonstrations, and contrasts Aristotle's use of such testimony with more recent practice. At 1318,34, Simplicius informs us that the fifth-century atomist Democritus and his followers called the atoms 'nature' (*phusis*), and at 1318,35 he attributes to them the view that atoms have weight. At 1319,20-3, he says that Thales, Anaximenes, Anaximander and Heraclitus explain generation and perishing in terms of condensation and rarefaction.

In Chapter 10 Aristotle proves that the first mover must have no parts or size; it is also indivisible. It is the cause of the eternal circular motion of the heavens, which in turn cause the changes in the sublunary world. The first mover causes motion without effort, and is located at the circumference of the sphere of the fixed stars. In this chapter Aristotle also discusses projectile motion, and states why a projectile does not stop moving when it loses contact with what throws it: the thrower imparts the ability to be a mover to the medium through which the projectile moves.

In his comments on this chapter Simplicius gives further insight into his views on kinematics and dynamics. He holds that in the case of motions that are bodily or due to force, a greater power causes motion over a greater length of time than a smaller power, and makes the same thing move in a shorter time than the smaller power does (1321,5-12). At 1322,8-35, he makes a specification which eliminates a loophole in Aristotle's argument that nothing finite can cause motion for an infinite time. At 1325,8-1326,27, he reports an objection which Alexander made to Aristotle's argument for the thesis that no finite magnitude can contain infinite power, and then defends Aristotle against it, and at 1326,28-37 he quotes an additional argument which Alexander supplied for the thesis. From 1326,38 to 1336,34 Simplicius subjects his Christian opponent Philoponus (whom he does not name but calls variously 'that Grammarian' and 'that man') to a bitter and sustained attack. He refers to Philoponus' arguments against Aristotle's view that the world, and also time and motion are eternal, which he refuted elsewhere (cf. n. 415), but here focuses his objections on Philoponus' assumption that if something possesses finite power it is perishable, and on his failure to distinguish between (a) causing motion eternally, where the mover possesses an infinite active motive power all at once together in actuality, and (b) being able to be moved eternally, which is a property of what is moved immediately by the first mover, and which is a passive capacity to be moved ad infinitum that exists potentially and not all at once together. Simplicius holds that while a finite body cannot have infinite power all at once together, it can be moved ad infinitum, and that the latter property rather than the former is what is required in order for the world to be eternal. Beginning at 1329,16 he presents and argues vigorously against several of Philoponus' own arguments which appeared in a lost work whose title is unknown, that the heavenly bodies are perishable, that they do not by their own nature possess infinite power, and that they possess finite power and hence are perishable. Simplicius relates Philoponus' belief in the perishability of the world to the tradition that the heavens were created 5500 years before the birth of Christ and that God will bring the world to an end in its six-thousandth year (cf. n. 455), so that in the early sixth century AD, when Philoponus and Simplicius were writing, the end of the world was nigh. In response, Simplicius marshals counter-evidence that tends to show that the world is not nearing its end (1335,3-16).

Immediately following his criticism of Philoponus, Simplicius argues at length that Plato's account of the world in the *Timaeus* (especially 41A-B, which Simplicius quotes and interprets) agrees with Aristotle's in several essential respects, in particular that the heaven is eternal and always moving both on account of its own nature and on account of the cause that creates and moves it (1336,35-1339,24). In this section Simplicius' Neoplatonism is evident in its terminology, its style of argumentation, its ideas, and its way of approaching Plato and Aristotle.

Next (1339,25-1340,8) follows a brief treatment of the question whether Aristotle's primary mover causes motion temporally or atemporally, the answer being that unlike things that cause motion being themselves moved, which cause motion temporally, it causes motion atemporally: 'it is by the agency of what is atemporal that time must exist in what is moved temporally'. At 1340,25-8, Simplicius supplies some surprising examples to illustrate Aristotle's claim that 'there can be more power in a smaller magnitude' (266b7-8), and disagrees with Alexander's objection to the usefulness of this claim (1340,32-1341,9). At 1341,20, he says that Aristotle desires to geometrize nature. At 1342,7, he uses the phrase 'indefinite and becoming smaller ad infinitum' to describe we would say is approaching zero as a limit. At 1342,39-1343,12, he generalizes Aristotle's proof at 266b8-20 (which is stated in terms of doubling the power and halving the time) so as to be applicable to other ratios.

At 1346,29-1348,5, Simplicius quotes with approval Alexander's discussion of a problem that arises in connection with Aristotle's account of projectile motion: Aristotle holds that projectile motion does not violate his principle that in every motion the mover is continuously in contact with the moved even though the projectile keeps moving after losing contact with the thrower, because the air receives from the thrower the power of causing motion in the projectile; but how can the air continue to move the projectile even after the thrower ceases to move *it*? Alexander's answer (which is different from Aristotle's) is that the thrower makes the air a self-mover for a limited time, and so the air can continue being moved, and hence move the projectile, even after the thrower stops moving it. Aristotle's solution is that the air is a not a self-mover (a view he criticized when Plato applied it to the soul) but no-longer-moved mover. In effect Simplicius interprets the principle 'if it is not moved it will not cause motion' more strongly than Aristotle. Where Aristotle takes it to mean 'if there is *no time* at which it is moved, it will not cause motion', Simplicius understands it as asserting, 'if it is not moved *at any given time* it will not cause motion *at that time*'.

At 1348,6-1349,10, Simplicius also considers a further problem about the nature of the air's self-movement: if air is like paradigm self-movers, viz. animals, it will be composed of two elements, corresponding to the soul and the body, one of which is an unmoved mover while the other is moved; but air is not composite in this way, so Alexander's attempt to rescue Aristotle's account fails. Simplicius' response is based on Alexander's claim that the air becomes 'in a way' a self-mover for a little while: namely, in a different (although not well explained) way from that in which an animal is a self-mover.

Simplicius next (1349,11-36) takes up an objection to the account he has given: why not simply say that the thrower makes the projectile, rather than the air, a self-mover for awhile? His answer is that the earthy nature of projectiles makes them unsuitable for either lateral or upwards motion, and that air (also water) as an intermediate element that is suitable for both upward and downward motion (and consequently for lateral motion as well) contributes to the persistence of upward and lateral motions of projectiles. At the end of this discussion (1349,36-1350,9) Simplicius expresses doubts about the adequacy of this solution, declaring that he has offered it in an attempt to deal with the problems in a way consistent with what Aristotle says, but inviting others to improve on his solution. In commenting on Aristotle's rejection of mutual replacement (*antiperistasis*) as the cause of projectile motion, Simplicius informs us (1351,28-1352,17) that (contrary to the opinion of commentators both ancient and modern) Plato did not hold that projectiles move by mutual replacement, but the cause of their motion is 'non-uniformity and inequality' (cf. *Tim*. 57E-59A).

At 1353,29-33, Simplicius gives a cosmological interpretation of Aristotle's conclusion that the cause of uniform eternal motion is unmoved and that the thing that this unmoved mover moves is eternal and is unchanging in its relation to the mover: the heaven and consequently the entire world are eternal, as is the primary mover of the heavens, and they stand in the same relation to one another unchangeably and forever. Aristotle's claim that the primary unmoved mover of the heavens is located at the most rapidly moving part of what it moves (267b7-9) gives rise to great difficulties, with which Simplicius wrestles at 1354,12-1355,38. If the mover is located on something that moves, how will it not be moved incidentally? Alexander's suggestion that it is not moved because is on the entire circumference, which does not move or change place as a whole, is rejected as inconsistent with Alexander's more correct claim that the mover is incorporeal and occupies no place, and also because the suggestion is not consistent with Aristotle's claim that the mover is located in what moves fastest. Simplicius' own proposal is that the mover is not strictly speaking in the heaven, but that the heaven is in it, since it surrounds the whole world by virtue of its infinite power. (Shortly below, he clarifies this proposal: the primary mover is not in the sphere of the fixed stars but in the entire body that is in circular motion, i.e. the entire world: 1357,13-17.)

At 1356,33-1357,17, he brings up a problem raised by Alexander: should we say that the motions of the spheres of the planets as they are carried along with the sphere of the fixed stars are continuous, eternal and uniform, even though the cause of these motions is the sphere of the fixed stars, which is moved? While Alexander's answer is that the primary unmoved mover, not the sphere of the fixed stars, is the cause of the motion of the planetary spheres, Simplicius suggests that the motion of the entire heavens including the planets is a single motion. At 1357,11-29, he quotes Eudemus' account of various ways in which things that are unmoved can cause motion. At 1358,18-1359,4, he praises Alexander for recognizing that in the case of something that is moving ad infinitum (like the sphere of the fixed stars), when we speak of its power or capacity (*dunamis*) to be moved, we are using the word 'power' only homonymously. Unlike movers, where their power can grow weary, here 'power' merely means a suitability to be affected, which may persist and be activated as long as it exists. Thus (*pace* Philoponus), what is moved ad infinitum has by its own nature the ability to be moved, and it gets its motion through the agency of something else; hence, the sphere of the fixed stars, which is finite, does not possess all at once together an infinite power of causing motion, but it is subject to motion ad infinitum, so that, again, in a way the planets, which move together with the sphere of the fixed stars, are moved by the same unmoved mover.

At 1359,5-1360,23, Simplicius draws parallels between Aristotle's way of proceeding from the moving, changing, extended and limited existence of bodies to the unmoved, unchangeable, unextended and unintermittent cause, and Plato's progression from the changing, moving, corporeal world of generation to the unchangeable, unmoved, intellective, eternal creator (*Tim.* 27D-28A), and he praises Aristotle for taking care to avoid language that suggests that eternal things are generated – a mistake made in their cosmologies by others (including Plato) who did not believe that the world has a temporal origin. At 1360,24-1363,24, he argues that Aristotle's primary mover, like Plato's demiurge, is an efficient as well as a final cause of the world and in particular of the heaven. Moreover, at 1361,11-1362,12 he cites several arguments from a book by his teacher Ammonius designed to present Aristotle's God as being, like Plato's, a Creator, albeit of a beginningless universe. This inadvertently makes Aristotle's God closer to the Christian God.

Simplicius concludes his commentary on the *Physics* with a detailed summary of the argument of book 8 (1363,25-1366,22), which ends by indicating the relation between physics and first philosophy: 'the entire structure of nature depends on a cause that is above nature, and the study of nature depends on first philosophy' (1366,19-21, cf. 1359,5-8).

The translation

The initial obstacle that confronts translators of Simplicius' commentary on Aristotle's *Physics* is the large number of misprints and other errors (particularly of punctuation) in Diels' text (*CAG*, vols. 9-10), which in some places make it untranslatable and in others dislocate the train of argument. The opportunity of consulting the principal manuscript of the text of book 8 (Marcianus 226, Diels' A) in the Biblioteca Marciana (for assistance in which I express my heartfelt thanks to the librarian and staff of that library) both made this situation apparent and enabled me to correct at least some of the faults.

The goal of this translation is to render Simplicius' text faithfully into acceptable English, while striving for consistency in the translation of important terms. The Greek-English Index and the English-Greek Glossary provide complete information about how individual words have been translated. Simplicius provides lemmata which indicate the stretch of text he is discussing. The lemmata typically contain the first few words and the last few words of the passage. In conformity with the intention of making volumes in this series usable without constant reference to other books, I have translated the entire Aristotelian passage, a practice which results in a new translation of virtually all of the second half of *Physics* book 8. This translation differs from other translations of the *Physics* in two principal ways. First, as we can tell from the commentary, Simplicius' text of Aristotle differed in places from modern texts. Accordingly, I have translated Simplicius' text, recording the differences from Ross's edition in the notes (collected in the Appendix). Second, since Simplicius quotes and paraphrases the Aristotelian text throughout his commentary, I was forced to make a translation of Aristotle that is more 'literal' and less idiomatic in English than is perhaps desirable, in order to preserve the Aristotelian phraseology in translating Simplicius' text.

In fact, Simplicius quotes Aristotle far more frequently than Diels' text indicates, and the present translation makes an effort to reflect this fact by putting all the quoted material in inverted commas. However, the flexibility of Greek word-order raises a problem in this connection. As an example, consider 1294,32-3: *estai gar to auto hama on kai ouk on, kai hote gegone ti, tote ouk on*, which comments on 263b11-12: *estai hama to auto on kai ouk on, kai hote gegonen ouk on* ('it will follow that the same thing both is and is not at the same time, and that it is not when it has come to be'). Simplicius adds *gar* ('for'), because the context in his text demands this particle; he puts *hama* after *to auto* instead of before; he writes *gegone* instead of *gegonen*, because *ti* ('a thing'), which he has inserted, does not begin with a vowel; and he has added *tote* ('at the time'). I translate as follows, using the quotation marks to indicate the directly quoted phrases within the sentence: for 'it will follow that' 'the same thing' 'is and is not at the same time, and' a thing 'is not' at the time 'when it has come to be'. I consider *gegone* a quotation of *gegonen* despite the addition of the nu-movable. Likewise I ignore differences in elision for this purpose. Paraphrases are not put into inverted commas, even when the paraphrase in Greek is so close that the English translation of Simplicius is identical with that of Aristotle. This practice, although typographically inelegant, displays how closely Simplicius was working with the text of the *Physics*, and helps us see how he went about his task as a commentator. I should point out, however, that there are borderline cases, where it is not clear whether Simplicius is deliberately quoting Aristotle or simply using the word he would ordinarily use, and in such cases one reader's judgment may differ from another's.

Book 8 of the *Physics* is devoted to certain issues that arise in connection with motion and change, and so the vocabulary of motion and change is prominent. In conformity with tradition, *kinêsis*, which applies to changes in place, quality, and quantity, is consistently rendered 'movement' or 'motion' rather than 'change', the latter term being reserved for *metabolê*, which applies to all *kinêseis* and also to the changes of generation and destruction. And since 'move' is ambiguous between 'cause motion' and 'undergo motion', the active verb *kinein* is normally rendered 'cause motion' and the middle/passive *kineisthai* as 'undergo motion', 'be moved', or 'be in motion' except in cases where there is no chance of unclarity.

Simplicius' text has not been previously translated into any modern language. In translating the Aristotelian text, I found the analysis in Ross's edition (*Aristotle's 'Physics': A Revised Text with Introduction and Commentary*, Oxford, 1936, pp. 441-55) most helpful, but at a late stage I was able to check my version against Graham's recent translation (D. Graham, *Aristotle, 'Physics' Book VIII*, Clarendon Aristotle Series, Oxford, 1999), and make some improvements.

My greatest debt is to Richard Sorabji, who invited me to contribute this volume to the series, and who provided friendly support at every stage. In particular, the lengthy conversations I had with him in Oxford in summer 1998 helped immeasurably my understanding of the Neoplatonic passages of the commentary. Four anonymous readers and Don Morrison (who forewent his privilege of anonymity) took the trouble to read the translation and offer suggestions for improvement, and I am grateful to them for their help. The translation benefited from the Greek-English Indices of the previously published volumes in this series. In composing the Greek-

English Index of this volume and also in achieving a high degree of consistency in the translation I was assisted by a concordance generated by *TLG Workplace*, a product of Silver Mountain Software, Inc.

Much of the work on this volume was completed on a sabbatical granted to me for this purpose by Pomona College, during which I had the privilege of working at the American School of Classical Studies at Athens and also at Cambridge University, where St John's College welcomed me as an Overseas Visiting Scholar. I want to express my deep thanks to these institutions for their support and assistance. I dedicate the volume to Voula, my lovely wife, who helped me in matters of translation and Hellenistic terminology, and who invented the *mot juste* for the kind of work involved in such a project: *Simplikiazesthai*: 'to be affected (perhaps adversely) by Simplicius'.

Textual Emendations

1262,5	Reading *kineisthai* instead of *hêgeisthai*
1268,6	Closing the quotation here; Diels neglected to indicate where it ends
1268,10	Adding *mê* in accordance with *Cat.* 12, 14a30
1270,3	Placing the comma after *phoran* (with MS A)
1270,19	Placing a comma instead of a full stop after *topon*
1273,31	Reading *monêi* with the MSS, which Diels misprints as *monê*
1274,1-2	Following Diels' suggestion to add *ta gar* before *ex enantiôn* and to read *hêi* instead of *ê*
1275,11	Punctuating with a full stop after *anthrôpon*
1276,17	Reading *metabolêi* for Diels' *metabolês*
1277,14	Closing the quotation here; Diels neglected to indicate where it ends
1279,7	*di' hous*. The masculine pronoun refers to *topôn* ('places') (1279,5). I have placed *katôthen ... diaphorai* (1279,6-7) in parentheses to bring this out
1285,13	Following Diels' suggestion (*tên men EG suntomôs ekalesen tên to E, tên de ZH*) for filling the lacuna
1288,8	I take it that S.'s comment on Alexander's text stops here, and punctuate accordingly
1294,28	Following Diels' suggestion to read *khronon* for *on*
1298,32	Reading Δ instead of A
1296,15	Reading *genomenon* for *ginomenon*
1299,7	Ending the parenthesis after *ekeinôi* instead of *egeneto* (1299,8), with equal MS authority
1303,3	Placing the comma after *eleusetai*, not after *kinoumenon*, with equal MS authority
1303,25	Punctuating with a comma after *kineisthai*, not a full stop
1307,5	Punctuating with a question mark instead of a comma
1307,12	Closing quote here (Diels neglects to close the quotation)
1308,28	Omitting the comma after *anakamptonta*
1310,5	Placing comma after *palin* (with MS A) instead of *pollakis* (1310,6) (with Diels)
1317,6	Closing the quotation from Alexander after *autês*, where Diels has it go to the end of the paragraph

1318,14	Accepting Diels' suggestion to read *proterôn* instead of *neôterôn*
1321,24	Adding a full stop after *amegethes*, as in MS A
1321,29	Adding a full stop after *einai*, as in MS A
1330,16	Placing the comma after *ekeinôn*
1331,13	Closing the parenthesis after *esti* (with the authority of MS A), instead of *apeiron* (1331,15)
1332,9-10	Removing the parentheses in 1332,9-10
1333,30	Punctuating with a question mark instead of a full stop
1336,9	Omitting the full stop after *holên* and adding parentheses
1337,11	Reading *diestos* with MS A
1346,3-4	Omitting *apodidôsi pros* with MS a. Diels prints these words, which he labels a 'corrupt reading of manuscript A'
1347,13	I translate following Diels' tentative suggestion to emend the MS reading *tês tou* to *toiautês, hôste*
1353,13-14	Placing a comma instead of a full stop after *kinein* and a semicolon instead of a comma after *diamenein*
1353,16-17	Placing a comma instead of a full stop after *einai* and a full stop after *estai*
1354,13	Reading *toutôn* (with MS A) for Diels' *toutôi*

Simplicius
On Aristotle Physics 8.6-10

Translation

[Simplicius, On Aristotle *Physics* 8.6-10]

<CHAPTER 6>

258b10-16 Since there must always be motion without interruption, [there must be something eternal, either one or more, that are first movers, and the first mover must be unmoved. It is irrelevant to the present discussion whether each of the unmoved movers is eternal, but that there must be something that is itself unmoved and apart from all change, both without qualification and incidentally, but that causes motion in something else,] is clear from the following considerations.

After proving that the primary mover[1] for each motion[2] is unmoved,[3]
he now proves that the primary mover must also be eternal, so that 1250,35
it cannot even admit the sort of change involved in generation or
perishing. The first argument is common,[4] applying to motion without
qualification, and is brief and clear: since there must be eternal
motion (as was proved at the beginning of this book[5]) and since motion 1251,1
is in what is moved (as we learned in book three[6]) and everything that
is moved is moved by something (as he demonstrates both in the
previous book[7] and in this one[8]), 'there must be something' that is a
primary mover and is eternal. The addition of 'unmoved' is not
repetition, but means 'in fact entirely unmoved', so that it participates 1251,5
neither in the change involved in generation and perishing nor in that
involved in incidental motion.[9] But since there are many things that
are themselves unmoved but move other things (This class includes
all the souls of animals; this is why animals are self-movers, composed of bodies that are moved and souls, which are unmoved but
move the bodies, so that the animals are moved along with them.), he
says that 'it is irrelevant to the present discussion whether each of 1251,10
the unmoved movers is eternal'. For whether all souls are immortal
is not presently being investigated. (These are the elements in self-movers that are unmoved but impart motion,[10] but it belongs to
another treatise to consider the soul.[11]) Nor is his present task to
investigate without qualification whether all unmoved movers are 1251,15
eternal, but to prove that 'there must be something', either one or
more than one (this matter, which is not to the point, has not yet been
determined[12]), and that it 'is itself' always 'unmoved' 'and apart from
all change, both' per se[13] 'and incidentally, but causes motion in
something else'. He says 'apart from all change' and not 'from all

motion' in order to include generation and perishing. Further, he declares the primary mover to be unmoved even 'incidentally'.[14] In this it differs from the elements in self-movers that are unmoved movers since they are unmoved per se but are moved incidentally because the bodies in which they are are moved. But the primary mover, being in every way separate from bodies, is beyond not only per se change but also incidental change.

258b16-259a8 But[15] suppose, if you might like,[16] that some things can be [at one time and not be at another, without a process of generation or perishing. (It is perhaps even necessary, if something without parts is at one time and is not at another, that every such thing is at one time (b20) and is not at another without undergoing a process of change.) And suppose also that some of the principles that are unmoved movers can at one time be and at another not be. However, there is not any way that all of them can <have this property>. For it is clear that for things that move themselves there is a cause of being at one time and not at another, for everything that moves itself (b25) must have magnitude, if nothing that is without parts undergoes motion. But what we have said does not necessitate that <every> mover <is without parts>. The cause of the fact that some things are generated and others perish, and that this happens continuously is no single one of the things that are unmoved but do not always exist, nor is it any of those things[17] that move[18] particular things, while these[19] move others; for of the eternity and continuity[20] of the process it is not the case either (b30) that each of these or all of them together are the cause. For its being thus[21] is eternal of necessity,[22] whereas all of these things taken together are infinite and they do not all exist at the same time. It is clear, then, that even if it happens ten thousand times that principles[23] (259a1) of unmoved movers[24] and also many things that move themselves perish and others succeed them, and this unmoved thing moves that thing, and that moves something else, none the less there is some thing that comprehends them all, that is apart from each of them, and is the cause of the fact that some things exist (a5) and others do not, and of the continuous process of change. And this <is the cause of motion> for these, and these are the causes of the motion of the other things. Therefore, if motion is eternal, the first mover will be eternal too, if there is one,] but if there are more than one, the eternal things will be more than one.

He proves again that there is some primary mover that is not only unmoved but also eternal, hypothesizing that eternal motion and

change occur among existing things and that the unchangeable cause 1251,30
of change must pre-exist the things that are always undergoing
change. Given the objection[25] which states that the principles of
motion in self-movers (i.e. the souls in animals), even though not
eternal, satisfy the account of a principle of motion[26] since they are
unmoved (For it is not through generation or perishing or through
change in general that they sometimes are and sometimes are not. 1251,35
For if this[27] held of them on account of change or motion, what causes 1252,1
the motion would have to pre-exist them; but since it does not hold
through motion or through change in general, but because they are
without parts, they would satisfy the account of a principle,[28] and it
follows that there is no need for an eternal principle.), he refutes this
objection immediately at the outset, granting that something can 'be
at one time and not be at another, without a process of generation or 1252,5
perishing', but 'perhaps' it is not possible, but even 'necessary, if
something without parts is at one time and is not at another, that
every such thing is at one time and is not at another without
undergoing a process of change'. 'Suppose', then, he says, that some
'of the principles' of being moved 'can at one time be and at another
not be' without motion or change, nevertheless 'not' 'all' the principles
'can' be at one time and not be at another. For there must always be 1252,10
something that causes such principles to be at the times when they
are, since being at the time when they are holds always of successive
individuals.[29]

He proves still more evidently that 'for things that move themselves' too, if we hypothesize that even the element in them that
causes motion at one time is and at another time is not, there is some
eternal cause of their being at one time and not being at another time.
For even if the elements that cause motion in the self-movers, which 1252,15
are incorporeal and without parts and without any generation and
perishing, at one time are and at another time are not – nevertheless
if it undergoes motion at all, a self-mover 'must have magnitude',
because nothing 'that is without parts' 'undergoes motion', as has
been proved.[30] But it is not yet clear whether *the mover* must be
without parts. Now if[31] among self-movers the things that at one time
are and at another time are not are bodies, it must be through
generation and perishing that these at one time are and at another 1252,20
time are not.[32] But of every case of generation and perishing there is
some cause, and the cause of the bodies' generation would also be the
cause <of the generation> of the elements in them that cause motion,
if in fact it exists together with them. Therefore, the unmoved
elements in self-movers do not satisfy the account of principle of
motion.

Further, he proves that 'the cause of the fact that some things are
generated and others perish, and that this happens continuously', 1252,25

with some things being generated and others perishing, is neither as one 'of the things that are unmoved but do not always exist', as is being posited,[33] nor a plurality of them related in such a way that 'those things move particular things' and others move 'others'. 'For of the eternity and continuity of the process', there is no one cause (which he calls 'each'). For if the thing is not eternal but the change is eternal, when the thing is not, either the change will not be either,
1252,30 in which case it will not be eternal or continuous, or the change will have another cause, if indeed it takes place even if the thing is not. After saying that 'of the eternity and continuity of the process it is not the case either that each of these or all of them together are the cause', he briefly eliminates [the possibility that] that any single non-eternal thing is the cause by saying that which is, 'thus is eternal of necessity'. But how could what is non-eternal be a cause of what is
1252,35 necessary and eternal? He omits to draw this consequence as being evident. He proves that neither can all of them together be causes[34] by saying 'all of these things taken together are infinite and they do not all exist at the same time'. They are infinite in that they come to be ad infinitum, and therefore they do not all exist at the same time. How, then, is it possible to say that all things together are causes of eternal and continuous change when they do not all of them always exist and they do not all exist at the same time? How could ones that
1252,40 do not yet exist or that no longer exist be causes of what exists? He will prove later on[35] that neither are all of them together the cause in
1253,1 the sense that all of them together are the cause of the same single change, but different ones are causes of different parts of it, this particular one of this part and that one of that part, e.g. things that exist now are causes of things that exist now and things in the future of things in the future.

But I think that it is possible to deduce this result too from what has been stated, that neither a single one of the noneternal things
1253,5 nor all of them together are causes of continuous eternal motion. For he assumes that 'its being thus', i.e. being the cause of continuous eternal motion, is 'eternal of necessity', or as Alexander and Themistius write, 'eternal and of necessity' (for the cause of what is eternal must always exist too), and to this affirmative premise he adds the negative premise 'each individual that is subject to generation and
1253,10 perishing is not eternal, nor are all such things together',[36] from which <it can> be inferred that neither the individual that is subject to generation and perishing nor all such things together are causes of continuous eternal motion. It would be superfluous to prove that each individual that is subject to generation and perishing is not eternal. But that not all of them together are eternal he proves through the
1253,15 fact that they are infinite, and infinite in such a way that they are being generated ad infinitum, 'and they do not all exist at the same

time'. For since different ones are being generated at different times
they do not all exist simultaneously, so that not even all together are
they eternal. For if they were eternal, all of them would exist simul-
taneously, and not some now and others in the future. Now it is clear,
he says, that even if the 'principles of unmoved movers' that are of
this kind[37] <are generated and perish and others succeed them> 'ten
thousand times', i.e. ad infinitum, as is the case for the unmoved
principles in self-movers (i.e. the souls of animals), and even if in fact 1253,20
there are many self-movers (For all sublunary things are such that
some are being generated and others are perishing, the immediately
preceding one each time becoming the cause of generation for the
next.[38]), anyway 'none the less there is some' eternal cause as well as
each of these, 'that comprehends' and marks out the changes of these,
providing them with precisely that inexhaustibility which neither 1253,25
each of the things that are generated and perish nor all of them
together can provide. For what is consecutive is neither continuous
nor eternal. But motion has been demonstrated to be eternal and
continuous.[39]

After proving, then, that neither can each of the causes of motion
that are subject to generation and perishing be causes of eternal
motion, nor can all of them together, he reasonably infers that it is
clear that there is some eternal cause of eternal change. Then he adds 1253,30
that 'for these' things that cause motion, which have been posited to
be non-eternal, 'this' eternal cause is the cause both of existing and
of imparting motion, and that it is also the cause of the existence of
the self-movers, in which such things that cause motion[40] <are pre-
sent>, and 'these' self-movers are 'the causes of the motion of the other
things'[41] – the things that are moved, but do not in all cases move
other things – since in fact what is self-moved is the principle and the
first of things that both impart motion and are moved, to avoid going 1253,35
to infinity, putting something that is moved by something else ahead
of each thing that is moved by something else. On the other hand, he
can be saying that 'this' eternal thing causes motion 'for these' things
which are always moving continuously (the things that undergo
circular motion), and 'these' are the causes of both motion and
generation for things that are generated and perish. But perhaps it
is better to understand 'and this for these' as referring to the elements 1254,1
in self-movers that are subject to motion, since they[42] were the subject
of the immediately preceding discussion.

As a conclusion of all that has been said he infers a brief and sound
conditional. For, he says, 'if motion is eternal, the' moving cause 'will
be eternal too', that is to say, the primary mover. Now since he has 1254,5
the antecedent established, that there exists eternal motion, he
confidently infers the consequent, that the primary mover is eternal.
And 'if' what is moved eternally 'is one', there will be one eternal

mover, 'but if' the things that are moved with eternal motion are 'more than one, the eternal' movers are 'more than one' too, since there is one for each motion.

1254,10 **259a8-13** [But we should believe] that there is one <first mover> rather than many, [and a finite rather than an infinite number. If the same conclusions follow, (a10) we always do better to assume them finite. For among things constituted by nature, what is finite and what is better must rather be present, if possible. And in fact it is sufficient <to assume that there is> one, which, being primary among unmoved things, and being eternal,] will be the principle of motion for the others.

He declares it more reasonable for the primary mover to be 'one' cause 'rather than many'. For the rule of many is not a good thing, as he proclaims in Homeric language when treating this issue in the *Metaphysics*.[43] But if there are many, it is more reasonable to suppose
1254,15 them 'finite' rather 'than infinite'. In addition he gives the reason: 'if the same conclusions follow, we' 'do better to assume them finite'. The cause of this in turn is that in things constituted by nature, 'if possible', 'the finite' must 'rather' 'be present', because this is better than the infinite,[44] and one is better than many. For it is posited that
1254,20 in all cases nature chiefly makes the best possible things.[45] And for this reason in the first book of this treatise, since both Democritus and Anaxagoras posited elements that are infinite,[46] he prefers Empedocles who declared that they are finite and produce all the things that the others have produced through their infinite elements.[47] He says this because he is about to prove that since the many continuous
1254,25 motions are finite and not infinite, there is one pre-existent motion (the motion of the sphere of the fixed stars) that comprehends the rest, and one of the motion-imparting causes that is beyond the others. Thinking now of these points he infers that 'it is sufficient <to assume that there is> one, which, being primary among unmoved things, and being eternal, will be the principle of motion for the others'.

259a13-20 It is also evident from the following consideration that [the first mover] must be (a15) something single [and eternal. For it has been proved that there must always be motion, and if there always is, it must be continuous. For in fact, what is always[48] is continuous, and what is consecutive is not continuous. But further, if motion is continuous, it is one. But one <motion> is that where the mover and the moved are each

> one. For if what causes motion is one thing after another,] the entire motion (a20) is not continuous but consecutive. 1254,30

After proving that there is an eternal primary mover for each thing that is moved continuously and eternally, and that it is reasonable to suppose this to be one, he goes on to prove the same thesis through another argument, which is inferred in three or more steps in the hypothetical manner,[49] as follows. If 'there must always' 'be' 'motion', 'it must be continuous'; if it must be always and continuous, it must 1255,1
be one; if it is necessary for motion to be always, continuous, and one, it is necessary for it to be generated through one eternal thing that is moved and one eternal thing that causes motion. Therefore, if it is necessary for there always to be motion, it is necessary for the primary mover to be one and eternal. But in fact the antecedent has 1255,5
been proved; therefore, the consequent is true too. That if it is always it is also continuous he proves via the claim that what is always is also continuous. For if it is interrupted, it is not always, but if it is continuous it is one. For if it is more than one it is not continuous but consecutive. Therefore, if it is continuous it is not more than one, but one. It is clear that it is not one by being an eternal motion composed of things that are generated consecutively, since consecutive motion 1255,10
can in fact be interrupted. It is not necessary that there should be motion immediately after the first mover and that the next mover should be generated, unless there is some single eternal cause of necessary succession in the movers, which is precisely what the argument is now introducing. But unless there is a cause of this kind it is possible for there to be a gap and for the next mover not to exist immediately when the first mover perishes.[50] But if that[51] is inter- 1255,15
rupted, the motion too will be interrupted, which is impossible. But if it is to be continuous and one, it will also be numerically one. <Motion> that is numerically 'one', as was proved above,[52] is that in which both 'mover' 'and' 'moved' are each numerically 'one'. For when the things that are moved are many there is not a single motion but consecutive motions (if even that), and likewise when the movers are many. For in the succession of movers there occurs a halt, since one 1255,20
must stop and another start, and so the continuity is disrupted. It was also said[53] that there is no need for the second mover to exist immediately after the first mover, for since the several <movers> are not in fact eternal, they will cause eternal motion in succession. For these things, which grow weary in being active, require rest during the continuity of their activity; and things that grow weary are mortal, whereas what is eternal does not grow weary. Further, in 1255,25
order to change from not causing motion to causing motion they themselves will need to be moved; but it was supposed that they are unmoved, since they are primary. Through these arguments he

proves that not even all things together that cause motion but are not
eternal can be causes of eternal motion by causing motion in succes-
sion. And he proved earlier[54] that they do not all exist simultaneously,
1255,30 either. So in fact it is clear from what has been said that he wants the
eternal motion-imparting cause of each continuous motion to be one.
So also in the *Metaphysics* he attempts to discover the number of
unmoved primary moving causes from the continuously moving bod-
ies that undergo circular motion.[55]

Someone might reasonably be puzzled, I think, as to how from what
1255,35 was proved at the beginning[56] – 'that there must always be motion' –
it follows that it is continuous. 'For', he declares 'what is always is
continuous, and what is consecutive is not continuous'. And yet, the
arguments up to now have proved that motion that is always is not
continuous but consecutive, since in fact he proves[57] that there is
always a change prior to any change hypothesized to be first and a
change posterior to any change <hypothesized to be> last, and such
1256,1 changes are consecutive, not continuous. If, then, 'what is always is
continuous', as he says, 'and what is consecutive is not continuous',
he infers categorically in the second figure[58] that what is consecutive
is not always. Now if the motion that was the subject of the proof was
consecutive, that proof has not shown that the motion is always or
1256,5 that it is one. And so neither will it be proved that the mover is one,
as he just said, from its having been proved that there must always
be motion. For it was proved <to be> always as being consecutive, not
as being continuous. This would be the puzzle. Perhaps we should say
that at the beginning[59] it was proved that there is always motion
consecutive to motion, for this evidently holds among the sublunary
things that are near us, in which prior to every motion another motion
1256,10 must pre-exist and after every motion another motion must exist
afterwards. It was then assumed[60] on the basis of an hypothesis that
prior to consecutive motion that is always there is continuous motion
that is always, which is the cause of the inexhaustibility of the
consecutive motion. Now what is stated here, 'if there always is, it
must be continuous. For' 'what is always is continuous, and what is
consecutive is not continuous', reveals that the consecutive insofar as
it is consecutive, is not continuous nor does it necessarily have the
1256,15 property of being always. Now since the consecutivity of motion in
sublunary things has been proved inexhaustible, the property of
being inexhaustible and always must belong to what is consecutive
as deriving from something continuous. For if the motion of the
heavens were not continuous and eternal, the motion of sublunary
things, being consecutive, would not be inexhaustible and eternal. He
reveals further on that it was assumed on the basis of an hypothesis
1256,20 that if consecutive motion is proved inexhaustible then there exists
a continuous motion, when he says, 'At the same time what was

hypothesized both now and previously, that some kind of motion can be continuous and eternal, will be evident by means of the same method',[61] and he proves[62] that circular motion is the only motion that is such. If, then, (a) supposing that there is inexhaustible consecutive motion among sublunary things, there must pre-exist a motion that is prior in nature, which is one, continuous, and eternal, and which causes what is consecutive to have the property of being always; and (b) supposing that there exists one continuous eternal motion, the thing that causes it must also be one and eternal; he reasonably declares that from what was proved at the beginning about there always being consecutive motion, it is proved that there must always also be motion that is one and continuous, and that what causes it is one and eternal, since in fact 'one' motion 'is that where the mover and the moved are each one'.

259a20-b20 One might, then, be convinced [that there is some primary unmoved thing] both by these arguments, [and also in turn by attending to the origins of things that cause motion.[63] It is evident that there are some existing things that are sometimes in motion and sometimes at rest. And consequently it has become clear that neither are all things in motion, nor are all at rest, nor are some things (a25) always at rest while the remainder are always in motion. Things that do both and that have the capacity of being in motion and at rest give proof of these matters. Things of *these* kinds are clear to all, but we want to prove the nature of each of the other two kinds too – that there are some things that are always unmoved and others that are always in motion. On our way (a30) towards this conclusion, we posited that everything that is in motion is moved by something, that *this* is either unmoved or in motion, and that <if> it is in motion, it is moved either by itself or by something else at each stage. We proceeded to establish that the origin of things that are in motion is, among things in motion, something that moves (259b1) itself, while the origin of them all is the unmoved. And we see that things that move themselves evidently exist, for example the class of living creatures and that of animals. These cases are in fact what suggested the idea that perhaps motion can come to be[64] in things although it did not exist at all <before>, because we see (b5) this occurring in them: they are unmoved at one time and then again they are in motion, as it seems. Therefore we must grasp that there is just one kind of motion that they cause in themselves, and that they do not even cause *this*, strictly speaking. The cause does not stem from the animal itself, but there are other natural motions in animals,

> which are not due to their own agency, such as growth, decline, and breathing. (b10) Each animal undergoes these when at rest and not undergoing the motion that is caused by itself. The cause of this is the environment and many of the things that enter the animal, such as, in some cases, food. For while it is being digested they sleep, and when it is being distributed they wake up and move themselves, though the first origin comes from outside. This is why they are not always being moved by themselves continuously. (b15) For what causes their motion is something else, which is itself in motion and causes change in relation to each of the things that move themselves. In all these things the first mover and the cause of their moving themselves is moved by itself,[65] although incidentally. For the body changes place,] and so that which is in the body does too, and (b20) it moves itself by leverage.

After proving in many ways that there is a primary mover that is unmoved,[66] he says 'one might be convinced' of this also 'by attending
1256,35 to the origins of things that cause motion', that is to souls in animals, which seem to be themselves origins and indeed unmoved origins of motions. If they prove to be neither the origins of every motion of the animal (for animals' souls are not causes of growth or decrease or
1257,1 breathing, but if at all, only of motion in place due to impulse, and they are not in the strict sense origins even of this; for the origins of this too will be proved to be external[67]), and further if they are not unmoved in every way, but are moved incidentally, being moved along
1257,5 with the bodies that are moved by them,[68] it is clear that they are not in the strict sense origins of motion, but there is some other origin prior to them.

The train of thought of the present argument is as follows: after saying that both from the preceding arguments 'one might be convinced that there is some primary unmoved thing' 'and also in turn by attending to the origins of things that cause motion',[69] he diverts
1257,10 his discussion to discuss the approach by which the causes of motions were discovered. What is evident everywhere comes to be a principle of what is unevident. Now since it is manifest that some 'existing things' 'are sometimes in motion and sometimes at rest', as animals clearly are and natural bodies, which move naturally to their proper places and naturally are at rest in them, and since the same point is
1257,15 proved also from things that undergo motion contrary to nature (for motion contrary to nature is a departure from their natural state of rest, which occurs by force), 'and consequently', he says, 'it has become clear that neither are all things' that exist always 'in motion', as Heraclitus seemed to say,[70] 'nor are all' always 'at rest, nor are some things always at rest while the remainder are always in motion': the

things that are sometimes in motion and sometimes at rest refute these segments of the division. 1257,20

Since these things that are sometimes in motion and sometimes at rest exist and are manifest, the point of the argument must be to investigate whether, given that these intermediates which have both attributes are found among existing things, the extremes exist too, namely, 'things that are always unmoved and others that are always in motion'. And 'on our way towards' the proof of this we posited that 'everything that is in motion is moved by something', which in fact has been proved in the preceding book,[71] and that the mover must be 1257,25 'either unmoved or in motion, and that' if 'it is in motion, it is moved either by itself or by something else at each stage', so that at each stage the mover is moved by something external. Now since this is absurd, in order to avoid going to infinity, the origins of things that are moved turned out to be two: the proximate origin is something that moves 'itself', 'while the origin of them all is the unmoved'.[72] In fact, every self-mover turned out to have an unmoved mover.[73] 1257,30

Further it was found that not only according to the argument does a self-mover necessarily pre-exist things that move and are moved, in order to avoid going to infinity, but it is also clear from perception that there are some existing things that both undergo motion and come to a stop, which have an element that causes them to move and stop that is not external, but in themselves: animals and in general everything that has a soul.[74] Even plants are nourished, grow, and 1257,35 decrease from <something> within. For even if things constituted by nature have an origin of motion in themselves, they have this origin not in connection with producing motion but in connection with being of a nature to be moved, as has been said before. For it has been said that things constituted by nature 'have an origin of motion, not of imparting motion nor of causing it, but of undergoing it'.[75] As a sign that animals and ensouled things in general are thought to be 1258,1 self-movers, he introduces the fact that this makes people pose the puzzle whether their motion is subject to generation, since it evidently comes to be in them though it did not exist before. For it seems that 'they are unmoved' 'at one time' and change through their own agency into being in motion.

This puzzle, which objects to the view that their motion is ungenerated, he confronts at the beginning of the book,[76] proving that 1258,5 animals are certainly not unmoved when they are not undergoing motion that originates with themselves, i.e. motion from place to place, which arises from impulse. While at that point he deferred discussion of this subject, intending to speak more clearly about it later, he now speaks with another consideration in mind, proving that animals are not wholly at rest before they begin to impart to them- 1258,10 selves the motion that they cause themselves to have. Rather there

is some motion in them that occurs through the agency of something external, which is the origin and cause of the change that results in their self-motion. For as he said earlier too,[77] there is one motion that animals 'cause in themselves' – motion involving change of place – and not even 'this' originates with them 'strictly speaking' and in
1258,15 every way. For the first origin of such motion is not in them, but they naturally undergo some other motions too and not as originating with themselves, e.g. the motions involved in growth and decrease, as well as breathing, sleep and waking. It is not the animal itself that wholly causes itself to have such motions, but rather the body that undergoes circular motion. In fact, the cause of breathing is 'the environment',
1258,20 air, and nutriment is the cause of growth, sleep and waking. 'For while it is being digested', the animals sleep because their sense organs are overcome by the exhalations originating from the nutriment that plunge the animal into a deep sleep because of the heaviness of the moisture and do not permit it to engage in sensory activity.[78] Then later after digestion, when the excretions become thinned out and move, the animal wakes up. And then it moves itself,
1258,25 'though the first origin comes from outside'. For the cause of falling asleep is the digestion of nutriment and the movement of the exhalations, which does not originate from the animal itself without qualification, but from the nutriment and from natural activity, not the activity due to impulse. The fact that animals need nutriment and sleep, and that while asleep they cannot undergo the motion that
1258,30 originates from themselves, is the reason why they do not move themselves continuously.

After saying 'the first origin comes from outside', he continues, 'for what causes their motion is something else, which is itself in motion and causes change in relation to each of the things that move themselves'. By 'something else' he means the body that undergoes circular motion, for this is what moves and changes the things that impinge externally,[79] by its own continuous motion in relation to each, that is
1258,35 to say <it moves them> in a way appropriate to each of the sublunary self-movers. The things that impinge externally cause natural motions like those of alteration: when they are heated or cooled by that,[80] the things that impinge or that cause some other change appropriate to the self-movers, move them. And this is why animals do not move themselves continuously – because they are subject to motion imparting
1258,40 origins that impinge externally. And they require rest and sleep.
1259,1 But since it is not true that animals are able to move themselves being unmoved at any time, and since the first cause of this motion, which seems to originate with themselves, is external, and is moved and changed in relation to each of them through the agency of something that perpetually causes motion, the thesis that the generation of

motion in animals takes place where there was no prior motion at all is eliminated.

After thus proving that what causes motion in self-movers is not 1259,5
strictly speaking an origin of motion because it too has an external origin of motion,[81] he next proves that the <element> in self-movers that causes motion is not strictly speaking unmoved, either. For 'in all these things', 'the first mover' (i.e. the soul), which is 'the cause of their moving themselves', <is moved>. For it is because of the soul that the animal is a self-mover. Now this primary mover 'is moved' 1259,10
'by itself' via the body as an intermediate, 'although incidentally' it causes itself to undergo such motion, for by moving 'the body' it makes it change 'place' by its own leverage. And <the soul> itself, 'which is in the body', changes place together with it so that it moves 'itself by leverage', although incidentally, for <it does so> by moving the body in which it is. The motion of the body caused in animals by the soul 1259,15
he compares to 'leverage' either (a) because what causes the soul to move the body is external <to the soul>, just as levers too are instruments of the things that are primary movers <of what the levers move>, or (b) because just as levers always move the things they push, being applied to them and not separated from them, the soul too moves the body, being always joined together with it, or rather (c) 1259,20
because such motion is forcible and not natural to the body. For the natural motion of each of the elements in the body is different – for earth downwards, for fire upwards.[82] This is why animals grow weary engaging in the motion caused by the soul and they cannot continue for long, whereas they do not grow weary of undergoing the natural motions toward the natural places of the elements. A sign of this is 1259,25
that each one moves more quickly when it comes to be near its proper place.[83] But he says that it moves 'itself' 'by leverage' not directly but incidentally, for what directly causes motion by leverage is the body.

> **259b20-8** From this we may be convinced [that anything that is an unmoved mover that moves itself incidentally, cannot cause continuous motion. And so, since there must be motion continuously, there must be a first mover that is unmoved even incidentally, if, as we said, there is to be (b25) among existing things a motion that is unceasing and immortal, and if that which exists is going to be permanently in itself and in the same <place>. For if the origin is permanent, the universe too must be permanent,] since it is continuous with the origin. 1259,30

After saying that the motion-imparting element in self-movers is unmoved per se but moved incidentally, being moved along with the body, which requires nutriment, sleep and waking and therefore does not undergo continuous motion, he infers that it is not possible 'that

1259,35 anything that is an unmoved mover that moves itself incidentally', should 'cause continuous motion' even incidentally. He bases his argument on mortal animals, which require nutriment and sleep and
1260,1 therefore are unable to undergo continuously the motion proper to animals – motion involving change of place, which is due to impulse. For if bodies that undergo motion are not of a nature to do so continuously, neither does what causes motion continuously move either the bodies themselves or itself incidentally, by being in a body
1260,5 that is moved. To this conclusion, that the <elements> in self-movers that are unmoved but cause motion do not cause continuous motion, he joins another argument, which proves that prior to the motion-imparting elements in self-movers there must be something else that is a more basic cause of motion. For if 'there must' 'be' continuous 'motion', whereas the motion-imparting elements in self-movers cannot cause continuous motion since they are unmoved per se but moved
1260,10 incidentally, 'the first mover' 'must' not 'be' moved even 'incidentally'. This is what 'unmoved' 'incidentally' signifies. Therefore, he says, there must be such a cause (a) if in fact, 'as' has been proved,[84] there is to be 'among existing things a motion that is unceasing and immortal'. For this is the kind of motion that is continuous: if it is not
1260,15 continuous, it is interrupted. But what is moved incidentally, in that it moves from one place to another (since this is what the argument is about), does not cause continuous motion because not even what is moved by itself is of a nature to be in motion continuously. And further, <there must be such a cause> (b) if indeed, he says, 'that which exists' is to be permanently 'in itself, and in the same <place>'. For there is nothing apart from 'that which exists', where by 'which exists' he means the world. And so the world is 'in itself' 'and in the
1260,20 same', not moving from one place to another or coming to be one thing from another. And it is permanently 'in the same' with regard to both the species of its substance and the quality of its motion. For what is in circular motion is permanently 'in the same <place>' as it moves. 'And this is the only possible way', says Alexander, 'for what causes motion not to be moved even incidentally – <namely,> if what is moved by it is permanently in the same <place> while it is being moved. For if such a mover is located in something that is a whole,
1260,25 and this whole is permanently in the same <place> (for the motion that belongs to what is in circular motion is by parts[85]), what is in it[86] would no longer be moved by itself even incidentally, for the motion it causes in the body does not involve a change of the whole in which it is, as we saw takes place in the case of mortal animals, which are moved by the soul by means of leverage, changing places in their
1260,30 entirety. For we will not say that what causes motion is in any given part of what is in circular motion – why should it be in this part rather than another when all of them are moved by it in a similar way? If

someone were to demand <us to tell> in precisely what way, then, it
will be in the whole', he says, 'we will reply that <it is in the whole>
in whatever way a person who says it is in some part of what is moved
hypothesizes it to be <in that part>. For it is not a body, so as to occupy
some distinct place, and it is no wonder if an incorporeal nature and
substance is in some whole simultaneously.' The first mover is the 1260,35
cause of the eternity of what is moved – this is what 'for if the origin
is permanent, the universe too must be permanent, since it is con-
tinuous with the origin' signifies. The origin would be permanent if
it is not moved even incidentally. And because every origin in the
strict sense is eternal (for as Plato says, 'if the origin has perished 1261,1
neither will *it* ever be generated from anything nor will anything else
be generated from it'[87]), also the universe, 'since it is continuous with
the origin', i.e. since it is moved proximately and immediately by the
eternal origin, must itself too be permanent and eternal. For since
the origin is always causing motion in the same way, it is impossible 1261,5
for what is moved proximately not to be eternal too. For if *it* were to
perish, the former thing[88] would no longer be an origin or always
causing motion. For the origin is always together with that whose
origin it is, and what is causing motion is always together with what
is moved. He is not now saying specifically about the first heaven,
which is moved proximately by it,[89] that it is eternal; he will prove
this in what follows. Instead he is saying generally that the universe,
i.e. the world, would be eternal in this way, if the origin of motion of 1261,10
all the things in the world that undergo motion were to depend on
that.[90]

> **259b28-31** However, being moved[91] incidentally [by oneself] is not the same as <being moved> [by something else. The latter attribute belongs also to some origins of the things in the heaven that undergo a plurality of locomotions,] the[92] former attribute (b30) belongs only to perishables.

When he says that as for unmoved things that move themselves
incidentally, it is impossible for them to cause continuous motion, 1261,15
anyone might pose the puzzle how the heavens – which themselves
are self-moving living things with a motion-causing element that is
unmoved per se and causes itself to move incidentally, but is moved
incidentally by something else – similarly undergo a continuous
motion. (For the planetary heavens are moved by the fixed heaven
with its own motion.) He solves this puzzle, saying that 'being moved 1261,20
incidentally by oneself' is not 'the same' as being moved 'by something
else. The latter attribute belongs' he says, 'also' to the planetary
spheres,[93] which undergo both their proper motion <rotating> around
their own poles, and a different motion, that of the sphere of the fixed

stars, since their own poles move[94] around *its* poles.[95] And it is possible
1261,25 for them to undergo the continuous motion caused by the sphere of the fixed stars, which is moved and causes motion continuously. For they are not moved by themselves incidentally, because the mover is in a whole and the whole is not moved, since things that are in circular motion are in motion by parts and not as a whole.[96] The property of being moved by themselves incidentally belongs 'to perishables'
1261,30 alone. (He says 'only' instead of 'alone'.)

Alexander declares that the souls in the planetary spheres are moved incidentally – not by themselves but by the sphere that moves their bodies, because the bodies in which they are located[97] do not undergo motion in the same direction as that in which they are moved by the unmoved elements in them. 'The first cause', he says, 'which is the motion-imparting element of the sphere of the fixed stars, will
1261,35 be moved incidentally neither by itself nor by anything else, because
1262,1 the sphere of the fixed stars undergoes but a single motion, in which the poles are permanently in the same <place>, or because the origin is not the form of the moving body but some separate substance.' Alexander says this because he thinks that all the other souls are forms that are not separate from bodies. It is more appropriate to
1262,5 examine this elsewhere, but it is worth remarking that Aristotle never believes that being moved[98] incidentally is one thing when he considers it as applying to mortal animals and quite another thing when it is caused by something else and is an attribute of the spheres that are said to wander. What is incidental cannot be eternal because it is a by-product of that which is per se and a deviation of the principal
1262,10 kind of thing, and no such nature is eternal. But there can be an account of what is affected by something else and is a principal kind of thing, and this is why Aristotle says 'being moved incidentally by oneself is not the same as' being moved 'by something else', not thinking, I suppose, that we should supply 'incidentally' with 'by something else'.

> **259b32-260a10** But surely, if in fact the mover[99] is always some such thing [(something that is itself unmoved and eternal) the first thing that is moved by this must be (260a1) eternal too. This is also clear from the fact that there is no other way for generation, perishing, and change to exist in things unless there is going to be something that is in motion that moves them. For that which is unmoved[100] will cause[101] motion in the same way, and it will in fact cause a single motion, (a5) since it does not itself change at all in relation to what is moved. But because what is moved[102] by the unmoved or[103] by what is already moved, varies in its relations to things, it will not be the cause of the

> same motion, but because it is in contrary places or forms it will
> cause each of the other things to be moved (a10) in contrary
> ways, and will cause them] at some times to be at rest and at 1262,15
> others to be in motion.

He proved first that there is always motion and that it has not been
generated at any time, not existing previously, and it never perishes
with the result that there is no motion.[104] Next he proves that the
primary mover 'for each motion'[105] (as Eudemus adds) must be un-
moved both per se and incidentally, and must also be eternal as the 1262,20
cause of eternal motion.[106] He now says that the primary thing that
is moved by the unmoved eternal mover[107] must be eternal as well.
For the mover as mover and what is moved as moved coexist. He also
gives another demonstration of this in addition. Assuming as agreed
that generation and perishing always exist among things that are, he
infers the following conditional: if there is nothing eternal that both 1262,25
causes motion and is moved, there cannot always be generation and
perishing. But in fact these things <always> exist. Therefore there is
something eternal that both moves and is moved. The minor premise
stating that generation and perishing always exist, he has posited as
agreed, as I said, for there cannot be generation of generation. He
proves the conditional as follows. What is generated and perishes[108] 1262,30
certainly requires a mover. Now the mover must be either unmoved
or moved. But 'that which is unmoved' in that it is unchangeable, 'will
cause motion in the same way', so that there will be either generation
or perishing alone.[109] 'But because what is moved by the unmoved'[110]
comes to be 'in contrary places' and the generative element sometimes
approaches and sometimes withdraws, as is stated in *On Generation* 1262,35
and Corruption, 'it will not be the cause of the same motion[111] but' 1263,1
'will cause' 'each of the other things[112] to be moved' 'in contrary ways',
and <will cause> the same thing sometimes 'to be at rest' and
sometimes 'to be in motion'.

Now it is true that unless there is something that is moved and
causes motion[113] there will not always be generation and perishing.[114]
And it is clear that what causes such motions[115] is itself both eternal
and in motion.[116] For if these motions[117] always exist, what causes 1263,5
them[118] must always exist. For if the causes of these motions[119] prove
to be not eternal but things that are generated and perish,[120] some
other cause of *their* generation and perishing will be sought in turn.
And that in turn cannot be unmoved, but if it is moved it will either
be always and eternal or subject to generation and perishing. And
again the same inquiry will ensue. For if what is subject to generation 1263,10
needs to have a self-moving cause (if the progression is not to go ad
infinitum), there will be some eternal first self-mover[121] being moved
continuously by the eternal unmoved first mover. In this way motion

is continuous and eternal,[122] and generation and perishing and the
1263,15 change of bodies into one another are ceaseless.[123]

After saying 'what is moved by the unmoved' he adds 'or by what is already moved' because the sphere of the fixed stars is moved proximately by what is unmoved, but the planetary spheres are moved by what is fixed but already in motion. He says 'in contrary
1263,20 places or forms' because the sun and the other stars at times approach us on account of the inclination of the ecliptic, and at times withdraw from us, and they are arranged similarly with respect to the south: sometimes they come to be higher in the sky and sometimes nearer the earth. They are in opposite forms, colder or hotter, because they are productive of these and suchlike opposite forms according to their varying positions. Someone might say that Aristotle here is keeping
1263,25 close to the passage in the *Phaedrus* where Plato says 'all soul has the care of all that lacks a soul; it traverses the whole heaven, at different times coming to be in different forms'.[124] For in consequence of their[125] observation of the forms there,[126] which are different, their[127] motions come to be different too, and the variety that results from their motions exists here.[128]

1263,30 **260a11-19** What we have said has also cleared up the puzzle we raised at the beginning, [namely, why in the world is it the case not that all things are either in motion or at rest, or that some things are always in motion and the others are always at rest, but that some things are sometimes <in motion> and sometimes not. The cause of this is now clear, namely that some things (a15) are moved by an eternal unmoved <mover>, which is why they are always in motion, while other things <are moved> by something that is in motion and changing, so that they too must be changing. But what is unmoved, as we have said, since it persists in a way that is simple, unvarying, and in the same condition,] will cause a motion that is one and simple.

After saying that the fact that some existing things are generated and perish and sometimes are moved and sometimes are at rest, is caused by their varying relationship to the movements of the heavens,[129] he
1263,35 reminds us that what has just been said solves the earlier puzzle.[130] When he was refuting the arguments that seemed to establish that motion comes to be, not existing previously, he said that it was puzzling to those who hypothesize a generation of motion why 'is it
1264,1 the case not that all things are either' always 'in motion or' always 'at rest, or that some things are always in motion and the others are always at rest', but there also exist in the world things that are sometimes in motion and sometimes at rest. 'The cause of this', he says, has 'now' become 'clear'. Because 'some things are moved by an

eternal unmoved <mover>', these must always be in motion, and because 'other things <are moved> by something that is in motion 1264,5 and changing' always, these too must always 'be changing. But what is unmoved, as we have said', since it is always in a similar state, causes 'a motion' 'that is one and simple' and continuous and eternal. And so it is reasonable that some things are always unmoved – those that are primary movers[131] in the strict sense; while others are eternal and always being moved similarly – those that are moved proximately by the unmoved <movers>; and still others are generated and perish 1264,10 and are sometimes in motion and sometimes at rest – those that are moved by things that are always in motion, but which have different relationships at different times.

<CHAPTER 7>

> **260a20-6** [These matters will] nevertheless [be still more evident if] we make a new start. [We should consider whether it is possible for any kind of motion to be continuous, or not, and if it is possible, what this is, and what is the primary kind of motion. For it is clear that if indeed there must always be motion, and if this motion is primary and continuous, (a25) then the first mover causes this motion, which must be one and the same, and continuous and primary.]

After proving that, agreed that there is eternal motion, both the primary mover and the primary thing that is moved must be eter- 1264,15 nal,[132] he next proves what kind of motion can be eternal and continuous. It is clear that this is the motion that the eternal mover will cause and that the eternal thing that is moved will undergo. It is reasonable that the primary kind of motion should be eternal and continuous if in fact the motions that depend on it are. So he now investigates what the primary kind of motion is, and after distinguishing the species of 1264,20 motion,[133] he proves that motion in place is the primary kind of motion in accordance with all the significations of primary,[134] and that the primary kind of motion in place is circular motion, which alone can be eternal and continuous.[135] And so the body that undergoes circular motion will be the first of things that are moved, and it is eternal, and this is what the first mover, being eternal, will move proximately. Thus together with demonstrating that what undergoes circular 1264,25 motion is moved primarily and eternally, he demonstrates that the primary mover and the primary thing that is moved are eternal. This is why he says[136] that this becomes evident from another starting point too, and appears together with the fact that the body that is in circular motion is what is moved primarily and eternally. He says that the starting point of the consideration is 'whether it is possible

1264,30 for' 'motion to be' 'continuous' 'or not, and if it is possible, what this is'. This in fact is 'the primary kind of motion'. The first point, 'whether it is possible for' 'motion to be' 'continuous', he defers for the present and will investigate it a little below;[137] he now proves what 'the primary kind of motion' is. Together with this he also demonstrates what <kind of motion> is continuous, for continuous motion is prior to non-continuous motions and eternal motion <is prior> to non-eternal motions. After setting out the problems in
1264,35 order (for the investigation about *whether* something is is prior to
1265,1 that of *what* it is[138]), he next proves why it is useful for him to prove what the primary and continuous motion is, deducing it potentially as follows: 'if' 'there must always be motion', as was proved, and this eternal motion is 'primary' 'and continuous' (for eternal and continuous motion is prior to motions that are not such), if I prove
1265,5 what this motion is, it is clear that 'the first mover' 'causes' 'this' 'motion' proximately and correspondingly, and this is the only motion that is one and the same, and continuous and primary, because it arises through the agency of the single primary mover, which causes motion continuously.

> **260a26-b7** For[139] there being three <kinds of> motion: [motion in respect of magnitude, motion in respect of affection, and motion in respect of place, which we call locomotion – this <last kind> must be primary. Increase cannot occur unless alteration (a30) pre-exists it, since what increases, in one sense increases by means of its like and in another by means of its unlike (for contrary is said to be nourishment for contrary). But everything is added to something by becoming like it. Now the change to contraries must be alteration. (260b1) And further, if it is indeed altered, there must be something that alters it and makes it hot in actuality from being hot potentially. Now it is clear that what makes it move [i.e. alter] is not <always> related <to it> in the same way, but at one time is closer and at another time is farther from what is being altered. But this cannot happen (b5) unless there is locomotion. Therefore if there must always be motion, there must always be locomotion as the primary kind of motion,] and if one kind of locomotion is primary and another kind is posterior, the primary kind of locomotion <must be the primary kind of motion>.

1265,10 'There being three' kinds of 'motion: motion in respect of magnitude', i.e. increase and decrease, 'motion in respect of affection', i.e. alteration (this has been proved to occur with reference to the affective qualities), 'and motion in respect of place, which we call locomotion', he will prove through many arguments that 'this <last kind>' is

'prior'[140] to the other kinds of motion. The first argument[141] proves 1265,15
that 'increase' cannot occur without 'alteration', or alteration without locomotion, so that locomotion is prior in nature to the other kinds of motion. For if none of the others occurs without locomotion, but it occurs without them, as will be proved, it is clear that eliminating locomotion eliminates the others, but not vice versa. And things of this sort are primary in nature. He proves that alteration pre-exists 1265,20
increase from the fact that what grows grows because of nourishment and nourishment is originally dissimilar to what is nourished and is somehow contrary to it (which is precisely why 'contrary' 'is said to be' 'nourishment' 'for contrary'), and later changes into its contrary, and, being made similar to what is nourished, it is joined to it and nourishes it and makes it grow. Therefore, it requires change in the 1265,25
form of alteration, since 'change' 'to contraries' in quality 'must' 'be alteration'. In saying that what grows does so by the addition of nourishment, he is now talking about growth more like a natural scientist than in the *Categories* when he says that the square grows by placing a gnomon around it.[142] It is also clear that nothing can be altered without locomotion, if in fact what is altered, is altered by 1265,30
something that alters it and changes it 'from potentiality'[143] to actuality. It is clear that what changes and alters it 'is not related' to it 'in the same way' in position now, when it is changing it, and before, when it was not yet changing it. For if it were similarly related to it, why would it not have been changing it then too, and why would the other not have been changing? Therefore there must have occurred motion in place either of both what causes alteration and what is 1265,35
altered, or at least of one of the two, in order for the one to cause alteration and the other to be altered. For in fact the nourishment would not have been digested and altered if it remained in its own place, but it is when it approached <the place where digestion occurs> 1266,1
that on the one hand it was altered and on the other hand our digestive capacity altered it. Therefore, there is need of locomotion if alteration is to take place. And concluding the argument he says, 'therefore if there must always be motion, there must always be locomotion', which has been proved to be 'the primary kind of motion', for what is primary by nature would be eternal rather than the others. 1266,5
But 'if one kind of locomotion' is 'primary and another kind is posterior' (and it will be proved[144] that circular motion is prior to motion in a straight line), *this* would be by nature the first of all motions, if indeed circular motion is the primary kind of locomotion and locomotion is prior to the rest.

260b7-15 Further, the origin of all the affections [is condensa- 1266,10
tion and rarefaction: heavy and light, soft and hard, and hot

> (b10) and cold are thought to be forms of density and rarity. But condensation and rarefaction are combination and separation, and it is in virtue of these latter processes that the generation and perishing of substances are said to be. And in combining and separating things must change in respect of place. Moreover,] the magnitude [of that which increases or decreases also] changes (b15) in respect of place.

This is the second argument proving that locomotion pre-exists motions of alteration and those of increase and decrease, and not only
1266,15 these, but also the change of generation and perishing. He proves that it pre-exists motions of alteration as follows: 'the origin' 'of the affections', i.e. of the changes of alteration, is 'condensation and rarefaction', 'but condensation and rarefaction are combination and separation', condensation being combination, and rarefaction being separation. But when things are being combined and separated, the
1266,20 change involved is change of place. Therefore, change in place is the principle of the changes of alteration. He shows that condensation and rarefaction are the principle of all the affections through reference to each species of quality. For, he says, 'heavy and light, soft and hard, and hot and cold are thought to be forms of density and rarity',
1266,25 heavy, hard and cold being kinds of density, and their contraries kinds of rarity. In this way white and sweet would fall under the heading of rarity because like heat they seem to tend to separate the perceptions relative to them, and their opposites seem to be kinds of density. It is clear that 'condensation' 'and' 'rarefaction' are 'combination' 'and separation', being different names for the same thing. After
1266,30 saying that condensation and rarefaction are combination and separation, he adds 'it is in virtue of these latter processes that the generation and perishing of substances are said to be', proving that change in place is prior to generation and perishing as well as to alteration and growth. For if generation and perishing are of their own nature combination and separation, as Democritus thought, and
1266,35 also Anaxagoras, Empedocles, and all who posited impassive primary bodies and generated the rest wholly out of these,[145] or all those who posited a single element, like Anaximander's intermediate,[146] and say that this produces the rest when condensed or rarefied, or even if generation and perishing occur in virtue of alteration, in this way too
1267,1 the claim that generation and perishing take place because of combination and separation is sound, if in fact combination and separation govern (so to speak) every alteration. He says, 'it is in virtue of these latter processes that the generation and perishing of substances are said to be', since these[147] are generations and perishings in the strict
1267,5 sense. Now in fact it has been said that changes in quality and magnitude occur through combination and separation. But after

proving by reference to combination and separation that change in
place pre-exists alteration, generation and perishing, he proves that
'the magnitude' of things that increase or decrease 'changes in respect
of place'. Not only does what is growing or becoming smaller change
in place, with the former occupying a larger place and the latter a 1267,10
smaller one, but also the nutriment by which the former becomes
larger must change in place, and so the nutriment is moved simulta-
neously with what is being nourished and is growing. And in turn, if
decrease occurs, something must be separated from what is becoming
smaller.

It is worth remarking that according to the same method of arguing 1267,15
from considerations of combination and separation, change in place
may also be shown to pre-exist what grows and becomes smaller. For
in these cases too change in magnitude is a change in place, since the
nutriment of what grows is combined with it, and since some sub-
stance is being separated from what is becoming smaller. Here too,
Aristotle wants to agree with his teacher. For Plato too, in the tenth 1267,20
book of the *Laws*, states that motion in place is the first of all motions,
and from this there exist combination and separation, and from this
increase and decrease. For he says, when things moving in place 'meet
with things that are stationary, they split them apart, but when they 1267,25
meet with other things that are moving in the contrary direction and
are brought to the same place, they are combined and become inter-
mediate and in between such things.'[148] – 'And in fact, things being
combined increase in magnitude and things being separated de-
crease.'[149] He says that generation exists as a result of increase and
perishing as result of decrease.[150]

> **260b15-19** In addition, [that locomotion is primary] will be
> evident if we consider the matter from the following point of
> view. [As in other cases, with motion too, primary may be said
> in several ways. A thing is called prior <to other things> when
> if it does not exist the others will not exist either, while it will
> exist without the others, and <a thing can also be called prior>
> both in[151] time] and in being. 1267,30

After setting out to present other arguments to prove that locomotion
is the primary kind of motion, he makes a division of the significations
of prior, thus supplying what he left aside at the beginning, and he
proves that locomotion is prior to the other kinds of motion in all the 1267,35
ways in which priority is said in the strict sense. Here he sets out
three types of priority. For 'a thing is called prior <to other things>'
in nature, 'when if it does not exist the others will not exist either,
while it will exist without the others, and' <a thing can also be called>
prior 'in time' as last year <is prior> to this year, 'and in being', 1268,1

namely, that which is more complete, since the complete is prior to
the incomplete, and this[152] is that which has achieved its own proper
being. Alexander says, 'A motion might be called prior specifically in
respect of being if it does not touch the being of the thing that is moved
1268,5 or of any of its attributes, but whose occurrence leaves it in every way
the same – and only motion in place is like this.'[153] Here he [Aristotle]
recalls these significations of priority, but in the *Categories*, in addi-
tion to priority in respect of nature and time he posits other signifi-
cations of priority.[154] There he speaks of something being prior in
1268,10 order or position, and spoke of the more honoured <as prior>. And of
things that do not[155] mutually entail each other's existence, the one
that is in any way the cause of the other's existence <is prior>, as
father <is prior> to son, and generally the cause <is prior> to the
effect. In book five[156] of the *Metaphysics* he sets out more types of
priority and posteriority:[157] For what is 'nearer' the determinate
1268,15 'origin' in each kind or nature or relation is called prior, and what is
'farther' from it is called 'posterior'. For example, <something is called
prior> 'in respect of place by being' near to 'some place', either a place
'determined' by nature, 'like the middle or the extremity, or to some
chance thing'. In comparison with the present, what is prior 'in time'
in the direction of the past is what is 'farther from the present', as
'the Trojan War <is prior to> the Persian War', but in the direction of
1268,20 the future, what is 'nearer to the present' is prior, as tomorrow <is
prior> to the day after tomorrow, 'if we treat the present as the
beginning and first thing'. He speaks third of what is prior 'in respect
of movement. For example, a child' <is prior> 'to a man' because 'it is
nearer to the first thing that caused motion'. For 'this too is without
qualification a kind of' 'origin'. Fourth is what <is prior> 'in power',
for the ruler is 'prior in power and' 'more powerful. This sort of thing
1268,25 is that according to whose choice the posterior must follow' '(choice
being an origin)'. Fifth are 'things' that are <prior> 'in order. These
are all things that are distant in relation to one determinate thing
according to' the 'definition.[158] For example, the person who stands
second in the chorus is prior to the person who stands third, and the
second lowest string of a lyre <is prior> to the lowest.'[159] There is a
starting point in those cases too: in the former, the place where 'the
chorus leader' is, in the second, the place where 'the top string' is.
Sixth is 'what' is prior 'in knowledge'; in one way this is what <is
prior> 'in definition' and in another way it is what <is prior> 'in
1268,30 perception'. 'Universals are prior in definition, but particulars <are
prior> in perception. And in definition the accident is prior to the
whole, as musical <is prior> to musical man, since the definition as
a whole will not exist without the part'. Seventh are 'the affections of
things that are prior. For example, straightness, which is an affection
of line per se, <is prior> to smoothness, which is an affection of

surface'. Eighth are 'things' <that are prior> 'in nature and being – 1268,35
all that can exist without other things, while *they* cannot exist without
them'. He says that 'Plato' 'employed' this 'division'. Ninth, 'the
subject is prior' to what is in a subject; 'this is why substance is prior'. 1269,1
The tenth falls out 'one way' 'as regards potentiality' and another 'as
regards actuality. For example, in potentiality, the half line <is prior>
to the whole line, the part to the whole, and the matter to the
substance, while in actuality' they are posterior. 'For they will exist
when' what exists 'in actuality is broken up'. This more complete kind 1269,5
of division of priority and posteriority is what he presents in that
treatise. But here, when investigating priority and posteriority in
motion, he employs all the significations that are relevant to the
subject. Priority in position, if it were found at all in motion, would
be covered by priority in time, or perhaps the motion of bodies that
are prior in position would be called prior in position too. It is worth 1269,10
remarking, however, that in the *Metaphysics* he joins together prior-
ity in nature and in being,[160] whereas here he divides them.

> **260b19-29** And so, since (b20) there must be motion continuously, [and it may be continuously by being either continuous or consecutive, but more so by being continuous, and since it is better that it should be continuous than consecutive, and we always suppose that what is better is present in nature, if it is possible, and since it *is* possible for there to be continuous motion (this will be proved later; for the present let it be posited) and (b25) this <continuous motion> can be no other motion than locomotion, locomotion must be primary. For what is undergoing locomotion need not be increasing or undergoing alteration, or indeed being generated or perishing, but none of these is possible unless the continuous motion] that the first mover causes [exists.]

After proposing to prove that locomotion is the primary kind of motion 1269,15
according to all the significations of primary, he first proves that it is
primary in nature, assuming that there must always be motion. (This
is what 'be continuously' means.) In fact, it has already been demon-
strated[161] that motion is ungenerated. Taking it as evident that
eternal motion is prior to motion that at one time is and then in turn
is not, he deduces as follows: the motion that is primary in nature is 1269,20
eternal, since there must be eternal motion; eternal motion is con-
tinuous motion; continuous motion is motion that comes under the
heading of locomotion; therefore the motion that is primary in nature
is motion that comes under the heading of locomotion. He also proves
that it is primary in nature from the fact that it does not mutually
entail the existence <of any of the other kinds of motion>: it does not

1269,25 need the other kinds of motion in order to exist (what is causing locomotion need not be increasing or altering, or undergoing generation or perishing either), 'but none of these can' exist 'unless the continuous motion that the first mover causes exists'. For it has been proved[162] that the principle of all things that are moved is the self-mover, and the principle of this is the unmoved. Eternal motion
1269,30 must be either continuous 'or consecutive'. But 'it is better that' motion that is eternal in the strict sense 'should be' 'continuous' 'than consecutive', for this movement is one if the thing that causes motion is one, whereas what is consecutive is not one. And 'we always suppose that what is better' exists in things having to do with nature 'if it is possible'. And it will be proved below that it is possible for there to be a continuous motion[163] and that 'no other motion' can be
1269,35 continuous 'than' 'locomotion'.[164] But now these claims have been assumed as posited for the meantime.

1270,1 **260b29-261a12** Further, it [locomotion] must be primary in time. [For (b30) this is the only kind of motion that eternal things can undergo. But for any individual that is subject to generation, locomotion must be the last of its motions, for after generation, alteration and growth are its first motions, while locomotion is a motion characteristic of things that are already complete. (261a1) But something else, which is prior, must be moved in virtue of locomotion, and it will be also the cause of generation for things that are being generated, although it is not itself being generated – as the begetter <is the cause> of the begotten. And yet, generation might seem to be the primary kind of motion on account of the fact that the thing must be generated (a5) first. This is the case for any individual that is generated: something else, which is prior, must be in motion, itself being a thing that undergoes generation but which is not being generated, and prior to this there must be another. But since generation cannot be the primary kind of motion (since in that case all things that are moved would be perishable), clearly none of the subsequent kinds of motion is (a10) prior either. By subsequent I mean increase, then alteration, decrease and perishing. All these are posterior to generation, so that if not even generation is prior to locomotion,] neither is any of the other kinds of change.

After proving that locomotion is prior in nature to the other kinds of motion,[165] he next proves that it is primary 'in' 'time' too, deducing
1270,5 this too potentially as follows: the kind of motion that is primary in time is that which 'can' belong to 'eternal things'; the motion that can belong to eternal things is locomotion; therefore locomotion is the kind

of motion that is primary in time. Of these two premises, the minor, which states that 'the primary <kind of motion> is that which eternal things can have' he omits as obvious. For just as eternal things are prior in time to things that are subject to generation, so too the motion 1270,10
that eternal things undergo, being eternal, will be prior to the non-eternal motions that things that are subject to generation undergo. He posits the major premise, which states that the only <kind of motion> that eternal things 'can' 'undergo' is motion in place, and he will prove it a little later.[166]

But since things that are subject to generation undergo motion in place later (animals move in place when they have already reached completion; first they are generated, and are nourished and grow by 1270,15
means of alteration, and only then do they move in place), it will not seem truly said that motion in place is prior in time to the other kinds of motion. In refuting this objection he declares that in the case of each individual that is subject to generation, motion in place is the last kind of motion to occur (since they move in place only when they have already reached completion),[167] but although they are generated 1270,20
and undergo alteration and grow first, and only then move in place, there must be something else that moves in place which pre-exists them, which is 'also the cause of generation' for what is generated, and is itself ungenerated, or is not generated then but before, as the father is <the cause> of the son and generally 'the begetter <is the cause> of the begotten'. For this must move in place in order to 1270,25
generate what is being generated. But the body that is in circular motion, which moves in place, is the cause of generation and is not itself generated. As for individual cases of generation, 'generation' will seem 'to be the primary kind of motion' because 'the thing must be generated' previously, and only when its substance exists can its accidents belong to it: alteration, growth, and change in place. This 1270,30
is the case 'for' each 'individual'. But 'something' 'else' 'must' 'be in motion' in respect of place – the cause of the thing that is being generated, which is not itself 'being generated' at that time. 'And another' in turn pre-exists 'this' as the cause of its generation – not being generated, but existing at that time and moving in place. And so before every individual generation there is local motion, which the cause of generation (which is not being generated at that time) 1270,35
undergoes. And before this is another – that which has eternal motion in place and is not generated at all. 'For man generates man and also the sun does',[168] as has been stated earlier.

After proving that motion in place pre-exists all generation even 1271,1
though in the case of particulars generation seems to be prior to the other kinds of motion, he points out as well that the same claim also results if we assume the contrary. For if generation were the primary kind of motion, 'all things' 'that are moved would be perishable'. For

1271,5 they must first be generated and only then undergo whatever kind of motion they do, if in fact generation is the primary kind of motion. But what is generated perishes. If, then, not all of the things that are moved are subject to generation and perishing (which has already been proved[169] from the existence of eternal motion, and which will be proved still more conclusively[170]), again, generation will not be primary. For it will not pre-exist in the case of things that are eternal
1271,10 and are in motion. But if generation is not primary, 'clearly' none of the other 'kinds of motion' that must exist after generation <will be primary> either: alteration, increase, and decrease. For only things subject to generation increase and become smaller, and alteration belongs to these alone since every alteration occurs in connection with the affective qualities, and only things that are subject to generation
1271,15 have affective qualities. If, then, the other kinds of motion are posterior 'to generation' and 'generation' is posterior 'to locomotion', clearly the other kinds of motion are posterior in time to locomotion. In this way too he proves that locomotion is prior in time to the other kinds of motion and he refutes the objection based on generation.

> **261a13-23** In general, it is apparent[171] that what is coming to be is imperfect [and is proceeding towards its principle, so that what is posterior in generation is prior in nature. And the last thing (a15) to belong to all things that are subject to generation is locomotion. Consequently, some living things are entirely without motion because of a deficiency (such as plants and many kinds of animals), but others acquire it as they are reaching completion. And so, if locomotion belongs more to things that have more fully attained their nature, then also this kind of motion will be prior to the other <kinds of motion> in respect of (a20) being, both for this reason and also because in undergoing locomotion what is moved departs from its being less than <it does with other> motions. For this is the only kind of motion that makes no change in its being, in the way that what under-
> 1271,20 goes alteration changes in quality] and what increases or decreases changes in quantity.

He makes the refutation of the objection based on generation the starting point of the third argument, which proves that motion in respect of locomotion is prior to the others in being too. What is primary in being he also names primary in nature, consistently with
1271,25 what is said on these subjects in the *Metaphysics*.[172] He deduces and demonstrates this thesis categorically as follows: the last <kind of motion> 'to belong to all things that are subject to generation is locomotion'; 'what is posterior in generation is prior in nature' and being. Locomotion, therefore, is prior in both nature and being to the

other kinds of motion. That motion is the last to belong to things that are subject to generation he proves from the fact that on the one hand, 1271,30
things that are less perfect[173] in being, such as 'plants' and the clinging animals which we call zoophytes, do not move in place at all, while to animals that reach completion, this is the last to belong. That 'what is posterior in generation' is 'prior in nature' and being he proves from the fact that what is generated is incomplete when it is still being generated, and it proceeds towards its own state of completion as 'towards its principle' and something primary in nature. 1271,35
For the state of completion is a principle as that for the sake of which and the end.[174] What is complete is prior in nature and being to what is incomplete, as that for the sake of which <is prior in nature and being> to what is for its sake. For what is natural for each thing <is determined by> what is in accordance with its state of completion. 1272,1
And what is complete is <a thing's> being and principle, for what is complete is what generates.[175] So just as the state of completion is last in generation but primary in both being and nature, so also the last kind of motion that things that are generated undergo is primary in nature and being. But by putting the supporting argument for the 1272,5
major premise, which says 'what is posterior in generation is prior in nature', *before* it in the words 'it is apparent that what is coming to be is imperfect and is proceeding towards its principle', and putting the supporting argument for the minor premise, which says the last <kind of motion> 'to belong to all things that are subject to generation is locomotion', *after* it in the words 'some living things are entirely without motion' and following, he puts the two premises themselves 1272,10
in the middle, putting the major first in the words 'so that what is posterior in generation is prior in nature' and then the minor in the words 'the last thing to belong to all things that are subject to generation is locomotion'. By arranging the deduction in this way he makes its analysis difficult to perceive all at once. At the end he deduces as follows, according to the first mode of hypothetical syllo- 1272,15
gisms:[176] 'if locomotion belongs more to things that have more fully attained their nature', i.e. if motion in place belongs more to things that have reached completion in accordance with nature and their own form, 'also' 'motion' in place 'will' 'be' 'prior' in nature and 'in respect of being' 'to the other' kinds of motion. Then he adds still another demonstration that motion in place is more complete and consequently also prior in nature and being. Motion in place, he 1272,20
declares, 'makes no change' in the condition of what is moved – not in being, as generation and perishing do, nor in 'quality', as alteration does, nor in 'quantity', as increase and decrease do. And what preserves its nature and being is more complete in nature and being than 1272,25
what does not preserve them.

261a23-6 Most of all it is clear that, [strictly speaking,] what moves itself [causes this kind of motion above all – motion in respect of (a25) place; and further we declare that this] – that which moves itself – [is the origin of things that are moved and cause motion, and is the primary <cause> for things that are in motion.]

This argument can be used to prove that locomotion is prior to the other kinds of motion according to the three types of priority[177] that
1272,30 he gives. It goes like this: it was previously proved[178] that what moves itself, which pre-exists in both time and being the things that cause motion and are moved, is the origin of such things, and that the self-mover undergoes motion that comes under the heading of locomotion specifically and 'strictly speaking'.[179] If, then, motion in place is proper to what is the origin and cause of things that are moved, and this movement that is proper to the origin and cause of things
1272,35 that are moved will be prior in both nature and time to the other kinds of motion, the conclusion is clear, that motion in place is prior in nature, being and time to the other kinds of motion. And then the discussion concludes. It can be seen that here too Aristotle forms his
1273,1 demonstration following the tenth book of Plato's *Laws*, since after presenting the species of motion and setting out motion in place as the primary kind, he [Plato] declares it to be the cause of the others – proximately of separation and combination, i.e. of alteration, and,
1273,5 in consequence of this, the cause of increase and decrease, and in consequence of *this*, the cause of generation and perishing. For after setting out motion in place <as the primary kind>, he continues, 'Each time they meet with anything, if the things they meet are stationary, they are split apart, but when they meet with moving things coming from the contrary direction, they combine and become intermediates, which are in between their original states'.[180] – 'And in fact, things
1273,10 that are being combined increase and things being separated decrease, whenever the existing state of each persists. But if it persists, it is destroyed by both processes'.[181] He says that generation occurs as a result of increase, and perishing as a result of decrease.[182]

261a28-b22 We must now prove what kind of locomotion is primary. [At the same time what was hypothesized both now and previously, that some kind of motion (a30) can be continuous and eternal, will be evident by means of the same method. Now it is evident from the following considerations that no other kind of motion can be continuous. All kinds of motion and change are from opposites to opposites. For example, the limits for generation and perishing are what exists and does not exist,[183] for alteration (a35) they are the contrary affections, for increase and decrease they are either largeness and smallness or completeness and incompleteness of magnitude. And changes to contraries are (261b1) contrary <changes>. A thing that is not always undergoing a particular motion, but that existed previously, must previously have been at rest. It is evident, then, that a thing that changes will have been at rest in the contrary state, and likewise for changes too: perishing and generation are opposites without qualification, (b5) and a particular instance of the one is opposite to a particular instance of the other. And so if nothing can undergo opposite changes at the same time, the change will not be continuous, but there will be a time interval between them. It makes no difference whether the contradictory changes are contraries or not, as long as they cannot be present in the same thing at the same time (for (b10) this is not put to use in the argument), nor does it <make any difference> if there is no need <for it> to *be at rest* in the contradictory state or if a change is not contrary to a state of rest (for perhaps what does not exist is not at rest, and perishing <is a change> to what does not exist), as long as there comes to be a time interval in between. For in this way the change is not continuous. In the previous cases too, what was put to use (b15) was not the contrariety <of the two changes>, but the impossibility of their coexistence. We must not be bothered that more than one thing will turn out to have the same contrary, i.e. a motion is the contrary both of a stationary state and of the motion towards the contrary. We need only grasp that the contrary motion is somehow opposite both to the motion and to the state of rest, as the equal and moderate (b20) <is opposite> to what exceeds it and to what it exceeds,] and that opposite motions and changes cannot coexist. 1273,15

After proving that locomotion is prior to the other kinds of motion and change in respect of all the types of priority, he now proceeds in turn to investigate 'what' in the category of locomotion 'is primary' in respect of the significations of primary that have been stated. We know that one kind of locomotion is rectilinear, another is circular, and the third is combined <out of the two>, just as is true of the kind 1273,20
of line on which the thing that undergoes locomotion is moved. When the primary kind of locomotion is discovered, 'at the same time what was hypothesized both now and previously, that some kind of motion can be continuous and eternal, will be evident'. For this primary kind of locomotion will be found to be continuous and eternal.

But we should remark that until now the existence of continuous motion was being assumed as an hypothesis. For the motion that was 1273,25

originally proved ungenerated and indestructible was not demon-
strated as being a motion that is one and continuous, but instead as
something that always is, but that is now one motion among existing
things and now another. It is characteristic of an exact science to use
a single method to discover several theorems, which are general and
useful in this way.[184] In this case, the discovery proceeds by first
1273,30 proving that none of the other kinds of motion or change aside from
locomotion can be continuous and one, and of the kinds of locomotion
neither rectilinear <locomotion> nor <locomotion> combined <of rec-
tilinear and circular> can be so, but continuity and eternity can
belong to circular motion alone.[185] And so he proves on the basis of
this that of the kinds of locomotion, circular motion is prior to
rectilinear. That no other kind of motion can be continuous, he
1273,35 demonstrates as follows: all the other kinds of motion (for now he
includes generation and perishing; then later he separates them from
motions in the strict sense,[186] since they are changes, not motions, as
was proved in the fifth book of the present treatise[187]) – now all the
1274,1 other 'kinds of motion' aside from locomotion 'are' 'from opposites to
opposites' (for things that are from contraries to contraries are oppo-
sites in that they are contrary).[188] However, he bases his argument
on contraries, although generation and perishing are opposites but
not contraries. Further, it is impossible for motions that are contrary
1274,5 or opposite in any way to be one and continuous. From this he infers
that all the other kinds of motion aside from locomotion cannot be
continuous.

That each of them is from opposite to opposite he proves by setting
out the opposite limits towards which they move. For 'the limits for
generation and perishing are what exists and does not exist', which
1274,10 are opposed as affirmation and negation; 'for alteration they are the
contrary affections' heat and cold, whiteness and blackness, etc.; 'for
increase and decrease they are either largeness[189] and smallness or'
rather 'completeness and incompleteness of magnitude', because
magnitude is common both to the complete, towards which increase
aims, and the incomplete, towards which decrease does. It is obvious
1274,15 that 'changes to contraries' and opposites 'are contrary' and opposite,
for this is the definition of contrary motions. Next, it is clear that a
single continuous motion cannot arise from opposite and contrary
motions. For if it were one, what is becoming white would simultane-
ously be becoming black, what is changing towards health would
simultaneously be changing towards sickness, what is increasing
1274,20 would simultaneously be becoming smaller, and what is being gener-
ated would simultaneously be perishing, if one and the same motion
arises from contraries. And it is also immediately clear that a single
motion cannot arise from contrary motions. For how could a single
motion arise from upward motion and downward motion, both re-

maining unmixed? It is impossible for contrary motions to be natural,
nor can motion that is natural and unnatural turn out to be single.
(Lateral motion, which *is* a single motion, seems to be a combination 1274,25
of upward and downward motion, but not one in which they remain
unmixed as the argument is now assuming, but the same sort of thing
as happens when the elements of a body, being contraries, are mixed
together and make some single compound.) And so, a single motion
cannot arise from contrary motions nor can anything simultaneously
have contrary motions.

But if the motion that arises from contrary motions is not one, the
contrary motions are interrupted by a state of rest. He assumes this 1274,30
premise in addition to the ones previously stated, and he establishes
it by use of the axiom that 'a thing that is not always undergoing a
particular motion, but that existed previously, must previously have
been at rest'. For if what is becoming white were to be becoming black
without having been at rest or having ceased to become white, it
would simultaneously be becoming white and becoming black, so that 1274,35
it will become white and black simultaneously when it has come to
the end of the motion.[190] But if it becomes white after having ceased
to be becoming white,[191] and from white it changes to black, it will
certainly be at rest for some time interval when it has come to be in
the form of white. For this is how the change to becoming black will
occur, <starting> from being white and no longer becoming white, for
it is not white at the same time as it is becoming white, nor is it purely 1275,1
white at the same time as it is becoming black. For <if it is,> it will
simultaneously be and not be at the limit. But if it is not the case that
it simultaneously is <at the limit> and departs <from it>, there will
be some time interval during which the moving thing was at rest at
the limit.

After proving first for motions from opposites to opposites, i.e. for 1275,5
motions in the strict sense, that they cannot be continuous, he
generalizes the argument to cover changes too, i.e. generation and
perishing (for these are the only kinds of change that are not kinds
of motion too), proving that the same thing that was proved for
motions holds for these too. (For something that is not coming to be
a man up to now but that will come to be a man, must first be at rest
for some time interval in the form from which it comes to be a man; 1275,10
a seed changes into a man only after remaining a seed for some time
interval.[192]) For opposite changes, like contrary motions, cannot occur
simultaneously. For before it has been generated, what is being
generated will not be changing into ceasing to be, if ceasing to be is
the change from being. But what is being generated is not yet. If this 1275,15
is so, then these kinds of change will not be continuous either. For
they both have starting points and ends to which they change, and
these are opposite to one another. 'But', he says, 'there will be a time

interval between them'; he does not say 'a state of rest', guarding against the objection that he will refute a little below. For while in the case of opposite motions, what is moved is at rest in between, it
1275,20 is different for things changing to the opposites being and not being. For what is at rest is an existing thing and is at rest in something that exists, but what is not an existing thing cannot be at rest, nor can anything be at rest in what does not exist. However, for these things[193] too there must be some time interval in between the changes from either extreme to the other.

1275,25 After proving for contrary motions that they must be at rest in between, he says that also in the case of contradictory opposites such as generation and perishing, even though the change is not from contraries to contraries, nothing prevents there from being a time interval in between, even though there is not a state of rest, with the result that changes from opposites are neither continuous nor single. For in the case of contraries too we saw that it happens that a state
1275,30 of rest occurs in between them, not in that they are contraries, but in so far as they cannot occur simultaneously, which holds no less for contradictory opposites. For in both cases the reason why they cannot be continuous is the same: in neither case can they coexist with one other – neither contrary motions nor motions that are contradictory opposites. This is why both are interrupted by a time interval.
1275,35 Likewise, he declares, even if 'there is no need <for it> to be at rest in the contradictory state' (just as it was shown <in the other case that it must be at rest> in the contrary state), this 'makes no difference' with regard to the motion from contradictory opposites not being single and continuous. 'For' 'what does not exist is not at rest', nor is anything at rest in what does not exist, 'and perishing <is a change> to what does not exist', and generation is a process from what is not the relevant kind of thing. But it is sufficient
1275,40 simply if for this kind of change too there occurs a time interval 'in between'. 'For' 'in this way' 'the change' is interrupted and 'is not' 'continuous'. For in the case of contraries 'too' it 'was not the
1276,1 contrariety <of the two changes>', that was put to use 'but the impossibility of their coexistence', and this belongs no less and perhaps even more to contradictory opposites, as we were taught in the *De Interpretatione*.[194]

After saying that as regards the non-existence of continuous mo-
1276,5 tion there is no difference between contradictory opposites and contraries, even if 'there is no need <for it> to be at rest in the contradictory state', he continues 'or if a change is not contrary to a state of rest', as in the case of motions some state of rest *is* contrary to a motion. For the state of rest in the condition contrary to that towards which the thing in motion is moving was posited[195] as contrary to the motion that it is undergoing towards the contrary of

that in which it is at rest. The state of rest *beneath* was posited as 1276,10
contrary to motion upwards. So even if in the case of change, i.e. of generation and perishing, there is no change opposite to the state of rest, since what has perished is not in a state of rest, this fact too has nothing to do with the argument that proves that just as motions that are contrary are not continuous, neither are changes that are opposite. For even if there is no state of rest in between, still, they too are 1276,15
interrupted by a time interval.

After saying 'or if a change is not contrary to a state of rest' on the grounds that there is a state of rest contrary to a motion, but not so for change,[196] he says 'we must not be bothered' by this as if the axiom 'one thing has one contrary' were being eliminated if we say that both a motion and a state of rest are contrary to a motion, but we should be certain of this point, 'that' in a certain way 'the motion' 'is opposite' 1276,20
'both to the state of rest' 'and to the' opposite 'motion'. And we should keep in view that this type of opposition, in which one thing is opposed to more than one, is also found in other cases; for instance, 'the equal and moderate' is opposite both 'to what exceeds it and to what it exceeds', i.e. to both the greater and the smaller. This is also how the virtues, which are in due proportion, are opposite to both excesses 1276,25
and deficiencies. So we should not be pedantic about whether they are <opposite> as contraries, but should make use both of the fact that a state of rest is opposite to a motion if they are opposed as contraries, and also of the fact that nothing can undergo contrary or opposite motions or changes at the same time. And he cautiously says, 'the contrary motion is somehow opposite both to the motion and to 1276,30
the state of rest'. For in the previous books, where his task was to discuss contrary motions, he declares[197] that the state of rest in a given condition is opposite to the change from that condition as a privation, while the contrary motion is opposite <to it> as a contrary – the equal being opposite to the greater and less as unequal, since one thing has one opposite, and the virtues <are opposite to the vices of excess and deficiency> as what is in due proportion[198] to what is disproportional.

'It is not off the point', declares Alexander, 'but[199] useful for the 1276,35
present topic that he safeguards the argument stating that one thing has one contrary. For since he has made use of the principle that motions toward contraries are contrary, in order to stop someone from objecting to this on the grounds that one thing has one contrary but the contrary of a motion is a state of rest, and therefore a motion 1277,1
cannot be contrary to it too; whereas on the other hand if changes from contraries are not contrary, nothing prevents their changes towards one another from turning out to be identical[200] – this objection is the reason why he introduces <talk of> the state of rest at this point, recalling what was proved about it, namely, that a state of rest is not 1277,5

strictly contrary to a motion but is opposed to it in the same way as a privation is opposed to a state, whereas what is contrary to a motion is a motion. This is why contrary motions cannot be one and also why the state of rest in the contrary condition cannot occur simultaneously with the change from that condition towards its contrary. For even if the state of rest beneath is not contrary to motion from beneath, yet
1277,10 is opposed to it in the way a privation is opposed to a state, for this reason too it cannot coexist with it. For non-coexistence does not hold because of contrariety, but because of opposition, which belongs not only to contraries but also to other things. So it remains that one thing has one contrary, even if it sometimes happens that one thing is opposed to many things according to different forms of opposition.'

1277,15 **261b22-6** Further, in the case of generation and perishing [it would seem to be completely absurd if a thing that is generated must perish immediately, and persist for no time interval. And so from these cases (b25) we may obtain confirmation for the rest,] since it is in accordance with nature that the case is similar for all.

That changes from opposites to opposites are not continuous but are interrupted by a time interval he proves from 'generation and perishing', deducing it potentially as follows. If generation and perishing
1277,20 are not continuous but are interrupted by a time interval, it would be credible that the other changes from opposites to opposites are not continuous either. But in fact the antecedent <is the case>. Therefore the consequent <is so too>. He proves the conditional premise on the basis of the claim that 'it is in accordance with nature that the case is similar for all' similar changes, when the conclusion follows by virtue of the similarity. For if the fact that the changes are not
1277,25 continuous holds because of their opposition,[201] it is necessary for all opposed changes not to be continuous. But perhaps the argument here is also a fortiori: what holds in the case of generation and perishing by virtue of the fact that things that are opposites in any way whatsoever do not coexist, will hold still more of changes that are opposite by way of contrariety. That generation and perishing are not
1277,30 continuous he proves by pointing out that it is 'absurd' that 'a thing that is generated' necessarily perishes 'immediately' 'and' persists 'for no time interval' in the form for the sake of which it was generated. For even if sublunary things are continuously in flux, there is in fact a kind of stability of which things in flux are capable, so that there may be a way of separating the forms, and a limit and arrangement and cognitive comprehension of them.

<CHAPTER 8>

261b27-262a6 [Let us now state] that there can be an infinite motion [that is single and continuous, and that this is circular motion. Everything that undergoes locomotion is moved either in a circle or on a straight line or with a combination <of those two motions>, so that (b30) if either of the first two is not continuous, that which is composed of both cannot be continuous either. Further, it is clear that what undergoes locomotion on a finite straight line does not do so continuously; it turns back, and what turns back on a straight line undergoes contrary motions, since motion upwards is contrary in respect of place to motion downwards, motion (b35) forwards is contrary to motion backwards, and motion to the left is contrary to motion to the right; these are the contrarieties of place. We have previously defined (262a1) what a single continuous motion is: the motion of a single thing in a single time interval and in respect of something that is undifferentiated in species. (For there were three things: (a) the thing that is moved, such as a man or a god, (b) when <the motion takes place>, for example, a time interval, and (c) third, that in which this is;[202] a place or an affection or a magnitude or (a5) a form.[203]) Contraries are specifically different and are not one,] and the differentiae of place have been stated. 1277,35

After proving that none of the other kinds of motions or changes can be continuous, and that if in fact locomotion is continuous it must be 1278,1
the only kind of motion that is,[204] he then sets out to prove what kind of locomotion can be single and continuous. 'Infinite' signifies what is unbounded and eternal. Motions that have limits, when they turn back, are in a state of rest in between.[205] He proves that no motion in place except circular motion can be continuous on the assumption 1278,5
that 'everything that undergoes locomotion' is moved 'either in a circle or' in a straight line 'or with a combination <of those two motions>'. And by proving that if it is impossible for either of the components of the combined motion.[206] For the composite locomotion participates in each of its components. So by proving that rectilinear motion is not continuous, he demonstrates as well that combined motion is not continuous either. For even if someone were to say that motion 1278,10
composed of both a continuous and non-continuous motion can be continuous, it is clear that he will grant that it gets this property from the continuous motion, and that the motion from which it gets the property of being continuous is continuous primarily and more strictly.

He proves that rectilinear motion is not continuous on the assumption that the straight line on which the thing that is in rectilinear

1278,15 locomotion is moved is finite, which he demonstrates in the third book
of this treatise through the fact that no magnitude is infinite in
actuality.[207] And so, after making this assumption he deduces as
follows: 'what is in locomotion on a straight line' undergoes a locomo-
tion on a finite straight line; what undergoes locomotion on a 'finite'
line 'turns back'; 'what turns back' on a straight line 'undergoes
1278,20 contrary motions', and contrary motions are not single or continuous;
therefore, what is in rectilinear locomotion does not undergo a motion
that is single or continuous. Further, it is clear that what is in
locomotion on a finite line must turn back if it is going to be always
in motion and not move once over the straight line and cease from its
motion. 'And what turns back' on a straight line 'undergoes contrary
motions'. For motions from contraries to contraries are contrary, and
1278,25 in respect of place, up is contrary to down, front is contrary to back,
and right to left,[208] for these are the turning points of things that are
in rectilinear motion. But contrary motions, which are interrupted by
a time interval, are not single or continuous, as has been proved
previously.[209] That is why neither alteration, nor increase and de-
crease nor generation and perishing, which are opposite to one
another, are able to be continuous and single.

1278,30 He recalls the distinctions that were previously employed in deter-
mining a 'motion' that is 'single' 'and' 'continuous',[210] and he proves
via this consideration too the claim whose proof is currently in
question, that motion on a straight line that involves turning back is
neither continuous nor single. 'For' there are 'three things' of which
motion is constituted: 'the thing that is moved such as a man or a god'
1278,35 (of which god signifies what is eternally in motion, since this is
immortal and divine, while man is an example of things that are in
motion in a part of time). The second thing constituting motion is the
'time interval' of the motion, and 'third, that in which this is', in
respect of which the motion takes place, i.e. 'a place' if the motion is
motion in place, 'or an affection' if it is in respect of an affection and
1279,1 alteration, or generally in respect of quality, 'or a magnitude' if what
is in motion is moved in respect of quantity, 'or a form' if <it is
changed> in respect of substance, as in cases of generation and
perishing. Therefore, 'a single' 'continuous' 'motion' is 'the motion of
a single thing in a single time interval' and 'in respect of' one
'species',[211] for example, if a man should be becoming white for one
1279,5 hour or if the heaven should be revolving eternally. However, things
that are in rectilinear motion and turn back move from contrary
places to contrary places. (They move upwards from below and
downwards from above,[212] and these are differentiae of place distin-
guished in terms of contrariety.) And on account of these[213] the
motions toward them are contrary too. Of things that fall under the
same genus, contraries differ the most.

Now if the contrary motions do not turn out to fall under one
undifferentiated species, the motion of things that turn back on a 1279,10
straight line can be neither continuous nor single.

But in fact the antecedent <is the case>; therefore the consequent
<is so too>. But it is also possible that the entire demonstration
contained in this passage is just one demonstration, which goes as
follows: rectilinear motion is composed of things that are contraries
and differ in species; motion <composed of> things that are contraries
and differ in species is not single or continuous. The conclusion is 1279,15
clear. The first section of the passage[214] establishes the minor premise, while the last section,[215] which determines what a single motion is, establishes the major premise.

> **262a6-12** A sign that motion from A [towards B] is contrary to motion [from B towards A is the fact that they stop one another and make each other come to a standstill if they occur at the same time. Likewise in the case of a circle, the motion from A towards B is contrary to the motion from (a10) A towards C. (They bring each other to a standstill even if they are continuous and there is no turning back, because contraries destroy and block one another.)] But lateral[216] motion is not contrary to upwards motion.

He also brings this demonstration that motions on a straight line are
contrary to one another since what is in motion turns back, using as 1279,20
evidence[217] the fact that things that move oppositely in this way bring
each other to a standstill and are destructive of one another's motions,
which is an attribute of contraries. He says 'sign' in the strict sense,
since arguments from consequences are ones that infer from effects[218]
and are not demonstrative, as when we deduce that someone has
given birth from the fact that she has milk. For having milk is a 1279,25
consequence of giving birth. By contrast, demonstrative arguments
deduce the posterior from the prior and the effects from the causes.
At any rate, from the fact of having given birth we deduce demonstratively that she has milk. In this case mutual elimination is not prior,
but is a consequence of the contrary motions, since contraries eliminate one another. However, not all things that eliminate <one another> are contraries. A negation eliminates an affirmation, and 1279,30
things that are opposites by way of the other types of opposition[219]
eliminate one another even though they are not contraries. For
<mutual> elimination is not the cause of being contrary, but a
consequence of it. This is precisely why it is a proof from a sign.[220] In
fact not only do things that move oppositely on a straight line make
each other cease from their motion; things moving oppositely to one
another on a circular line make one another cease from their motion

1279,35 too, though they do not move oppositely in the way that things
<moving oppositely> on a straight line do, beginning from contraries,
one of them from A, which is up, say, and the other from B, which is
down. In this case they begin from the same place and are not moving
1280,1 towards the same parts of the circle, for example, if both were to begin
from point A, which is on the circle, and one of them were to move
towards the part of the circle labelled B and the other towards the
part labelled C.[221] Things that move in this way make each other come
to a standstill when they meet.[222] On straight lines the limits are
1280,5 contraries, but not so on a circle, because there are no limits on it.
However, the things that move oppositely on this too come to a stop
'even if' the motions 'are continuous' 'and' there occurs 'no' 'turning
back' as on a straight line. For the things moving oppositely do not
stop one another on account of the turning back, but because of their
opposite motion. After using the fact that reverse motions eliminate
1280,10 one another as a sign that motions on a straight line are contrary, he
confirms this from the fact that motions that are not opposite and do
not move oppositely do not eliminate one another, but are merely
different from one another, as motion that takes place from right to
left and lateral motion in general <is not opposite> to movement from
down to up, although they are different and are not continuous or of
the same species, but since they are not opposites they can coexist.
1280,15 But reverse motions on a straight line do eliminate one another,
which is more than just differing in species.

262a12-b8 [But that motion on a straight line] cannot [be continuous is] above all evident [because when it turns back it must stop – not only on a straight line (a15) but also if it is undergoing locomotion over a circle. For undergoing circular locomotion and undergoing locomotion over a circle are not the same thing; for there can be cases where a thing continues being in motion and others where it turns back again when it comes to the same <point> from which it set out. That it must come to a stop is confirmed not only by perception but also by argument. This is the starting point: there being three things, (a20) beginning, middle and end, the middle is both in relation to each of the other two and in number it is one but in definition two. Further, there is the distinction between being potentially and being actually, so that any point on a straight line between the two extremities is a middle potentially but not actually, unless <the thing that is in motion> divides <the line> there and after stopping there begins (a25) to move again. In this way the middle becomes a beginning and an end – the beginning of the latter <motion> and the end of the first. For example, if A is in locomotion and comes to a stop at B and again undergoes locomotion towards C. But when it undergoes locomotion continuously, A can neither have come to be at point B nor have departed <from it>, but it can only be there at an instant, not in any time interval, except that of which the instant is a dividing point, (a30) namely, the whole time interval. (If anyone will suppose that it has come to be there and has departed, A will always be coming to a stop when it is undergoing locomotion, for (262b1) A cannot have come to be at B and have departed at the same time. Therefore <it must have done so> at a different[223] point of time. Therefore what is in between will be a time interval. And so, A will be at rest at B. And likewise at the other points too, since the same argument holds for them all. (b5) But[224] when A, the thing undergoing locomotion, uses B, the middle, as both an end and a beginning, A must stop because it makes <the point> two, just as if one were to do so in thought.) But it has departed from point A as its starting point] and has come to be at C when it finishes <its motion> and stops.

He supplies another demonstration too, from which he declares that it is not only more <evident> than from the earlier arguments, but 'above all evident' that 'motion' that takes place over and over 'on a straight line' 'cannot' 'be' 'continuous' since what is in motion turns back. He proves this from the fact that everything that 'turns back' must first come to a stop. 'Not only' is this necessary for things that turn back 'on' the 'straight line', but also if something that is moving on a circle comes to some point and turns back from there, it too must first stop because what is in motion is unable to turn back without first stopping. After saying that 'not only on' the 'straight line but also if it is undergoing locomotion over a circle' (i.e. on a circle) the same thing happens if it turns back as does on a straight line, he shows what the difference is between 'undergoing circular locomotion and undergoing locomotion over a circle'. What continues undergoing locomotion always in the same direction, and has its circular form in virtue of the motion itself, undergoes 'circular' locomotion, while what undergoes motion on a circle from a given point and then turns back and goes back 'to the same <point>', undergoes locomotion 'over a circle'. This makes it clear that circularity is not a feature of the motion, but of the path of the motion.

That what turns back 'must' first 'come to a stop' he proves in two ways: both from 'perception' and from 'argument'. Each person must become a judge of his own perception, but after proposing to prove it by argument, he first distinguishes ('there being' these 'three things, beginning, middle, and end') the status of 'the middle' 'in relation to each of the other two' extremities (that it is the beginning and end,

<respectively>), and what sort of middle is a middle in actuality and
1281,1 what sort <is a middle> potentially, and <he proves> that what is moving from the beginning towards the end comes to a stop at what is in actuality a middle by coming to be at it and departing from it. For then strictly speaking it is in actuality a middle, and an end in relation to one of the extremities and a beginning in relation to the other. These things being assumed, he proves generally that what is
1281,5 in motion must come to a stop at a point at which it comes to be and departs, using that point as both beginning and end. And what turns back does precisely this. For the point at which it turns back is a point in actuality, not potentially.[225] And so it comes to a stop at the point at which it turns back. For if it comes to a stop also at the middle points when they are occupied actually and not potentially, this is all
1281,10 the more reason to think that it comes to a stop at what is a point in actuality alone. It necessarily has to stop at a point which it must use as end and beginning. And such is the end of the straight line, where it turns back.

He first proves that what is in motion on a straight line does not come to a stop at every point on the straight line. This is the best way for him to prove that being at rest is a property[226] of things in
1281,15 rectilinear motion when they turn back. For if what undergoes locomotion comes to a stop at every point, in the first place it will not come to a stop at the point of turning back any more than anywhere else on the straight line, since there is a point everywhere. Also, the same thing would hold for circular motion, since there are points everywhere there too. In this way no locomotion would be continuous. And so, what undergoes locomotion does not come to a stop at every
1281,20 point, since in what is continuous the points do not even exist in actuality. For this reason, they are not middles in actuality either. He proves that the points in what is continuous are not in actuality middles of the extremities as long as it is taken as continuous, by specifying what are, strictly speaking, middles of what. For things that have a beginning, end and middle, 'the middle' becomes the other <extremity> 'in relation to each' of the parts: the end in relation to
1281,25 the beginning, and the beginning in relation to the end. 'In this way' it 'becomes a beginning and an end' according to the argument. 'In relation to each of the other two' <extremities> it becomes 'both', when each of the extremities is also taken in both ways, as both beginning and end.[227] For then the middle becomes in one way the end in relation to the beginning, and in another way the beginning in relation to the end, e.g. C in relation to A, and likewise in relation to B.[228]

1281,30 Alexander does not think we should understand 'the middle is both relatively to each' in this way. He thinks it false that it becomes both in relation to each, but holds that because when the middle is opposed

to each of the two it is the other, it becomes both. But if each of the extremities is taken as both – as both beginning and end (for example, on a straight line from up to down, the top is the beginning in relation to what moves from it, and the end in relation to 1281,35
what moves towards it, and the same goes for the bottom) – the middle too will be in definition both in relation to each, for 'in number it is one'.[229] As long as a straight line is taken as one and continuous, the points on it between the extremities exist only potentially, and since they exist potentially they will not be middles strictly speaking. For what exists in actuality is all that exists 1281,40
strictly speaking. The middle comes to be in actuality when what is in motion on a straight line divides it at one of the points on it by 1282,1
stopping there.[230] The point at which the moving thing comes to a stop comes to be in actuality both a point and a middle, and 'in relation to each' of the extremities it comes to be 'both' 'in definition'. So as soon as what is undergoing locomotion stops there, there is also a middle in actuality, which is a beginning and an end, and as soon as there is a middle in actuality, there is also a stopping at it. The same thing would not be a beginning and an end in relation to each of the extremities, that is, a middle in actuality for what is in motion, if there did not come to be a stopping at it and if the continuous were not divided by the stopping, nor would there come to be a stopping at 1282,10
it if it were not taken as a beginning and an end, i.e. a middle in actuality. And so <being> a middle in actuality and the stopping at it are consequences of one another, for the stopping of the thing that is in motion proves that what is 'one' 'in number' comes to be 'two' 'in definition', whereas what is in motion continuously and stops nowhere would not be said to have come to be at anything or to have departed from it (for it did not use anything as an end and a beginning), nor would it make anything a middle in actuality as long as the straight line remains continuous and undivided. For what 1282,15
comes to be at something must be at rest at it for some time interval, since it is not possible to come to be at something and to depart from it at the same instant (as he will say shortly below[231]), or both come and go away. It would both be and not be at the same thing at the same time. And so, <it does this> at different instants. But if this is so, and if between any two instants there is a time interval, during 1282,20
this <time interval> it will be at rest at that at which it has come to be and has not departed.

But what is in motion 'continuously' 'can' not 'have come to be' 'at' any of the points on the straight line; it would be said to be at them not 'in' a 'time interval', but at the instants in the time interval, and these are not a time interval but limits of a time interval. Just because something is at something it does not need to be there in a time interval too. For if it were at it in some time interval, it would have 1282,25

to be at rest there. So for each instant there is some corresponding point on the straight line at which it is. And so, 'A', which is undergoing locomotion, <can> 'neither have come to be' at B 'nor have departed' <from B>, 'but' <it can> 'only be' at it 'at an instant', not 'in' a 'time interval'. For the intermediate instants in a time interval are analogous to the intermediate points on a straight line. And so, what
1282,30 is in motion continuously is only potentially at the intermediate points and the intermediate instants.

After saying that there is no time interval in which what is in motion continuously on a straight line is at any of the points on it, he adds 'except that of which the instant is a dividing point'. This is equivalent to saying that even though what is in motion continuously
1282,35 on a straight line is not in the strict sense and per se at any of the points on it during a time interval, however, more generally and incidentally, what is in motion on a straight line can be said to be at each of the points during the time interval <containing> each of the
1283,1 instants at which it is at any of the points – that it is within the time interval during which it is going completely through the straight line. For by being at the instant, which is a division of the time interval, it could also be said (though not strictly speaking) to be in the time interval within which the instant is. For the instant is not a part of
1283,5 the time interval, but is a cut of it, at which the time interval is cut and has the capacity to be cut. For just as a line is divided at one of its points, which is without parts, so also a time interval <is divided> at an instant.[232]

After saying that what undergoes locomotion continuously neither comes to be at nor departs from any of the points on the line, he proves this from the principle that whatever comes to be at or departs from
1283,10 any point must come to a stop, and so, what is moving continuously will always be stopping at all the points, which is impossible. That what comes to be somewhere and departs comes to a stop is clear because it is impossible for anything both to come to be in something and depart <from it> at the same instant, for it would simultaneously both be and not be in the same thing. Therefore <it must do so> at different instants. But if this is so, and there is a time interval between all instants[233] (since instants are not next to one another[234]),
1283,15 then during the time interval in between it would be at rest at the point at which it is posited to have come to be and departed. But if it does this at each of the points on a straight line (for why should it do so more at one point than at another?) what is in motion continuously on a straight line would be at rest on it – and what could be more irrational than this? He says 'therefore at a different point of time',
1283,20 meaning the same as 'in a different instant'. For an instant in a time interval is analogous to a point on a line.

After discussing the situation of what is in motion continuously on

a straight line and proving that it does not come to be at and depart
from any of the intermediate points but is at them only at an instant,
he says,[235] 'but when' 'the thing undergoing locomotion', 'A', 'uses' one 1283,25
of the intermediate points, 'B', as a 'middle', i.e. as 'a beginning' and
'an end', that is when 'it must stop' at it. But it *must* use the point at
which it turns back in this way. It 'must' therefore 'stop' at the point
at which it turns back. And this is what was proposed at the beginning
to prove. Further, 'because it makes <the point> two, just as if one
were to do so in thought' is equivalent to saying 'for this is how it
makes what is numerically one two, by its stopping'. It does not divide 1283,30
the point, which is indivisible, but uses it twice, once as an end and
once as a beginning, by coming to a stop at it. While showing how it
makes one two, he adds 'just as if one were to do so in thought'. What
becomes two in this way was posited to become two in definition. For
he says 'and in number it is one, but in definition two', i.e. in thought. 1283,35
For just as the person who uses the same thing as two notionally,
taking it once as an end and once as a beginning, has not divided what
he has used by using it in both ways, but in a way has made one thing
two notionally – the same holds also for what stops at it and uses it
both as beginning and as end. After showing how what is in motion
continuously makes use of the points between the extremities – not 1284,1
using the same point as an end and a beginning, but being at each of
them at an instant and not in a time interval, he adds that it does not
use the point at the beginning of the straight line or the point at the
end in the same way. For the intermediate points on the line are
potentially, and, as we saw, in order to come to be in actuality they
need something that will divide the straight line by its stopping. But 1284,5
the points at the extremities are in actuality. This is in fact why what
is in motion on a straight line must depart from the beginning and,
when it has moved over the whole line, come to be at the end. And if
it turns back it will have to depart from the point at which it has come
to be. But what has come to be at something and has departed from
it must have been at rest at it for the time interval between the instant 1284,10
at which it came to be at it and the instant at which it departed from
it. And so everything that turns back must be at rest at the endpoint
at which it turns back. Now this is what is being proved. But while
bringing the argument towards the conclusion that what is in motion
on a straight line departs from the beginning and comes to be at the
end, before positing the consequent of this, that what turns back from 1284,15
the end must first be at rest there, he makes use of what he has just
said and proved to solve a certain puzzle that seems to arise.

262b8-17 Therefore this is the response that should be given to the puzzle as well. [For there is the following puzzle: (b10) if line

E is equal to line F and A undergoes locomotion continuously from the extremity towards C, and if A is at point B at the same time as D is undergoing locomotion from the extremity F towards G uniformly and with the same speed as A, then D will have come to G before A does to C. For what sets out and departs (b15) earlier must arrive earlier. Therefore[236] A has not come to be at B and departed from it at the same time, which is why it is delayed. For if <it it has come to be and departed> at the same time, it will not be delayed,] but <if this is to happen> it will have to come to a stop.

1284,20 After proving that anything that comes to be at something and departs from it must stop at it, he brings up a puzzle that is directed against this view and he solves the puzzle by making determinate what was assumed indeterminately in it. He puts the puzzle briefly and unclearly; this is its main point. If what comes to be at something
1284,25 and departs from it must be at rest at it for some time interval, it will follow that things that are in motion with equal speed on equal straight lines and that began to be in motion at the same time, do not go through the equal straight lines in the same time, which is absurd. And so the antecedent is absurd too – that what comes to be at something and departs is at rest for some time interval. The aforementioned absurdity follows if one of the things that are in motion with equal speed is assumed to have come to be at one of the
1284,30 intermediate points and to have departed <from it> and the other is assumed not <to have done so>. For if there are two things, A and D, that are in motion with equal speed on two equal lines, E and F, and A, <which is in motion> on EC, is assumed not only to be at one of the points between E and C, e.g. B, but also to come to be at it and depart <from it>, then if what has come to be at something and departed <from it> must be at rest for some time interval between
1284,35 the two instants – that in which it comes to be and that in which it departs – it will follow that D moves over FG more quickly although
1285,1 it has equal speed to A and has begun its motion at the same time. This followed as the result of assuming in the case of A both that it has come to be at one of the points between E and C and that it has departed <from it>, assuming moreover that no such thing happens with D, but allowing its motion to be continuous. And so in fact the points between F and G will be potentially, and D, which is in motion
1285,5 continuously, is only *at* the points but does not come to be <at them> or depart <from them>. Now then, if also A, which has equal speed to D, is assumed to be in motion continuously, it no longer follows that D gets through all of FG faster than A does through EC. For both of them get through all <of their respective lines> similarly, being at the intermediate points, which are potentially, and it is not the case

that either of them comes to be at any of the points and departs <from it>. For it is not necessary that what is at something also comes to be at it and departs <from it>, as the argument that poses the puzzle 1285,10
assumed for one of the moving things, A, and inferred that it had not been truly demonstrated that what comes to be and departs is at rest for some time interval in between. Aristotle, taking lines EC and FG, <calls EC 'E' for short, and FG>[237] 'F'.

Alexander does a good job of clearing up the unclear claim about D, 'for what sets out and departs earlier must arrive earlier': 'This 1285,15
does not mean, as one might suppose, that A and D did not begin to move at the same time. For, in fact, not only are they posited as beginning at the same time, but also if they did not begin at the same time the conclusion would not be at all absurd. What this means is that A and D, which began together from E and F, undergoing locomotion with equal speed, proceed, A towards B and D towards 1285,20
some other point analogous to B on the straight line FG and equally distant from F as B is distant from E. But from these points (B and the analogous point on the straight line FG) A and D no longer set out at the same time, because A, which was hypothesized to come to be at B and depart <from it>, must also have been at rest at it, while 1285,25
D, which was not hypothesized to come to be at the point analogous to B, but only to be at the points that are potentially, while it moves continuously on the straight line, need not be at rest. And so, after arriving at the analogous points at the same time, the one that is at rest and the one that is not at rest do not set out at the same time, but the one that is not at rest <sets out> first. And since they have equal speed, the one that departs earlier from the middle (if that 1285,30
happens <to be the point in question>), must arrive at the end earlier.' Aristotle himself states the reason in the words, 'therefore A has not come to be at B and departed from it at the same time'. For 'A has not come to be at B and departed from it at the same time' because it in fact was at rest there. Therefore it must be delayed. For 'if <it has come to be and departed> at the same time, it will not be delayed'. But it must be delayed because of being at rest at B. The preceding 1285,35
statement, 'and if A is at point B at the same time as D is undergoing locomotion from the extremity F towards G', also contains a certain unclarity. How could A be at B at the same time as D is undergoing locomotion from the extremity F, if in fact A began from E and D from F at the same time? But it seems that he 1286,1
means: at the time when A is at rest at point B, D, which is moving continuously from the extremity F towards G, is undergoing locomotion and is not at rest anywhere as A is according to the hypothesis.

262b17-22 Therefore we must not suppose that when A came to be at B, [D was simultaneously moving from the extremity F (for if A will have come to be at B, (b20) its departure will also occur, but not at the same time); rather, it was there at a cut of time and not in a time interval.] Now in this case, when the motion is continuous, it is impossible to speak in this way.

At this point begins the solution, which argues from previous results. He says that we should not posit that A, which is in motion on EC, has come to be at B at all, which would result in saying, along with the person who posed the puzzle, that 'when A came to be at B, D was simultaneously moving from the extremity F'. 'For if' 'A' had come to be 'at B', it would also have really departed <from it> 'but' <would> 'not' <have done> both 'at the same time', but there would have been some time interval and state of rest in between, and the argument that posed the puzzle would be strong. But if the motion of A were continuous just as that of D is, A will not be at B in a time interval nor will it come to be at B at all, but instead of coming to be <at B> it will be <at B>, and instead of being <at B> in a time interval, it will be <at B> at an instant. For what moves continuously is at each of the points (that for the time being are potentially), and is there not in a time interval but at an instant, which is the limit of a time interval. This is why it neither comes to be at any such thing nor departs <from it>, so that in the case of continuous motion that takes place on a continuous magnitude, it is impossible to say that what moves on it comes to be at any of the points on the continuous <magnitude>, because it is not in actuality, but only potentially, that any of these points is.

262b22-8 But in the case of something that turns back, [we must speak in this way. For if G is undergoing locomotion towards D and then turns back and undergoes locomotion downwards again, it has used the extremity D as an end (b25) and a beginning – the one point as two. This is why it must have stopped. It has not come to be at D and departed from D at the same time,] for in that case it would simultaneously be and not be there at the same instant.

After proving how in the case of what moves continuously it is not possible to speak of coming to be and departing, he infers that in the case of what comes to the end and turns back it is necessary to say that it has come to be at the end and has departed <from it>. For the end of the straight line is no longer potentially, as we saw is the case for the intermediate points, but is in actuality. This is why it must come to be at the end and depart from it. He proves this again through

an illustration with letters, changing the letters (according to Alex-
ander), now making G the thing that moves, where this was pre- 1286,30
viously D, and D the end of the straight line, where this was
previously G. He says that when after reaching the end, D, G turns
back again from it, 'it has used' 'the one point as two' (as 'a beginning'
and an end), and we saw that this is what it is to come to be and
depart. This is why it must necessarily be at rest for the time interval
between the instants in which it came to be at the end, D, and in which 1286,35
it departed, turning back from it. This is how Alexander understood 1287,1
the change of letters. But perhaps just as he [Aristotle] previously
has placed D, the thing that moves, at the extremity of the straight
line FG in order to take it as moving from the extremity, so now too,
keeping it at the extremity, he employs <that letter> for the extremity
of the straight line, while G, which was formerly the end of the 1287,5
straight line, he uses for the beginning of the straight line, that is, he
uses it for the moving thing D, in order that by making the turning
back <be> at the beginning he might take the beginning simultane-
ously as an end too, since if in fact he had wanted to take the end
simultaneously as a beginning too, he would have moved D towards
G. That it is not possible for it to have come to be at D and to have
departed from it at the same time he proves by saying, 'for in that 1287,10
case it would simultaneously be and not be there at the same instant.'
For as for what has come to be at something and departed <from it>,
when it has come to be at it, it is at it, and when it has departed and
gone away from it, it is no longer at it. So if someone were to say that
something has both come to be at something and has departed <from
it> at an indivisible instant, he would be saying that simultaneously
and at the same instant something both is and is not at the same 1287,15
thing – it is in that it has come to be <at it>, and it is not in that it
has departed <from it>.

> **262b28-263a3** But in fact we must not apply the solution of the previous puzzle. [We cannot say that G is at D at (b30) a cut of time, but that it has not come to be or departed <from D>. For it must reach an endpoint that exists actually, not potentially. Now although the points in the middle exist potentially, this one does so actually; it is an end when considered from below (263a1) and a beginning when considered from above. And therefore it is related in the same way to the motions. Therefore a thing that turns back on a straight line must stop.] Therefore there cannot be continuous eternal motion on a straight line.

While solving the puzzle that states that if A will have come to be at
B and have departed <from B>, and if what comes to be and departs 1287,20
comes to a stop in between, it follows that D, which has equal speed

to A, goes through the straight line FG, which is equal to EC, more quickly than A <goes through> EC, which is absurd (for things with equal speed go through equal lines in an equal time interval) – while solving this puzzle he says that A does not come to be at B and depart
1287,25 <from B> in a time interval, but 'it was' at it 'at a cut of time and not in a time interval', because in what is continuous the points do not exist actually but potentially. Therefore in the case of things that turn back we should not, he declares, say the same thing, that it is 'at D' at an instant, and yet has 'not come to be' at it or 'departed' from it as well. For this <point> exists already 'actually' and not 'potentially', like things in between the beginning and the end. He brings as
1287,30 confirmation of this the fact that what is moving to something must reach it at some time, and that it[238] is already an end actually, not potentially, as we saw that the intermediate points are[239] while the thing in motion is moving to the end. They are ends potentially, because the thing in motion could have made them too ends actually by stopping at them and dividing the continuity. For what is in motion
1287,35 continuously over a finite continuous straight line uses none of the intermediate points as if they existed actually. But at some time it must in fact arrive at the end and come to be at it actually, the end too existing actually, and not potentially, like the points before it. For
1288,1 as the points in the substrate are, so also does what moves continuously on the substrate come to be in them: potentially in those that are potentially, and actually in those that are actually.

Alexander knows another reading as well, which he in fact puts down as his preferred reading, and which is found in the majority of
1288,5 manuscripts. I shall stop after expounding it, since I am commenting on a different one. The other one[240] goes as follows: 'For it must reach an end that exists actually, potentially'.[241] This would mean that what is in motion, being potentially at the end as long as it is in motion, must reach an end that is already an end in actuality, when it comes to be at it.[242]

But what is coming to be at what is an end in actuality, is coming
1288,10 to be at an actual point during a time interval and not in a cut of time. For while a thing in motion that turns back from it[243] is <still> moving toward it, it must use it as an end, but when it is moving away from it, it must use it as a beginning, and there must come to be a state of rest between the end and the beginning. But, he says, as the straight line on which the moving thing moves has the same <point> as end
1288,15 and beginning, so too must motion up 'from below' and motion down 'from above' <have the same point as end and beginning>. For of this too the top is end and beginning.

After proving in many ways what he proposed, that things that turn back must come to a stop in between, and that things moving on a straight line must turn back unless they stop when they come to be

at the limits, in concluding his remarks he briefly sets out the whole 1288,20
deduction as follows: things in motion permanently on a straight line
turn back; things that turn back come to a stop in between; things
that come to a stop in between do not undergo continuous motion;
therefore, things in motion on a straight line do not undergo continu-
ous or eternal motion. The two premises saying that what is in motion
on a straight line turns back and that what turns back comes to a 1288,25
stop, he includes in the phrase 'therefore a thing that turns back on
a straight line must stop', and skipping over as obvious the premise
that things that come to a stop do not undergo continuous motion, he
infers the conclusion, which says 'therefore' 'motion' 'on a straight
line' 'cannot be' 'continuous' and 'eternal'.

263a4-b3 We must respond in the same way also against those 1288,30
who propound (a5) Zeno's argument,[244] [that if we must always cross half the distance, but there are an infinite number of these, and it is impossible to get entirely through an infinite number of things. Some put this same argument differently, holding that at the same time as the thing is moving we can first count the half motion as it comes to be at each halfway point, so that when it has got through the whole motion (a10) it follows that we have counted an infinite number, which is admittedly impossible. Now in our first discussion of motion we put forward a solution that depended on the fact that a time interval contains in itself an infinite number of things. For it is not at all absurd if someone traverses an infinite number of things in an infinite time interval; and in fact the infinite belongs to both the length and (a15) the time interval in the same way. This solution is adequate for the person who propounded the argument, since he was asking whether it is possible to get entirely through or to count an infinite number of things in a finite <time interval>.[245] But with regard to the matter and the truth, it is inadequate. For if someone leaves aside the length and the question whether it is possible to get entirely through an infinite number of things in a finite (a20) time interval, and inquires this with regard to the time interval itself (for the time interval contains an infinite number of divisions), this solution will no longer be adequate; rather we must state the very truth that we asserted in the previous discussion. For if someone divides a continuous motion into two halves, he uses one point as two. For he makes it a beginning (a25) and an end. But this is exactly what both the person who counts and the person who divides it into halves are doing. However, if he divides it in this way, neither the line nor the motion will be continuous. For a continuous motion is over

> something continuous, and in what is continuous there are an infinite number of halves, although they exist potentially and not in actuality. But if a person makes them exist in actuality, he will not make (a30) the motion continuous, but will make it stop, which is evidently what happens when someone counts the halves: for he must be counting (263b1) the one point as two. For it will be the end of the one half and the beginning of the other,] if he is not counting the continuous motion as one, but as two halves.

After proving that the points between the extremities on what is continuous exist potentially and not actually, and consequently that 1288,35 what is in motion continuously on what is continuous neither comes to be at any of them nor departs from them, since if it does this it has divided the continuous, he says that we must use this consideration 'also against those who propound Zeno's argument'. Concerning the arguments Zeno of Elea employed in training his interlocutors, be- 1289,1 lieving that he was proving that there is no motion, in the sixth book of the present treatise he [Aristotle] said that they are four, stated them, and raised an appropriate objection to each.[246] Having taken up one of those four arguments, he now reminds us of the refutation that was advanced then, and says that it will be refuted better, in a 1289,5 more naturally fitting way, on the basis of what has just been proved. The argument stated by Zeno which he now recalls is the following. 'If there is motion, there will be something that has traversed an infinite number of things in a finite time interval. Because dichotomy ad infinitum is possible in everything that is continuous, there will be an infinite number of halves because every part of it has a half. So what has moved over a finite line will have traversed an infinite 1289,10 number of halves in the finite time interval in which it traversed the finite line. But assuming in addition the opposite of the consequent in the conditional premise, that it is not possible for anything to get entirely through any infinite number of things in a finite time interval because it is not possible to get entirely through an infinite number of things at all, he eliminates the existence of motion.'[247]

Thus Zeno. But he [Aristotle] says that some put it differently, 1289,15 saying, 'If there is motion, since there are an infinite number of halves in everything that is continuous, what is moving over what is continuous can count each individual half. But if this occurs, when the moving thing has traversed the finite magnitude, the person who is counting will have counted an infinite number of halves. So if this – counting an infinite number – is impossible, what this is a consequence of is also impossible, and we have seen that it is a consequence 1289,20 of the existence of motion.'

On the previous occasion he objected to Zeno's argument on the

grounds that it infers from the existence of motion the absurd conse-
quence that what is moving goes completely through an infinite
number of halves in a finite time interval; he said that it does not go
completely through the infinite number <of halves> in a finite time
interval, but if it does go completely through them, the time interval
is infinite too. As the magnitude has an infinite number of halves, the
time interval does too: both of them are similarly continuous and 1289,25
divisible ad infinitum. Therefore it will not traverse the infinite
number of things in a finite <time interval> but in an infinite number
<of times>. Now this refutation, which he employed on that occasion,
he says is sufficient against someone who asserts as an absurd result
that it goes through an infinite number of things in a finite time
interval (for it is not in a finite time interval <that it goes through>
the infinite number of things, but, if it does so at all, in an infinite
number <of time intervals>, since 'the infinite' occurs in both the 1289,30
magnitude 'and the time interval' 'in the same way', so that 'noth-
ing'[248] 'absurd' follows). 'But' in fact 'with regard to the matter' itself
'and the truth' according to the matter, it is not yet sufficient, and it
does not truly confront <the argument>. For objecting to a false
argument and establishing how the truth is are not the same thing.
We can even hinder the progress of a false argument by the use of
some falsehoods. But that this reply is insufficient for the solution of 1289,35
the puzzle that arises from the facts of the matter, he establishes after
changing the question. 'For if someone', he declares, 'leaves aside the
length and the question whether it is possible to get entirely through
an infinite number of things in a finite time interval' or to count them,
and were to ask the same things 'with regard to the time interval
itself' on its own, (for if he takes some finite time interval in which
the moving thing has moved, since it contains an infinite number of 1289,40
halves – because it is continuous and divisible ad infinitum – he will 1290,1
be postulating that it goes through the halves and counts them; if it
goes through the whole, <it> also <goes through> each of its parts, so
that if there is an infinite number of halves, when the moving thing
has traversed the whole it will have traversed and counted an infinite
number of parts of a time interval in a finite whole), the same puzzle
arises even if someone only thinks that the halves of the magnitude 1290,5
itself are being counted when the moving thing is traversing it, even
if we do not make use of the time interval as Zeno did. This is why
Aristotle seems to have added to Zeno's argument the person who
puts the same argument differently, from considerations of counting,
because the refutation advanced at that time, which proves that the
time interval is infinite in the same way as the magnitude, was 1290,10
inadequate against such a question, just as it was inadequate against
a person who puts it from considerations of time or magnitude alone.
For, he says, the refutation previously advanced, which says that both

magnitude and time are infinite in the same way, will no longer apply to a person who poses the puzzle in this way: anyone who states this <refutation> will be saying nothing against the claim. For the one
1290,15 who is posing the puzzle made use of the infinity of the segments and the claim that an infinite number of things can be got completely through, even though what is moving is going through them and counting them. This is why, he says, that refutation should be considered adequate against Zeno and those who have put the argument in that way (for it does not go through the infinite number of things in a finite <time interval> but in an infinite <time interval>), but for the argument on the same topic that can be put differently,
1290,20 and for the truth, and for proving the nature of the matter <it should be considered> inadequate.

Rather, this puzzle too must be solved by what was said shortly above. It is this: everything that is continuous contains in itself potentially, not in actuality, the things at which it can be cut – the line <contains> points and the time interval <contains> instants. A person who divides them into halves and counts them takes them in
1290,25 actuality. This is how he divides the continuous and 'uses one point as two. For he makes it' both 'a beginning' and an end, just as we previously saw that the person moving on the straight line and turning back does. And the way in which that person no longer moves as on a continuum or undergoes a continuous motion, but uses the same point as both a beginning and an end, coming to be at it and in turn departing from it, is also the way in which the person acts who
1290,30 counts each half and always says that he is moving over the first half; he is dividing the continuous and taking certain middles on it in actuality. But being divided it no longer remains continuous, for neither the magnitude nor the motion nor the time interval is still continuous when divided. So the one who makes such a division no longer proves that there are an infinite number of things in what is continuous and finite, for what has been actually divided does not remain continuous.
1290,35 And so those who attempt in this way to prove – either by dividing or by counting – that there are an infinite number of halves in what is continuous, through the very proof eliminate the continuity in which they choose to establish the infinite number of things. For nothing continuous is still continuous when it has been divided.

Someone will say that if everything continuous is divisible ad
1290,40 infinitum and every part of the continuous is continuous, then by
1291,1 dividing the continuous a person does not eliminate the existence of the continuous. In refuting this objection he [Aristotle] states how continuous things are divisible ad infinitum: it is because they contain the infinite number of things potentially, not in actuality, because they can accept a cut at any one of their parts whatsoever. Now what has undergone a continuous motion has moved over things that

are infinite potentially and not in actuality, and 'in the continuous' time interval too there are 'an infinite number of halves' 'potentially' and not in actuality. But if potentially, then not as things that are being counted or partitioned; these are features of something that is already being actually divided and is no longer continuous. So the person who is doing the dividing will stop the cuts and will take a finite number of them. For the halves in the continuous thing were seen to be infinite potentially but finite in actuality; the infinite number of them in what is continuous is not potentially in such a way that they can be taken as infinite in actuality too. For the infinite number of them is not in actuality but in such a way that they can be cut ad infinitum. But the existence of cutting ad infinitum is in always being potentially and in coming to be and in being cut, not in having been cut. As with an athletic contest and a day,[249] it is not possible to take <the infinite things> all together, but their existence is in coming to be. That the one who is counting the halves divides what is continuous into a finite number of things and does not preserve it as one and continuous (although it was because it was assumed to be such[250] that it was potentially divisible ad infinitum), he proves through the fact that a person who does this is 'counting' 'the' 'one point' 'as two' – as the end of the former part and as the beginning of the latter one. A person who divides something that is whole and continuous into halves takes the same point in both halves. But in fact, a person who does this and uses the one point as two is dividing what is continuous, as has been proved for things in motion, and therefore the one who counts the halves is dividing what is continuous. But if he does not preserve the continuous as continuous but is dividing it into halves, he is not taking the motion as continuous or one either.

> **263b3-9** Therefore to a person who asks [whether it is possible to get entirely through an infinite number of things either in time or in length, we should say that in a way it is (b5) and in a way it is not: if they exist in actuality it is not possible, but if they exist potentially it is possible. For a person who moves continuously has incidentally traversed an infinite number of things, but not without qualification. For it is an accidental attribute of a line to be an infinite number of halves,] but its essence and being are different.

He then teaches us generally how to respond 'to a person who asks whether it is' at all 'possible to get entirely through an infinite number of things either in time' – when traversing something in a finite time interval a thing may be said to be going completely through the infinite number of instants in it – 'or in length' – when having traversed a finite length it may be said to have traversed the infinite number of points on it. 'To a person who asks' this he says we must answer that in one way it is possible and in another it is not; for things that are infinite potentially it is possible, but for things that are infinite in actuality it is impossible because in the first place there do not even exist things that are infinite in this way.

1291,35 'How it is possible to traverse things that are potentially infinite he indicates', says Alexander, 'in the words "for a person who moves continuously" and following. For since the line over which the thing that has moved has moved, has as an accident the attribute of being potentially divisible ad infinitum and of having an infinite number of halves potentially, that which has got entirely through the line "has"
1292,1 "incidentally" "traversed" the potentially "infinite number" of halves, i.e. it has traversed the line to which the attribute of having "an infinite number of halves" potentially belongs incidentally. For as we have seen, the "essence" "and being" of the line do not consist in the attribute of having an infinite number of halves, since it would no longer be the case that what has moved over the line has traversed
1292,5 an infinite number of halves incidentally, but <it would have done so> per se, if in fact it has moved per se over a line whose essence consists in having an infinite number of halves. But since this belongs to a line not as the essence of line, but is an accidental attribute, while breadthless length belongs to it as the essence of line, that which has traversed the line has traversed breadthless length per se, and has traversed incidentally all the accidents of the line, e.g. if the line were
1292,10 white, <it would have traversed> the white thing <incidentally>, and likewise for each of the other <accidents>, which as we have seen include the attribute of having an infinite number of halves potentially.' This is how Alexander expounds the present passage, in these very words. But in the first place one might notice, I think, why he adds 'potentially' to 'it is an accidental attribute of line to be an infinite
1292,15 number of halves'. And not only did he write the passage that way,[251] although the manuscripts that have come to me do not have 'potentially' added, but he also expounded it as follows. For, he says, 'after getting entirely through the line it has incidentally traversed the potentially infinite number of halves, i.e. it has traversed the line to which the attribute of having an infinite number of halves potentially belongs incidentally.' But how does he mean that the attribute of
1292,20 being potentially divisible ad infinitum is an accident of the line, if in fact the line remains continuous and being divisible ad infinitum is the definition of everything that is continuous? We should notice too that Aristotle does not say as Alexander does that the line has the attribute of being divisible potentially ad infinitum as an accident, but that 'it is an accidental attribute of a line to be an infinite number of halves'.

But with reference to the passage of Aristotle we need to pose the

puzzle how a person in motion per se continuously over a finite line 1292,25
has traversed an infinite number of things incidentally. For what is
incidental, even if it is incidental, nevertheless is and has existence.
For what is in motion incidentally or is being heated incidentally,
exists too, and the one is moving and the other is being heated even
if incidentally. A sailor in a boat is being moved by the wind inciden- 1292,30
tally, because he happens <to be in> what is being moved per se, i.e.
the boat, but nevertheless the man in the boat is being moved too even
if he is being moved incidentally. But the infinite number <of halves>
do not exist, <which would be necessary> in order for what is under-
going motion per se continuously over a finite line to have traversed
the infinite number <of halves> incidentally, and for it to be an
accident of the line to be an infinite number of halves. Perhaps
Aristotle is now presenting us a different kind of accident, calling 1292,35
what belongs to something potentially an accident of it. For this is
the way in which he says that the attribute of being an infinite
number of halves is an accidental attribute of a line – they are in it
potentially, even if the attribute of being able to be divided ad
infinitum is the definition of the continuous. But since each thing is
most of all what it is in virtue of its actuality, and not in virtue of the
potentiality in it, he says that the being and essence of the line is not 1293,1
in virtue of the infinite number <of halves> in it, because they are in
it potentially, but in virtue of being one and continuous, which belong
to the line in actuality. And even if we define the continuous as that
which can be divided ad infinitum, we are defining what is in actuality
in terms of what is in it potentially, as if someone were to define a 1293,5
bull as what can become a bee.[252] But if Aristotle says 'accident'
instead of 'potentially' in this way, Alexander's addition of 'poten-
tially' in the claim 'it is an accidental attribute of a line to be an
infinite number of halves' is superfluous. For then it would be the
same as if someone were to say 'for it would belong potentially to line
to be an infinite number of halves potentially'.

But how is the line potentially an infinite number of halves? 1293,10
Everything that is potentially can also come to be in actuality, and
one who hypothesizes what can come to be as having come to be may
be hypothesizing something false, but not something impossible. So
there will be an infinite number of halves in actuality. But it was
proved in the third book of this treatise that neither magnitude nor
number can be infinite.[253] So we must understand the claim that it is 1293,15
an infinite number of halves as being used instead of the claim that
it is being divided into halves ad infinitum. For the line is capable of
this, but not of being an infinite number of halves – unless someone
were to call what is being divided into halves ad infinitum an infinite
number <of halves>, on the grounds that the dichotomy never has a
limit – and this is precisely how the argument that poses the puzzle

took the halves as infinite, not as coming to be ad infinitum as the
1293,20 truth really has it. Aristotle grants this much to the argument that poses the puzzle, that in no way is there an infinite number <of halves> at the same time, and he makes the refutation depend on what is potentially and in actuality, calling them infinite potentially, as he presents it in the third book of the present treatise. Not thus: 'just as if it is possible for this to be a statue, this will in fact be a statue, so too something is infinite that will be in actuality';[254] but
1293,25 'this is how the infinite exists', he says, 'by one thing always being taken after another and what is being taken is always finite but always different';[255] 'just as a day or an athletic contest <exists> by one thing always occurring after another, the infinite <exists> in this way too';[256] and, he says, 'this infinite', i.e. what is ad infinitum, 'is clear in the case of time and among humans and in the division of
1293,30 magnitudes'.[257] In that passage Aristotle adds that what can be infinite, i.e. what is ad infinitum, 'we must not take like a particular such as a man or a house, but the way a day or an athletic contest is
1294,1 spoken of, whose existence has not come to be like a substance', i.e. it does not persist, 'but as always something finite that is subject to generation or perishing, but is always one <finite thing> after another.'[258]

263b9-26 It is also clear that if[259] anyone does not always put [the point] of time (b10) [that divides earlier and later with the later state of the thing, it will follow that the same thing both is and is not at the same time, and that it is not when it has come to be. Now the point is common to both <time intervals>, the earlier and the later, and is the same and one in number, but it is not the same in definition (for it is the end of the one and the beginning of the other). But so far as the (b15) thing is concerned, it always belongs to the later affection. Let ACB be the time interval and D the thing. This is white in time interval A and not white in B. Therefore at C it is white and not white. For if it was white during all this time interval [A], is true to say that it is white at any moment of A whatsoever, and not white in B. But C is in both. (b20) Therefore it must be granted <that it is white> not in the whole <of A>, but <in all of A> except the final instant, C. This already belongs to the later affection.[260] And if it was becoming white[261] and if[262] it was ceasing to be white[263] in all of A, it has come to be or has ceased to be <white> at C. And so that is the instant at which it is first true to say that it is white or not white – or else when it has come to be it will not be, and when it has perished it will be, or alternatively, (b25) it must be white and not white,] and in general must [be] and not be, [at the same time].

After recalling that the instant is the beginning of one thing and the 1294,5
end of another, that it is a division of time, not a time interval, that
it is without parts, and that it is not possible for anything to come to
be in something and depart from it at the same instant (for something
would be and not be in the same thing at the same time), he next
proves that as regards the time interval that is being divided, the
instant at which the division of the time interval occurs is equally in 1294,10
both parts of it, being the end of one 'and the beginning of the other'
(for this is how the time interval is continuous – by virtue of the fact
that its parts have the instant as a common boundary), but as regards
the thing that is in the time interval, it[264] is always to be assigned to
the later time interval and to the later thing or affection (the one that
comes to be in the later time interval), and is the beginning of it. For
example, if something changes in some time interval from white to 1294,15
black, as regards the time, the first instant at which it has changed
from white to black is in both the earlier time interval and the later
equally, being the end of the one and the beginning of the other; but
as regards the thing, it is attributed to its being black, no longer to
the white thing from which it changed,[265] so that the thing that is
being taken in the later time interval (e.g. what is becoming black or 1294,20
the black spot) has taken the beginning of its being such[266] from this
instant, and this instant is no longer being taken as common to the
things as it was to the time intervals, but is being assigned to the
later thing (the thing that exists in the later time interval, of which
it is the beginning), and is being taken away from the thing that exists
in the earlier time interval – the time interval of which this instant 1294,25
is the end, being numerically one but not one in definition, when the
same <instant> is taken as end and beginning.

That this is the case he proves by making evident the absurd
consequence of its not being the case. For, he says, if 'anyone' does
'not' 'always' rank 'the' 'point', i.e. the instant, 'that divides' the 'earlier
and later' time interval[267] with 'the' later 'thing', but takes it as
'common' 'to the earlier and' the 'later' with things as with time 1294,30
intervals, so as to be in both things as we saw it is in the time
intervals, something absurd will follow, for 'it will follow that' 'the
same thing' 'is and is not at the same time, and' a thing 'is not' at the
time 'when it has come to be'. That this is so, he will demonstrate.
The claim, 'but so far as the thing is concerned, it[268] always belongs
to the later affection' is equivalent to 'regarding the things that are 1294,35
taken in the time interval, it always belongs to the affection that came
to be later'. For affections too are things. Alternatively,[269] 'but so far
as the thing is concerned, it always belongs to the later affection' 1295,1

means 'regarding the thing in the time interval, it always belongs to the affection that comes to be in it later', and he is calling the affection into which the change has occurred the later one, unusually saying 'affection' in connection with all kinds of motion.[270]

1295,5 The following is the demonstration by which he proves that if the instant is also taken as common to the things the way <it is common> to the time intervals, it will follow that 'the same thing' both is and is not 'at the same time' and that something is not 'when it has come to be'. He takes D for the thing and he takes two time intervals, A the earlier and B the later, divided from one another at C. And he
1295,10 hypothesizes that D 'is white' 'in' the earlier time interval 'A', and 'not white' 'in' the later <time interval> 'B', after the division at C, while C is the common boundary of both time intervals, being the end of A and the beginning of B. And he says that if for the thing[271] too C is hypothesized to be the end of the one and the beginning of the other, D, which is white in time interval A, will also be white at the end of
1295,15 A, i.e. C. And again if the non-white thing exists in the whole of B, it is clear that it will be non-white at the beginning of it too. But as we saw, the beginning is C, which is the end of A. 'Therefore' 'at C', which is without parts, it both is and is not 'white'. This is the absurd consequence of C's being the beginning of the later thing and the end of the earlier, just as it is of the earlier and later time intervals.

1295,20 He says that we must not grant that D is white in the entire first time interval so that it is still white at the end; 'it must be granted' to be so 'in the whole' time interval but not at its end too. Rather, the end must be taken away from the first thing, viz., the white one. On the other hand, C must be made the beginning of the non-white thing
1295,25 that exists in the second time interval. It becomes clear also from consideration of coming to be and ceasing to be that C must be assigned to the latter. If something 'was becoming white' 'in all of' the time interval 'A' or if 'it was ceasing to be white' in it, 'it has come to be or has ceased to be <white> at C', if C is the end of A. And so if something is becoming white in time interval A, it is white first at C, and if something is ceasing to be white in time interval A, at C, which
1295,30 is the end of A, it will first have ceased to be <white> and will be not white. So if that which is coming to be or ceasing to be in time interval A, in time interval B has either come to be or ceased to be as also at C, it is clear that C must be assigned to B as its beginning,[272] since it has the same nature, but not to A, if indeed it is coming to be in A but at C it has come to be. But if someone will say that it is common to
1295,35 both A and B just as the intermediate instant is common to the time intervals on either side, it will follow that 'when it has come to be' it is not and that when 'it has perished' it is – results that are evidently absurd. That these things follow is clear from the following. If what is coming to be white in time interval A were posited as still coming

to be also at its end, C, at which <point[273]> it has changed, it would
not be <white> at C. For what is coming to be is not yet. But in fact, 1295,40
in the entire time interval B it is posited as having come to be, and
so it will have come to be also at the beginning of it [viz. of B], i.e. at 1296,1
C. Therefore it has come to be at C too (because C is the beginning of
B in which it has come to be) and it is not yet (because C is the end
of A, in which it is still coming to be). And what is coming to be white
in A is posited as still coming to be at the end of it [viz. of A]. The
same argument also holds for what has ceased to be: if what is ceasing 1296,5
to be (whether this is the white thing or whatever it may be), is until
it ceases to be and D was perishing in the whole of time interval A,
then if it were posited as ceasing to be at its[274] end, C, too, it would
still be at C. But if in all of B it has ceased to be and is not and C is
the beginning of B, it follows that it still is at C (because C is the end
of A, in which it was ceasing to be), but because it is the beginning of 1296,10
B, in which it had ceased to be, it will have ceased to be. Therefore
when it has ceased to be it still is, so that both impossibilities will
follow if we say that C is common – that what has come to be is not
when it has come to be and is, and that what has ceased to be is when
it has ceased to be and is not. Further, the same thing is both white
and not white at the same time, as was proved, and the same thing
both has come to be[275] and is not at the same time. So we should assign 1296,15
the end of the thing in time interval A to the thing that exists in time
interval B in order to avoid falling into the contradiction.

'Also the puzzle', says Alexander, 'about when Socrates died might
be solved in this way. The puzzle goes as follows. When did Socrates
die? In the time interval in which he was dying or in that in which he 1296,20
was dead? It is impossible <for him to have died> in any time interval
between these, since there is no <time interval> in between. But it is
not true to say that he died in that <time interval> in which he was
dying (he was alive in every part of the time interval in which he was
dying) or in that <time interval> in which he was dead (one who is
already dead cannot die), since in fact we say "he died" in the period
after his dying, and since his being dead comes after "he died", <which 1296,25
serves> as a beginning <for it>.[276] In fact, "he died" <did not occur>
in a time interval, but at the end of the time interval in which he was
dying, whereas "he is dead" <is true> at every future time interval 1296,30
after "he died". So this is solved by the previous remarks. For he
changed from living and he died at the first[277] limit of the time interval
in which, while alive, he was dying, not in the future time interval,
in which he was no longer living. So he died neither in the time
interval in which he was living nor in that <time interval> in which
he was dead, for he did not die in a time interval at all, but at the end
of the time interval in which he was living.' It is like asking when a
child was born – when it was being born or when after being born it

is being raised. In fact it is clear that neither of these <is the correct answer>, but <that it was born> at the end of <the time interval> of
1296,35 being born; neither in the time interval before that nor in that <time interval> after that, and not in a time interval at all.

263b26-264a4 If anything that is after previously not being, [must come to be a thing that is, and is not when it is coming to be, then the time interval cannot be divided into indivisible times. For if D was becoming white at A, but it has simultaneously come to be and is <white> at another indivisible (b30) time, B, which is next <to A> – if at A it was coming to be <white>, it was not <white>, and at B it is <white> – then between <A and B> there must be some process of coming to be, and consequently (264a1) a time interval in which it was coming to be. The same argument does not hold for those who deny that there are indivisibles. Instead, it has come to be and is <white> at the last point of the very time interval in which it was coming to be <white>, and nothing is next or consecutive to this,] whereas indivisible times are consecutive.

When he was proving that what turns back must be at rest at the
1297,1 <point> from which it turns back, he made use in the proof of the claim that nothing can come to be at anything and depart from it at the same instant (which is a limit of a time interval, not a time interval). This was seen to follow from the facts that a time interval is not composed of things without parts and that instants cannot be
1297,5 next to one another. If this is posited it follows that between the two instants (the one in which the moving thing came to be <at the given thing>, which is an end of its coming to be <there>, and the instant at which it departed) some time interval has occurred in which it must be at rest, since instants are not adjacent to one another. He now proves this very point, that a time interval is not composed of or divided into things without parts. In fact he proved this result
1297,10 through several <arguments> in book six of this treatise,[278] where he put it forward more generally, that nothing continuous is composed of things without parts. Now he proves the same thing specifically for time, confirming the argument through the results he has just proved. He makes use of the fact that the primary change in virtue of which something has changed, does not take place in a time interval but at the end of the time interval during which it was changing, and from
1297,15 this result he proves the present point. He assumes as an evident axiom that 'anything that is' later 'after previously not being, must' 'come to be' before it is, 'and when it is coming to be' it is not yet. It is clear that the argument is true for things that pass into existence by way of generation, but not for the things of which he was saying

earlier that at one time they are and at one time they are not, without generation or perishing.[279] When this axiom is posited he declares 1297,20
that it is impossible for 'the time interval' to 'be divided' into instants as 'into indivisible times', or be composed of instants. For if there were a thing, e.g. D, that is coming to be white, and there were some time interval, A, composed of indivisibles, then 'D', the thing that is coming to be white, is not yet white at an indivisible time at which it is coming to be 'white', for it is posited that what is coming to be, when it is 1297,25
coming to be, is not yet what it is coming to be. So D is not yet white in time interval A. 'But it has simultaneously come to be'[280] (instead of 'already is'[281]) white 'at' the 'indivisible' 'time' 'B', 'which is next to' A. Therefore there must be some time interval between A and B in which it has come to be white. For it cannot be said to have come to be <white> at a limit of time interval A which is between the times A and B, since what is without parts does not have a limit. But if this 1297,30
is the case, then A and B, which are indivisible,[282] cannot be next to one another. For if every coming to be is in a time interval (for it cannot be at a limit of a time interval, since what is indivisible has no limit), and if D must be coming to be <white> between the time interval in which it is <white> and the time interval in which it is not yet <white>, there will also be some time interval in between that <time interval> in which it is <white> and that <time interval> in which it is not yet <white> (i.e. A and B).

And if between times A and B some indivisible time is taken, in 1297,35
which one will say that D has come to be <white>, just like the person who says that B <is the indivisible time at which D has come to be white>, the same argument holds again <that was used> for times A and B. For since D is <white> at B, but was not <white> at A (for it was coming to be <white> at A), again there must be some time interval between these in which D will have come to be <white>. He says 'it has simultaneously come to be and is <white> at another 1298,1
indivisible time, B, which is next <to A>', instead of 'it is already a thing that has come to be and is at another indivisible time',[283] for he is not assuming that it has come to be at B, since he is investigating at what <instant> it has come to be.

He then refutes the objection <that claims> that the same puzzle can arise also for those who hold that time is not composed of 1298,5
indivisibles. For it seems reasonable that in this case as in the former there is some time interval between that in which it was coming to be and that in which it is.[284] Again, similarly, it will not be in A, since it was coming to be in A; but it will be in B; therefore, between the time without parts[285] and B, which as we saw are next to one another, there will be some time interval. Now if the puzzling feature is 1298,10
common,[286] it will not yet have been proved that time is not composed of indivisibles. So he says that those who do not make time out of

indivisibles are not also subject to the same puzzle. And he adds the reason: for 'it has come to be and is at the last point of the very time interval in which it was coming to be, and nothing is next or consecu-
1298,15 tive to this' point, 'whereas indivisible times are consecutive'.[287] This is why those who say that the time A is indivisible cannot say that it was coming to be in some <part> of A, but that at the limit <of A> it is; for the indivisible has no limit. And so, if it was coming to be in A and A is without parts, and if it cannot have come to be at any <part> of A, but at B, which is a time without parts too, it already is, then
1298,20 there must clearly be something between <A and B> in which it came to be. For between that in which it was coming to be and that in which it is, is that in which it has come to be, and if this cannot be a limit[288] (since A and B are without parts), it will be a time interval. And so A and B will no longer be next to one another as was posited. And if the time between were said to be without parts too, the same absurdity follows again. But according to those who say that A is not a time
1298,25 without parts, what is coming to be in time A is able to have come to be at the limit of A, and already to be in B. Thus it will no longer be necessary to look for any time interval between A and B in which D will have come to be. For in time A it was coming to be, and it has come to be at the limit of A, which is between the time A, in which it was not yet (since it was coming to be) and B, in which it already is.
1298,30 For what is coming to be must have come to be in something between the time in which it was coming to be and that in which it already is – something different from either, which cannot be ascribed to A if A is without parts. For there cannot be one <part> of A in which D[289] was coming to be, and another in which it has come to be.

After saying 'it has come to be and is at the last point', he adds,
1298,35 'and nothing is next or consecutive to this', clearly <meaning> a point of the same kind as it is. He proves through this that instants, which are limits of a time interval, are not next to one another or consecutive to one another in such a way that the time interval is composed of them. But everything that is taken after the limit of time A is a time interval. He adds 'consecutive', indicating the way in which indivisi-
1299,1 bles could be said to be next to one another: by being in sequence just like things that are called consecutive in the strict sense; not because their parts are joined together, since they do not have parts.

1299,5 Next after this Alexander poses a puzzle that is not very cogent, I think. 'If someone were to say', he says, 'that at the limit of time interval A, at which it has already changed, D is, but that in time interval A it is not (for it was coming to be in that),[290] and if he were to use the same argument and demand there to be some time interval in between time interval A and its limit, in which it came to be, the same puzzle seems still to remain. For at the first <time> at which it
1299,10 has changed into being,[291] it is true to predicate being of it.' Even so,

he does a good job of solving it, saying that if what it had come to be
in was a time interval, it would already be in this as in a time interval,
and there would have to be something earlier than this, in between
this and that in which it was coming to be. But if that in which it has
come to be is not a time interval but the limit of the time interval in
which it was coming to be, it will no longer follow that there needs to
be something between the limit and the thing whose limit it is. For 1299,15
that in which it has come to be must be something between the time
interval in which it is not yet but is coming to be and that in which it
already is. But if someone were to say that at the limit of A it has not
only come to be but also is, it is clear that he will also say that at the
beginning of A it is not, even though, in a way, what is already coming
to be is.

> **264a4-6** It is evident that if (a5) it was coming to be in the whole 1299,20
> time interval A, [the time in which has come to be plus the time
> in which it was coming to be is no more than] the entire time in
> which it was only coming to be.

After proving that D was coming to be in time interval A, but that it
has come to be at the limit of A, he proves very cleverly that that at
which it has come to be is not a time interval but a limit of a time
interval. For 'if', he says, 'it was coming to be in the whole time
interval A' (and at its limit it has come to be), and 'the time' 'in which' 1299,25
'it was coming to be' were to be proved equal to the sum of the <time>
'in which' 'it was coming to be' 'plus' that 'in which it has come to be',
it is clear that when that in which it has come to be is added to the
time interval in which it was coming to be it does not make the time
interval longer. And so that in which it has come to be was not a time
interval. For when a time interval is added to a time interval, it makes
the whole longer. That it does not make it longer is clear since it has 1299,30
come to be at the limit of time interval A, and the limit is without
parts. Indeed, on the one hand it was immediately clear that the limit
of the time interval is not a time interval. And on the other, the
seeming paradox that the time in which it was coming to be together
with that <time> in which it has come to be is no longer than that
<time> in which it was coming to be alone, produces great clarity. It
seems to me that he has added 'entire' to 'than the time in which it
was only coming to be' not without purpose, but in order to indicate 1299,35
that it is necessary to take it together with its limit.

'Setting out from these arguments', declares Alexander, 'it is
possible to prove that the Stoic propositions which some say change
truth value indeterminately are not really such.[292] They are of the
following sort: "If Dion is alive, Dion will be alive". Even if this is now 1300,1
true, starting with a true statement, "Dion is alive" and concluding

with a true statement, "He will be alive," still there will be a time when the additional premise,[293] "But in fact Dion is alive" is true, but the conditional will change truth value to false, because there will be
1300,5 a time when, "Dion is alive" is still true but it will not be true that "He will be alive too". But unless this is true, the whole conditional would change truth value and become false, for it is not always the case that when "He is alive" is true "He will be alive" <is true too>, since in that case Dion would be immortal. However, it will not be possible to specify and state when, while he is alive, "He will be alive" will not be true. This is why they say that the change in truth value
1300,10 of such propositions occurs at an indeterminate and indefinite time.' This is the kind of proposition that is said to change truth value indeterminately. The distinctions Aristotle has just drawn make it possible to see that the claim that it changes truth value is not sound. If everything that changes has changed at the limit of the time interval in which it is changing and at this limit it is no longer in that
1300,15 <state> from which it was changing, but is then in that <state> into which it both was changing and has changed, it would also be possible for the fact that Dion is alive, if it changes, to change at some extremity of the time interval in which he is alive, and this is a limit of a time interval and not a time interval. At this <limit> it is no longer true that Dion is alive, since at this <limit> he has already died. Therefore if 'Dion is alive', which is being said at the present instant, were to be said at the instant at which he has changed from living, it
1300,20 will not be true. But if 'Dion is alive' <is said> at another instant before this one, it is true. But between the instants in which Dion has already changed and is no longer alive, and in which he is still alive, there is a time interval during which he is clearly alive, if in fact it was at the limit of that <time interval> that he has changed from being alive. And it will follow that at any instant at which it is true to say 'Dion is alive', 'Dion will be alive' will be true. And so 'He will
1300,25 be alive' does not change truth value before 'He is alive', for it would change truth value before it only if the instant at which 'Dion is alive' is true and the instant at which he has changed from being alive, were next to one another. But since between any instant in which it is assumed truly that Dion is alive and the first instant at which he has changed from being alive there is a time interval during which he is still alive, it is clear that whenever 'Dion is alive' is true, 'He will be
1300,30 alive' will be true too, and that this will not make Dion immortal. But if 'Dion is alive' is not said at an instant but in a time interval whose limit will be the first <instant> at which he is dead, having changed from being alive, then that is when 'Dion will be alive' will change truth value to false, because the limit of the time interval is added to the thing that is in the time interval whose beginning this limit is, as
1300,35 was proved earlier.[294] For the instant at which he first changed from

being alive is the end of the time interval in which he was alive and the beginning of that in which he is already dead.

> **264a7-9** [These and suchlike arguments[295] are] what we may place confidence in as appropriate <to the topic>. [But if we look at the matter dialectically,] this same conclusion would seem to result [from the following considerations as well.]

After proposing to show what is the continuous and naturally primary kind of locomotion that is an attribute of the first thing that undergoes motion in place, and after dividing the kinds of locomotion into circular, rectilinear, and combined, and after specifying in advance that if either simple motion is not continuous, the combined motion that is made up of both will not be continuous either, he proves through several arguments that rectilinear motion is not continuous in two ways: (a) that what undergoes locomotion on a finite straight line turns back, and what turns back undergoes contrary motions, while contrary motions are different in species and are not one or continuous; and (b) in turn <he proved> that things that turn back must be at rest in between, and by being at rest in between a thing undoes the continuity of the motion.[296] After proving these results for rectilinear locomotion on the basis of the kind of motion that has contrariety and the magnitude on which the motion that turns back takes place – features that are appropriate to motion in that they belong to it per se – he says 'these and suchlike arguments are what we may place confidence in as being appropriate' to what is being proved, calling 'appropriate' the deductions that come from <premises> that are appropriate to the present topic. (It is his practice to call these 'demonstrative' as well.[297]) He next announces that he will prove this same point 'dialectically'.[298] Dialectical deductions are those that are not based on terms that are appropriate and near to the thing at hand, like the earlier ones, that were based on the differentiae of rectilinear motion and on the magnitude that underlies such motion, but are those based on terms that are more common and more general and capable of applying to other things too. It is his practice to call <such arguments> dialectical, on the grounds that they arise through accepted arguments. For he will next put forward arguments that do not concern things that move only in place and that are not based on attributes that belong per se just to these things, but are based <on attributes that belong per se> to all things that change from opposites to opposites and that are based on opposites, which belong strictly speaking not only to things that undergo rectilinear motion – <they belong> not only to these, but also to all things that change.

1301,30 **264a9-21** Everything that is in motion (a10) continuously [and arrives at a certain thing in its locomotion, if it is not knocked out of its way by anything, was also undergoing locomotion to that thing before. For example, if it arrived at B, it was also undergoing locomotion to B – not <only> when it was near, but right from when it started its motion. For why now rather than earlier? And likewise for the other <kinds of motion>. So, then, when a thing that (a15) is in locomotion from A reaches C, it will come back to A, being in motion continuously. Therefore, when it is in locomotion from A towards C, it is then also in locomotion to A in respect of its motion from C, and so <it is undergoing> contrary <locomotions> at the same time. For the rectilinear <locomotions> are contrary. And at the same time it is changing from something in which it is not. Therefore, if this is impossible it must (a20) come to a stop at C. Therefore the motion is not single,] for motion that is interrupted by a stationary state is not single.

He proves that no motion from contrary to contrary can be continuous, assuming a general axiom stating that everything that 'arrives at a certain thing' via a continuous motion without being impeded, had its
1302,1 impulse towards that very thing it came to in completing the motion, even at the time when it was starting its motion. For everything that undergoes motion, just as it moves from something determinate, also moves towards something determinate, unless it is impeded by something. For example, if something that moved downwards[299] has undergone locomotion without being impeded, it had its impulse
1302,5 towards the bottom when it was moving too, and 'not' only 'when it was' already 'near' the bottom but even from where it was starting the motion through which it has undergone locomotion continuously to this <place>, as if it was starting to move in order to come to this. 'For why' is it coming to be in this 'rather' 'than' in this? This is what 'for why now rather than earlier?' means. 'And likewise', he says, 'for
1302,10 the other' kinds of motion. For, he says, the argument is true not only for locomotion, as in the example he gives to put the axiom to use, but it is likewise true for the other kinds of motions too – alteration, and increase and decrease, and generally for those that change towards contraries. It does not hold for generation and perishing, as we will learn.[300] Now once this is postulated truly, the present point is proved
1302,15 through a conditional of the following sort, and the argument takes place as it did in the example of locomotion. If what is undergoing locomotion upwards from below on a straight line does not slow down and is not at rest when it comes to be up, but continuously with the same impulse undergoes locomotion downwards, it is clear through the axiom that also when it was starting its motion from below it had

its impulse towards the bottom, not only when it has come to be up
and is moving downwards. So, if what is moving from the bottom 1302,20
moves continuously, it has its impulse simultaneously to move both
to the top and to the bottom. But in fact, motion to the top is contrary
to motion to the bottom, since <they proceed> from contraries. There-
fore, if what is moving on a straight line moves continuously, it is
undergoing 'contrary' locomotions. But 'this' is 'impossible'. And so
what this is a consequence of is impossible too, namely, that what
moves on a straight line both moves continuously and turns back. 1302,25
After taking A as the beginning of the straight line and C as the end,
and positing that what moves continuously on the straight line from
A to C will also move from C to A likewise, and inferring that what
moves in this way will be undergoing 'contrary' locomotions 'at the
same time', he recalls why this follows, saying 'for the rectilinear
<locomotions> are contrary'. And this <is the case> since <the mo- 1302,30
tions proceed> from contraries. For the limits of a straight line are
contrary to one another: the top <is contrary> to the bottom, the right
to the left, the front to the back. Not only is this how these are[301] in
virtue of their relation to us; they are also in virtue of their own nature
proved to exist in the world prior <to their relation to us>, as we are
taught in the second book of *De Caelo*.[302] Another absurdity too is a
consequence of the claim that what is in motion on a straight line 1302,35
moves continuously and also turns back. If what moves from below
will be undergoing locomotion continuously and with its own impulse
from A back towards the bottom, A, then it will move towards coming
to be in that in <place> which it is. But this is absurd; nothing moves
through coming to be in that <place> in which it is. What moves,
moves from something[303] to something <else>. He next draws yet a
third absurd consequence of the claim that things that turn back on 1303,1
a straight line move continuously – that something is changing 'from
something' 'in which' 'it is' not yet. For if right from when it starts
moving, a thing that is undergoing locomotion to some place is moving
to that to which it will come,[304] that which moves from below would
right away be moving no more to the top than to the bottom. But the
downwards motion starts from the top. Therefore what is moving 1303,5
from below will also be moving from the top in which it is not yet. For
locomotion to the bottom does not start from anything else. But in
fact what is moving from somewhere cannot move from there if it is
not in that <place> from which it will undergo that motion. For if
anyone were to say that it is when it comes to be at the top that it
undergoes the downwards motion, he will agree that it first is coming 1303,10
to be at the top and since it is coming to be at the top, it can also depart
from it, but in this way it would no longer be moving continuously.

But perhaps someone will object and say that according to this
argument neither will it move continuously in moving upwards from

below. What starts to move from below to the top must first undergo locomotion to the middle, and thus, *then* from the middle to the top, with the result that it is moving from that in which it is not yet. Therefore this will not move continuously either, even if it does not turn back. But it must be said in the first place that motion upwards is not said to originate at the halfway point or the middle, but from below, and further that what is moving continuously to the top must not come to be at the halfway point. For if it comes to be there, it must also depart <from there> and be at rest in between, dividing[305] the halfway point in actuality; but what is moving continuously to the top will be said to be at the halfway point, but not to come to be <at it>. However, what is going to move to the bottom must first come to be at the top. For the motion to the bottom is not <part> of something simple,[306] so that it will already be moving from a point at which it must come to be but has not yet come to be. Now if these things are impossible: (a) to undergo contrary motions at the same time, (b) to move towards coming to be in that in which it is,[307] and, (c) third, for something to move from a <place> in which it is not yet – and if these <consequences> follow if we say that motion on a straight line is continuous, it is impossible for it to be continuous. To the aforementioned arguments Alexander adds the following one too, though it does not set out from the same axiom. 'If it is impossible', he says, 'for unnatural motion to be continuous and one with a natural motion, and, for something that moves and turns back on a straight line, if one of the motions it undergoes is natural, the other must be unnatural, then the motion composed of the natural and the unnatural motions is not continuous.'

264a21-b1 Further, the following considerations as well make this evident more generally [for every kind of motion. If every thing in motion undergoes one of the stated kinds of motion and undergoes <one> of the states of rest opposite <to these> (for as we have seen there are no others beside these), and further if a thing that is not (a25) always undergoing a particular motion (I mean motions that are specifically different, not where one motion will be[308] a part of a whole motion) must previously have been in the opposite state of rest (for the state of rest is a privation of motion) – then if the rectilinear motions are contrary and nothing can undergo contrary motions at the same time, it follows that the thing undergoing locomotion from A (a30) towards C could not be undergoing locomotion from C towards A at the same time. But since it is not undergoing the locomotions at the same time, and since it will undergo the latter motion, it must first be at rest at C. For as we saw, this is the state of rest opposite to the motion from C. So it is clear (b1) from what has been said] that the motion will not be continuous. 1303,35

He first proceeded to formulate his argument for the case of rectilinear locomotion, and then for the case of all motion in the strict sense, in which the change takes place from contrary to contrary, although 1304,1 even then the argument proceeded using a straight line as an example. Now, however, the demonstration becomes more general, covering all change, including both generation and perishing. It makes use of privation, since all things that change, whether in respect of place or of quality or quantity or of being, come to be <what they change 1304,5 into> from their proper privation. So it was reasonable for him to say that he will speak 'more generally' 'for every kind of motion', i.e. for all change. For everything that changes to something which it was not previously changes from not <being> this, i.e. from its proper privation, to being this kind of thing, as was proved in the first book of this treatise.[309] His argument goes as follows. After assuming that 1304,10 things that do not move eternally, but only for a time, 'undergo[310] one of the stated kinds of motion', which proceed from opposites to opposites, (<they change> either in virtue of generation and perishing or of increase and decrease, or of alteration or of rectilinear locomotion; but it is clear that if they are not always moving they at some time also undergo one of the states of rest that occur at the opposites 1304,15 of the things to which the moving things are moving; for there is no other kind of motion aside from those, or any other kind of rest), he then deduces that what does not move eternally either 'undergoes' 'one of the stated' 'kinds of motion' or 'undergoes' one 'of the states of rest' 'opposite' to these. But before undergoing that motion, 'that which undergoes[311] one of the stated kinds of motion' that are not eternal, 'must' 'have been in' 'the state of rest' that is between this 1304,20 motion and the motion opposite to it. But what is at rest in between does not undergo a continuous motion. In this way it has been demonstrated generally for every change that originates from a privation. He then applies the argument to backward-turning rectilinear locomotion, which was his task to prove not continuous. For 'if' 'the rectilinear' 'motions' 'are contrary' – the one from the top to the 1304,25 bottom <being contrary> to that from the bottom to the top, e.g. if the motion 'from A' to 'C' is contrary to the motion 'from C' to 'A', and for this reason what is moving from C to A cannot undergo both motions at the same time, 'it must first be at rest at C. For as we saw, this is the state of rest opposite to the motion from C', since it is between the two motions and interrupts their continuity. Therefore it is clear that 1304,30 rectilinear motion is not continuous. This is the entire demonstration of the argument.

After saying 'if a thing that is not always undergoing a particular

motion', he explains how to take 'a particular motion' in the words 'I
1304,35 mean motions that are specifically different, not where one motion
will be a part of a whole motion.' For what is undergoing motion to
the top (motion that, since it differs in species from motion to the
bottom and is opposite to it, cannot make one continuous motion with
it), before its motion to the top, 'must' 'have been in' 'the state of rest'
that is contrary to this motion, namely, <the state of rest> at the
1304,40 bottom. However, as regards a part of the motion to the top, what is
moving towards the top need not be at rest before undergoing *that*
1305,1 motion. For what is moving over the part before that one was poten-
tially moving over that part too, and since both belong to a single
species, nothing prevents the motion made up of both from being
continuous and therefore <nothing prevents the thing that is moving>
from *not* being at rest first (the way we saw that what undergoes
motion that is different in species must be at rest before undergoing
1305,5 this motion, because it is not possible to make the opposite motions
a single continuous motion as <it is possible with> motions that differ
in respect of parts;[312] for these are not interrupted by a state of rest).
This is precisely why things that undergo a continuous eternal motion
are always in motion at continually different parts of it, though since
the species is the same they are not compelled before undergoing that
motion to be in the state of rest opposite to the motion they undergo,
1305,10 the way things that do not undergo motions that differ in respect of
parts, but that undergo motions that differ in species <do>. If it were
subject to this kind of motion and were not undergoing it, it would be
in the privation of it. And the privation is the state of rest, from which
<originates> the change to that motion. For everything that changes
to something which it was not previously, changes from not being this.
And therefore also what changes to <be undergoing> this kind of
1305,15 motion which it was not undergoing previously will change to it from
its privation. And the state of rest in the opposite <condition> is the
privation. For what changes to white from black, before changing to
white must be at rest in <the condition of being> black, from which
it changes to white. For even though black is contrary[313] to white, in
fact, in the case of contraries the worse always has the definition of
1305,20 the privation.[314]

264b1-6 Further, the following argument is more proper <to the subject> [than the previous ones. The non-white has perished and has become white at the same time. Therefore, if the alteration to white and from white is continuous and does not remain stationary for any time interval, (b5) the non-white has perished and has become white and has become not white at the same time.] For the time of the three will be the same.

After proving through more common arguments, which he also calls more dialectical, that rectilinear motion is not continuous, he now professes to prove the same result again through more proper ones, i.e. ones based on per se attributes. For the per se attributes of something are proper to it.[315] Among the per se attributes of motion are (a) that it is in a time interval[316] and (b) that what has changed to something has at the same time both changed to what it changed to and left that from which it changed.[317] (c) That there is a time interval between instants is also <a> per se <attribute>.[318] Now since he is going to base his proof on these <attributes>, he reasonably says that the present proof will be based on ones that are more proper. His preceding arguments were based on contraries and opposites, which are common to more things, which is why he calls them more general. And indeed he bases this argument on attributes that belong per se to motion, and it does apply to all motions. This is why he constructs it as applying to alteration. What he says is the following: if motion from contraries to contraries is continuous, then 'at the same time' that from which the change <begins> 'has perished', and that to which the change <proceeds> has come to be at the same instant. And so 'the non-white has perished[319] and has become white at the same time. Therefore, if' the turning back 'is continuous', it will have changed both to white and from white to non-white at the same time. And so these three things will be at the same instant, which he calls time, when he says 'the time of the three will be the same' – the perishing of the non-white, the generation of the white, and the change from white to non-white. And so the same thing will both be white and have already lost something of its whiteness, for this is <what> changing from white <is>. But if in order to avoid this absurdity we were to say that it has changed to white and from white not at the same instant but at different instants, there will be some time interval and state of rest in between, and the motion will no longer be continuous. So if some things are of necessity at the same instant – the perishing of the non-white and the generation of the white – but others will be at the same time because of the hypothesis that posits that motion is continuous – the white and the change from the white to the non-white – the three things will be at the same instant: perishing of non-white, generation of white, change from white to non-white. This is how the change from the non-white would turn out to take place to the non-white, the motion being one and continuous, and what has perished would turn out to be changing into itself. For the non-white having perished in the non-white[320] changes back into the non-white, e.g. the black <changes back> into black. But nothing changes into that from which it is changing. This proof is more proper than the previous ones, because in them the absurdity that was inferred concerned the fact that the motions are contrary, whereas here it depends on making something both having changed from something to something and simultaneously at the same instant having changed back from this to that from which it changed to this. So the more proper time is to change than contraries are, the more <proper> this proof is than the preceding ones.

264b6-9 Further, it is not true that if the time is continuous so is the motion; [the motion can be consecutive. How could the
1306,25 extremity of contraries,] for example of whiteness and blackness, [be the same?]

He adds this to sway our imagination,[321] which on the basis of the continuity of time makes motions, which are in time, continuous too. For if a thing is in something continuous, there is no need for it too to be continuous. If that were so, all things that are in time will be
1306,30 continuous even if they are very far from one another and entirely dissimilar. (For the whole of time is continuous.) But it is not necessary. Things in a jar do not have to be continuous even though the jar is. So 'time' is 'continuous', but the motions in it which are motions towards contraries and from contraries will not be continuous with one another, but 'consecutive'. For as we saw, things are consecutive if there is nothing of the same kind in between.[322] And between
1306,35 motions towards opposites there is no motion, but a state of rest, which is not of the same kind as the motion. And so, motions that take place in this way are not continuous, but are consecutive to one another. That contrary motions cannot be one continuous motion he proves from the definition of things that are continuous. As we saw,
1307,1 things are continuous that are joined at a single common boundary and whose limits are one.[323] But contraries do not have a common boundary and it is not possible for the extremities at which they will be joined together to be one. For the extremities of contraries are contrary to one another, and contraries are not one. For 'how', he says,
1307,5 'could' the extremities 'of whiteness and blackness' 'be' one and 'the same?'[324] And so, contraries, whose limits are contrary, will not be continuous with one another. 'But', says Alexander, 'someone can understand the extremity of the contraries <to mean> the end of the continuous whole. For of a single continuous motion there will be a single end,[325] and so what is becoming white and what is becoming black will have one end. But what would this turn out to be? Not the
1307,10 black, because this is not the end of <the process of> becoming white, nor the white for this cannot be the end of <the process of> becoming black either, nor any mixture of both, because it does not undergo both motions at the same time.'[326]

264b9-18 But locomotion on a circular line will be single and continuous, [since nothing (b10) impossible follows. What is in motion from A will at the same time be moving towards A in virtue of the same propulsion (for it is in motion towards that at which it will arrive), but it will not undergo contrary or opposite motions at the same time. For not every motion to a given point is contrary or opposite to motion from that point. It is contrary when the motion is on a straight line[327] (b15) (for on it there are contraries in respect of place, e.g. the points on[328] a diameter, since they are farthest apart) and it is opposite when the motion is on the same length. Therefore nothing prevents it from undergoing motion continuously] and not being interrupted for any time interval.

After proving that neither any other kind of change than locomotion 1307,15
nor rectilinear locomotion can be continuous because of the impossi-
bilities that follow if any of these is hypothesized to be continuous, he
says that only circular motion can be 'single and continuous', since
nothing 'impossible' follows if it is posited [to be continuous]. For if
we proved that the other kinds of motion are not continuous because
impossibilities follow if we posit any of them as continuous, then if 1307,20
nothing impossible follows when circular motion is posited to be
continuous, we should confidently consider it continuous. For a thing
that is in motion in a circle, having started from A and being brought
back to A in virtue of the same impulse and tendency, does not come
first to anything contrary to A, as was proved to happen in the case
of rectilinear motion, which was the reason why motion composed of
contrary motions could not be single and continuous in that case. But 1307,25
what moves in a circle does not in its rotation undergo motions that
are either 'contrary' or at all 'opposite'. For neither is any part of the
motion from A to A contrary to any other part, nor is the whole motion
from A to A 'contrary or' at all 'opposite' to the whole motion from A
to A. He shows what the contrary and opposite motions are in the 1307,30
case of locomotion. The motion from A is not without qualification
'contrary' or 'opposite' to the motion to A (for in fact this is how circular
motion takes place), but the motion from A is contrary to the motion
to A when it takes place on a straight line. 'For' 'the points on a
diameter[329]' are 'contrary' 'since' they are the 'farthest' from each
other, and things that move from the points on a diameter and back
to them[330] undergo contrary motions since <they are moving> from 1307,35
contraries to contraries. If A and B are the ends of the diameter, the
motion from A to B is contrary to that from B to A. He calls the straight 1308,1
line of the circle the diameter, showing in addition that the ends of
the diameter are the 'farthest' distant from one another as is possible
on a circle. This is precisely why reverse motions that take place on

the diameter are contrary. He says ‘when the motion is on the same length’ ‘it is opposite’, since the motions on the diameter are contrary because they[331] are the ‘farthest’ distant from each other, and likewise the motions that mark out the limits up and down, right and left, and front and back – the ones according to nature – are contrary since they take place between things that are contraries in place. But things that turn back on any other straight line will undergo opposite motions even if not contrary ones, for <they undergo motions> as from a beginning to an end or from an end to a beginning, either because one of them is up in relation to us and the other is down in relation to us, or one is right in relation to us and the other is left, or one is in front in relation to us and the other is behind. These are opposite in relation to *something*, and so all motions that turn back on straight lines of these kinds, even if they are not contrary, are at least opposite. For we say that only motions from contraries are contrary, and ‘the points on a diameter’ are ‘contraries’ ‘as farthest distant from one another’, says Alexander. But perhaps in fact it is the things that mark out contrary places that are the contrary ends of a straight line, as I said before.[332] For also the ends of the radius, which is half the diameter, are contrary because one of them is up and the other is down. ‘It is possible’, says Alexander, ‘to speak of the motion on the same length as opposite even if it does not take place on a straight line. For if several lines are joined from one point to another, supposing that they are not straight (since there is only one straight line from one point to another[333]), the things that move on these lines in reverse directions move oppositely not on a straight line, but on the same length. So the opposite motions on such lines are not contrary, because the contrary motions are on a straight line, but they are opposite, because they oppose and eliminate each other.’ Now if things that move on a straight line and turn back[334] undergo either contrary or opposite motions, and neither contrary nor opposite motions can be continuous with one another, rectilinear motions that turn back will not be continuous, but ‘nothing prevents’ circular motion from taking place ‘continuously’ on the same <magnitude> ‘and not’ ‘being interrupted’ ‘for any time interval’.

264b18-19 For circular motion is motion from itself[335] to itself,[336] whereas rectilinear motion is motion[337] to another <thing>.

After proving in the previous argument that circular motion is not prevented (like rectilinear motion) from being continuous by being composed of contrary or opposite motions, he now proves that there is no actual point at which it will have to come to be and <from which it will have to> depart and <at which it will have to> be at rest in between, as was proved <to happen> with backward turning rectilin-

ear motions, which is one cause of their not being in motion continu-
ously. Circular motion that takes place from the same point to the
same point does not have any actual point on the circle towards which
the moving thing first set out and which it will use as both an end
and a beginning and <at which it> will be at rest in between. If it does
have some such point, it would not be said to move from the same
point to the same point, but <it would be said to move> to the first
point at which it had to come to be. But motion on a straight line is 1309,10
from the same point to a different one – to the opposite end of the
straight line, at which it first had to come to be as at an end. And
turning back from that as from its beginning, it moves back to the
moving thing's original place. 'From the same' means from the place
in which it originally was, 'to it' means to that place, and 'to another' 1309,15
means to another place than where it originally was.

> **264b19-28** Further, (b20) motion on the circle <is> never at the same <points>, [whereas rectilinear motion <comes to be> at the same <points> repeatedly. Now what is always coming to be at different points can be in motion continuously, but what <is coming to be> at the same points repeatedly cannot, for it must be undergoing opposite <motions> at the same time. And so there cannot be continuous motion on a semicircle (b25) or on any other circular line, for it will have to be in motion over the same points repeatedly and to undergo contrary changes. For it does not join the end to the beginning. However, circular motion does join them] and is the only complete <kind of motion>.

From this argument too he proves the difference between circular and 1309,20
rectilinear motion in virtue of which circular motion can be continu-
ous while rectilinear motion cannot. For 'motion on the circle', he says,
i.e. circular motion, '<is> never at the same <points>'. Alexander
supplies 'comes to be', in order that it may be: 'never comes to be at
the same points', because there is not in actuality any point on the
circumference at which it will come to be and stop, or at which it comes 1309,25
to be and departs <from> at all, while 'rectilinear motion', which is
on a straight line, must come to be 'at the same <points> repeatedly'
and depart <from them repeatedly>. And from the fact that some-
thing never comes to be at the same <points> it follows that it has no
need to stop, whereas from the fact that something comes to be at the
same <points> repeatedly it follows that it must stop again and again
at the <point> where it comes to be. For it departs from the <point>
at which it actually comes to be, and between coming to be and 1309,30
departing there must be a stopping as was proved earlier.[338] Now
things that turn back on a straight line come to be at the same point

twice – when they stop and when they start again – whenever it moves twice consecutively over the same distance, stopping at the point at
1309,35 which it turns back and starting again from it.[339] And it does this as many times as it turns back. This is why motion that is always coming to be at different points can be continuous, while that which comes to be 'repeatedly' at the same <point> cannot.[340] 'For' things that turn back 'must' stop in between coming to be <at a point> and departing <from it>, and indeed <must> 'be undergoing' 'opposite' motions. But
1310,1 neither what stops in between nor what undergoes opposite motions can be undergoing a continuous motion. What undergoes a continuous motion undergoes a single motion once, while what stops and then moves undergoes as many motions and as many times as it changes from a stationary state. So whenever something stops and, turning
1310,5 back at the same <point>, moves back – because it stops, it moves repeatedly and, because (since it turns back) it moves back on the same line,[341] it turns out to undergo the same motion[342] repeatedly, in fact as many times as it turns back.

How then will circular motion too not be at the same <points> repeatedly if in fact it rotates repeatedly? Rather, it is not the case that what comes to be at the same <point> several times has thereby also undergone this motion repeatedly; but if, moving continuously
1310,10 with a single impulse, force and tendency, it comes to be several times at the same <point>, which exists potentially but not actually (in such a way that the same <point> comes to be beginning and end), and if it is not at rest in between, the motion will be continuous. But if it is at rest in between and again undergoes the same motion in reverse, as happens with things that turn back, whether they are on a straight line or on a circumference, it will come to be at the same <point>
1310,15 several times because each one of the motions that are interrupted by the states of rest exists separately, and thus what moves undergoes many motions and moves repeatedly. It <undergoes> each of the parts of the motion as many times as <it undergoes> the entire <motion>, once for each time it moves over the substrate. If it were not moving continuously it would not be moving as over a single thing or be undergoing a single motion. But anyone who said that in the case of continuous motions too what has come to be at the same <point>
1310,20 several times has moved over it repeatedly, would not be saying it strictly. It is not repeatedly strictly speaking, because it is not many motions but a single continuous motion over a single continuous <line>.[343] For the parts of continuous things are potentially in the continuous things and are not determinate per se. This is why they do not <undergo> many <motions> or <move> repeatedly, whereas things that turn back do not move continuously, but consecutively.
1310,25 What turns back starts again from where it stopped, and it is

impossible for anything that moves in this way to move continuously, as has been proved.

After proving that what undergoes the same <motion> repeatedly
must undergo opposite motions, turning back, and be at rest in
between, he does well to remark that the cause of this is not the fact
that the line is straight, but that it is something finite that has limits 1310,30
in actuality. For indeed things that move permanently on a semicircle
or any other arc of a circle that has a beginning and an end that are
distinct in actuality, must turn back. Circular motion alone can be
continuous, since it does not have a beginning and an end that are
distinct in actuality. For indeed rectilinear motion too <could be
continuous>: if the straight line were infinite, nothing would prevent
something from moving continuously on it. But because what <is 1310,35
moving> on a finite line must turn back, and it must both be at rest
in between and also undergo opposite motions – for both reasons such
motion cannot be continuous. He himself briefly adds the reason why
motion on a finite straight line is not continuous. 'For it does not join', 1311,1
he says, 'the end' on it 'to the beginning', as happens with things that
are in circular motion, where every point that is taken is both end
and beginning, since at every <point> the end and the beginning are
joined because of the circularity and are not distinct as they are on a
straight line. For even if on a straight line the end also becomes a 1311,5
beginning for a thing that turns back, and the beginning becomes an
end, still the beginning is not joined to the end on the straight line.
E.g. A <is not joined> to B, if these are the limits of the straight line.
However on a circle the beginning and the end are joined to each
other, because wherever the thing that moves on a circle starts, it
ends up at the same <point> after traversing the line of the circle. For
this reason also, circular motion is the 'only complete <kind of 1311,10
motion>', because it leaves nothing outside itself, as a finite straight
line does, that can be increased since it has something outside itself,
whereas a circle does not admit additional increase.

264b28-265a2 It is evident from this division [that none of the other kinds of motion can (b30) be continuous either, since in all of them it happens that motion takes place over the same things repeatedly. In the case of alteration, it is the intermediate affections, in change involving quantity it is the magnitudes in between, and likewise in the cases of generation and perishing. It makes no difference whether we make the stages over which the change occurs (265a1) few or many, or whether we add or take away one of the intermediates,] since either way it happens that motion takes place over the same things repeatedly.

After proving that rectilinear motion cannot be continuous but that 1311,15

circular motion can, and basing this proof on the claim that rectilinear
motion comes to be at the same <points> repeatedly, but circular
motion never <is> at the same <points> in actuality – he says 'from
this division', which proves that what turns back cannot move con-
tinuously (since it comes to be at the same <points> repeatedly and
1311,20 uses the same <point> in actuality as both a beginning and an end,
and therefore is at rest in between and in its motion undergoes a
plurality of motions that are opposite), but what does not turn back
can <move continuously> – through this distinction, he claims to
prove (a) that not only do motions on a straight line, motions on a
1311,25 semicircle, and generally motions on a finite line involve the things
moving on them turning back, and therefore <such motions> cannot
be undergone continuously, but also (b) that neither can anything
that undergoes any of the other species of motion or change (i.e.
increase, decrease, alteration, generation, perishing) be in motion
continuously. For in all these <kinds of change> too, he says, 'it
happens that motion takes place over the same things repeatedly' –
1311,30 in fact, as many times as they are marked out by a state of rest. Even
in the case of alteration, when something changes from black to white
and back from white to black and does this repeatedly, since 'the
intermediate' periods of rest between becoming white and becoming
black are many and mark out the motions, they make them many too.
The motion from the beginning to the end of becoming white is one
motion, and that from the beginning to the end of becoming black is
another, because the state of rest in the condition of having become
1311,35 white is one and that in the condition of having become black is
another. These are what he calls 'the intermediate' <things> in both
1312,1 alteration and motion in respect of quantity, <where it is> 'the
magnitudes' 'in between' the motions of increase and decrease, at
which it must be at rest. The fact that these are many makes the
motions many <too>. 'And likewise in the cases of generation' 'and
1312,5 perishing.' Since all these motions are finite, what moves must turn
back and undergo contrary or opposite motions. Therefore they must
undergo not one but many motions, which are contrary or opposite,
whose number must be calculated. 'It makes' 'no' 'difference' whether
the change takes place through more or fewer intermediates,[344] e.g.
'it makes' 'no' 'difference' whether becoming white comes to be through
1312,10 more changes and becoming black through fewer; the motions are two
and so are the states of rest that mark them out. We should regard
the intermediates, I suppose, not as if <the things in motion> are
arriving at a complete form, so that they are also interrupted by a
stopping, as if something were to go from white to red, from red to
gray and from gray to black (these are forms and what is in motion
has a stopping at them), but as if the motion is proceeding through
1312,15 things that are incomplete. For then it will be one and continuous up

to <the point of being> black, since its impulse is towards that, and that is <the point> from where its turning back occurs. And likewise for the other kinds of change. For the turning back evidently proves the interruption of the continuity. If the intermediates were infinite in number, it would be possible to say that it does not turn back because it never comes to be at the limit.[345] But since in the case of
quality the intermediate forms are finite, and in the case of quantity 1312,20
the intermediates are between determinate magnitudes in either direction according to the proper nature of each thing, and with generation and perishing the <states> at which the change <is bounded> are marked out either by the contradictory state or by privations, the issue whether the things that move towards contraries or opposites turn back through more or fewer <states> does not
trouble the previously stated argument at all. But things that turn 1312,25
back come to be repeatedly at the same <points> as at an end and as at a beginning in the way that has been stated, because the motions turn out to be many. For this reason such motion is not one or continuous either.

> **265a2-12** Therefore it is clear from these remarks that the natural philosophers [who declare that all sensible <things> are always in motion are incorrect. For they would have to undergo (a5) one of the kinds of motion that have been discussed, and particularly, according to them, alteration. For they say that things are always in flux and decay, and moreover they also call generation and perishing alteration. But the present discussion has said generally about every kind of motion that no kind of motion apart from circular motion can be undergone continuously, and consequently neither (a10) alteration nor increase can.] This is the end of our discussion [of the fact that apart from circular locomotion, no other kind of change is either infinite or continuous.]

After proving that none of the other kinds of motion or change except 1312,30
for circular motion can be continuous, belonging to the heavenly bodies alone, he criticizes the natural philosophers who claim that 'all sensible' bodies 'are always in motion'.[346] For if in fact everything were in circular motion like the heavenly bodies, it would be possible
for everything always to be moving. But since it is evidently clear that 1312,35
sublunary <bodies> undergo the other motions – altering, increasing, becoming smaller, moving in respect of place in a straight line, and being generated and perishing (for these are the changes of sublunary <bodies>) – moreover, in fact according to those people too these are
the motions found among perceptible <bodies>. They speak of gen- 1313,1
eration and perishing too, in saying that perceptible <bodies> 'are

always in flux and decay', and it is clear that they will say that alteration <exists>, and that 'they call generation and perishing alteration' because generation and decay take place when the bodies undergo alteration. Now if it has been proved that none of these kinds
1313,5 of motion can be engaged in continuously, and it has been proved 'generally' that 'no kind of motion can' be continuous except 'circular motion', it is clear that neither alteration nor increase and decrease, which they too saw among perceptible <bodies>, can be continuous. It is reasonable that the natural philosophers who follow Heraclitus, keeping in view the perpetual flux of generation and the fact that all
1313,10 corporeal things are coming to be and departing and never really are (as Timaeus said too[347]), claim that all things are always in flux and that you could not step twice into the same river.[348] For even if they had recognized this state of rest between the opposite motions, which Aristotle brings to light, they would have reckoned it as of no account in comparison to the entire flux, on the grounds that it is impercep-
1313,15 tible. For one swallow does not make a spring, as Aristotle himself nicely pronounces in the *Ethics*.[349] With these words he concludes the discussion and recalls 'that' 'no' 'change' 'is' 'infinite'. For even continuous change is not infinite, but ad infinitum: it is inexhaustible because of its repetition. But aside from circular motion no other motion is continuous in such a way as to be eternal.

<CHAPTER 9>

1313,20 **265a13-27** [It is clear] that circular motion is the primary kind of locomotion. [For as we said previously too, every locomotion is either circular, (a15) rectilinear, or a combination <of the two>. The former two must be prior to the last because it is composed of them. And circular motion is prior to rectilinear motion since it is more simple and complete. For it is impossible to undergo locomotion over an infinite straight line (for nothing is infinite in this way, and at the same time, even if something were, nothing could undergo motion <over it>, since what is impossible does not happen, and it is impossible (a20) to traverse an infinite line). On the other hand, a motion that turns back on a finite line is composite, i.e. is two motions, and if it does not turn back it is incomplete and perishable. But in nature, in definition, and in time, the complete is prior to the incomplete and the imperishable to the perishable. Moreover, <a kind of motion> that can be eternal is prior to one that (a25) cannot. Now circular <motion> can be eternal] but none of the others can, neither locomotion nor any other kind, [for a state of rest must occur, and if it does, the motion has been destroyed.[350]]

After proving that circular motion is the only kind of motion that can
be continuous and eternal (for it has been proved that none of the
others is continuous), he next proves that circular motion is also the
primary kind of motion. That locomotion is prior to the other kinds 1313,25
of motion has already been proved,[351] and it is also posited by him
that one kind of locomotion is 'rectilinear', one kind is 'circular', and
one is 'a combination' of these.[352] Assuming in addition as evident that
a motion that is a combination of certain <other motions> is posterior
to the simple motions out of which it is combined, he now proves that
circular motion is prior to rectilinear motion. When this is proved, it
will have been proved that of all kinds of motion 'circular motion' is 1313,30
'primary'. For the proof of this he assumes as agreed first, that the
complete is prior to the incomplete, and second, that the simple is
prior to what is not simple. After proving that circular motion is
complete and simple while rectilinear motion is neither of these, he
infers that circular motion is prior to rectilinear motion. First, that 1314,1
rectilinear motion is not complete he proves by making a division. A
straight line must be either infinite or finite, but first, it is impossible
for there to be an infinite straight line. For it has been proved in the
third book of the present treatise[353] that there is no magnitude that
is infinite in actuality and unable to be got completely through.
Further, even if someone were to grant that an infinite straight line 1314,5
exists, nothing could undergo motion over it. It would be 'impossible'
to get entirely through it. For if everything that comes to be, comes
to be by virtue of being able to come to be, that which is not able to
come to be will not come to be. But it is impossible to undergo an
infinite motion since the infinite is unable to be completely got
through and cannot be apprehended. Therefore, nothing undergoes
an infinite motion, so that even if there were an infinite straight line 1314,10
nothing would undergo motion on it.[354] Therefore rectilinear motion
must take place on a finite line. But if this is so, what moves on it
must either turn back or not turn back. But if it turns back, it will
undergo a composite motion that is composed of two motions contrary
to one another; 'if it does not turn back', it will be 'incomplete'. For it
will be 'perishable', and what is perishable is incomplete.[355] That 1314,15
rectilinear motion is incomplete he proves in this way. In view of the
fact that this motion is incomplete and perishable, he then adds that
complete motion must be prior to incomplete, and imperishable
motion must be prior to motion that is perishable, according to the
significations of prior in virtue of which he proved locomotion prior
to the other kinds of motion.[356] These were 'in nature, in definition,
and in time'. For complete and eternal motion (which is what circular 1314,20
motion is) is prior 'in nature' because if this is eliminated the other
motions, which are incomplete and perishable (e.g. rectilinear mo-
tions), will be eliminated too. But complete motion will not be elimi-

nated along with the incomplete ones. For since the cause of generation and perishing for things that undergo rectilinear motion is circular motion, it is clear that eliminating *it* eliminates the others,
1314,25 but not vice versa. 'In definition' means in respect of being. For what is eternal and complete is prior in being to what is perishable and incomplete. 'And in time', because what is eternal is prior to what is not eternal and the complete is prior to the incomplete.

But[357] since circular motion has not yet wholly been proved eternal – it was proved continuous not by virtue of being a motion (for every motion has the kind of continuity that both magnitude and time have)
1314,30 but because it does not need either to be terminated or to turn back like the motion on finite lines on which the beginning and the end are distinct from one another. What is continuous in this way can be eternal, if it lasts, but also can not be <eternal>, as with the circular motion of ordinary wheels moving in the same place. However,
1314,35 nothing prevents things that move on a finite line from moving continuously to the end, but no matter whether they stop at the end or turn back, their motion is followed by a stopping. This is why such motion cannot be eternal. Now if this was said truly, it is clear that 'circular <motion> can be eternal' but of the other kinds of motion,
1315,1 rectilinear locomotion cannot, much less the others, to which rectilinear locomotion was proved prior. So if according to all the significations of prior '<a kind of motion> that can be eternal' 'is prior' 'to one that cannot', circular motion will be prior to all the other kinds of motion, and this is precisely what it was his task to prove now.

1315,5 **265a27-b8** It is a reasonable conclusion that circular <motion> is single and continuous, [but rectilinear motion is not. For in rectilinear <motion> the beginning (a30) and end and middle are determined, and it has them all in itself, so that there exist points from which the thing in motion will start and where it will stop (for everything is at rest at at the limits, whether at the point from which <it sets out> or at the point where <it stops>), but in circular motion they are indeterminate. For why should any given point on the line be a limit more <than any other>? Each one is equally a beginning, a middle, and an end, so that <anything that undergoes circular motion> is both always (265b1) and never at a beginning and an end. This is why a sphere is in a way both in motion and at rest, since it occupies the same place. The reason is that these are all attributes of the centre: it is the beginning, and middle, and end of the magnitude, so that since it is outside (b5) the circumference, there is no place where the thing that undergoes locomotion can rest as having finished traversing its course (since it is always in locomotion around the middle and not towards the extremity), and because the centre remains fixed,] in a way the whole is always at rest and is also continuously in motion.

After proving that 'circular' motion can be 'singular', 'continuous' and eternal, but 'rectilinear motion' can be neither continuous nor eternal, he brings to light the most basic cause of each of these two facts in
the present passage. 'For', he says, 'in rectilinear' motion, the 'begin- 1315,10
ning and' the 'end and' the 'middle' are distinct in actuality, since these are also distinct on the straight line on which the motion <takes place>, and all of them exist in actuality. For a finite straight line has a beginning, an end and a middle, 'so that' what is moving on it too has the beginning of the straight line as the place 'from which' it 'will
start' its motion, and the end as where 'it will stop'. For this reason 1315,15
the motion too will be limited. For things that move from one place to another 'are[358] at rest at the limits' when the motion is concluded, and before the beginning. But circular motion does not have a beginning and an end distinct from each other, since a circle does not. For this particular point on it is no 'more' a beginning or a middle or a
'limit' than any other <point is>. And so if nothing prevents them from 1315,20
all being middles not in actuality but potentially, and none of them is compelled to be an end or a beginning in actuality as <happens> in the case of a straight line, but all of them <can be beginnings and ends> potentially, like the points on a straight line between the beginning and the end,[359] what moves can move continuously as on the points in between <the beginning and the end> on a straight line, because there is no beginning, middle or end in actuality. In this way
too, what moves over an entire circle can move continuously, because 1315,25
no matter what point of the circle the thing that is moving on it is at, it is at what is potentially but not actually a beginning and a middle and an end, because none of the points on a circle is determinately a beginning, a middle or an end in actuality, but any one that is taken
can be 'a beginning, a middle and' an end. This is how motion on a 1315,30
circle will be too – never coming to be at what is in actuality 'a beginning', middle or end. And this is why it is continuous, just as the motion on a straight line between the limits can be continuous too (at least this much of it), but when it reaches the actual end of the straight line, it is at rest, whereas motion on a circle can never be at
rest since no end exists in actuality. For it is always potentially but 1315,35
never in actuality 'at a beginning' and a middle 'and an end'. Therefore, he says, whereas what is in motion on a straight line is sometimes in motion and sometimes at rest, because the motion, being finite, is followed by a state of rest, 'a sphere' moving circularly at the
same time 'is' 'in motion and at rest'; <it is at rest> not because the 1316,1
motion is followed by a state of rest (for in actuality there is no end

of the motion), but because what is moving always occupies 'the same' 'place'. In the case of rectilinear motion, the cause both of the motion's being interrupted and of the moving thing's not being permanently in the same place is the fact that the beginning and the middle on the very straight line on which the motion takes place are distinct (for what moves comes from the beginning to the end, which is distinct from the beginning, and moves from place to place as it proceeds from the beginning to the end, and when it comes to be at the end, which exists in actuality, it terminates its motion). Correspondingly, with what moves circularly, since the beginning, the middle and the end are not distinct from one another and are not on the circle on which the motion takes place, but inside it, they do not interrupt the motion or make what moves move from place to place. But the centre is 'the beginning' 'and middle' 'and end' of the circle: the beginning because the circle exists at an equal distance from the centre, the end because all the radii terminate at it, and the middle because it is equally distant from the circle in all directions. So, since circular motion takes place not towards the end or towards the beginning or towards the middle of the circle (these being distinct as on a straight line) but 'around the' centre, i.e. around the beginning, the end and the middle, which are not distinct, for this reason it can be continuous and also occurs in the same place. For since what moves is always equally far from the end and does not approach it more and more, it is not compelled to go towards the end or to depart from the place around which the motion takes place, which always remains fixed.

265b8-11 Also, the next results follow reciprocally: [because rotation is (b10) the measure of motions, it must be the primary kind of motion, since all things are measured by what is primary.] And also because it is primary it is the measure of the others.

From the fact that circular motion is the measure of the other motions he proves that it is the primary <kind of motion>. That it is the measure is evidently clear, for day and night are the time of circular motion, and it is by this and its parts that we measure all motions, and it was also proved in the fourth book of the present treatise,[360] when he was discussing time, that circular motion is the measure of the other motions. That the measure is prior to what is measured is clear, if in fact everything that is measured is measured by the simplest thing in the same species: the <thing> of many cubits <being measured> by the cubit, that of many feet by the foot, that of fingers by the finger, and a number by the unit. And what is simpler is prior to what is more composite. He does well to remark that not only is its primacy proved from its being a measure, but also the fact that it is

the measure is inferred from the fact that circular motion is primary,
which has been demonstrated in many proofs. What is primary is
simplest, and among things that are measured, the simplest is the 1317,1
measure. For being primary and being a measure are mutually
implied by one another, which he calls following 'reciprocally'. Alex-
ander, who found the text reading: 'For because locomotion[361] is the
measure of motions, it must be primary', says 'He said "locomotion" 1317,5
instead of "circular motion," employing the term for the common
thing[362] to apply to it.'[363] However, in some manuscripts I have found
'rotation' instead of 'locomotion'.

> **265b11-16** Further, [only] circular motion can be uniform. [For things in rectilinear <motion> undergo locomotion nonuniformly from the beginning and towards the end. For the farther[364] things move away from what is at rest, the faster they move. But with (b15) circular motion alone neither beginning nor end is by nature located on it,[365]] but outside.

He says that the fact that the beginning and the end are not on the 1317,10
circle 'but outside', which is the reason why circular motion is con-
tinuous and occurs in the same place,[366] is also the reason why it is
the 'only' 'uniform' motion that can be undergone. This too proves that
it is more complete and primary in nature, for the uniform is more
complete than and prior to the non-uniform. He deduces the uniform- 1317,15
ity of circular motion in turn through a comparison with what is in
rectilinear motion and with its non-uniformity. 'For' 'things' that are
in 'rectilinear' motion whether natural or unnatural 'undergo locomo-
tion' 'from the beginning' 'towards the end' 'non-uniformly'. Things
that are moved unnaturally and by force, (a) if they are thrown they
move faster near the beginning and slower when they are moving
away from it as the power of the thrower is then being exhausted, and 1317,20
(b) if they are pulled they move slower at the beginning and faster
when they approach what pulls them.[367] On the other hand, it is a
matter of conviction that things that move naturally move faster
when they are already approaching their proper places. 'The farther
things' that move naturally, he says, 'move away from what is at rest',
that is to say, from that <place> in which they are at rest unnaturally,
i.e. before they undergo this natural motion, 'the faster' 'they move'. 1317,25
For after its period of rest at the top, which is unnatural, a dirt clod
moving naturally to the bottom moves faster near the bottom, as fire
does near the top, and this happens because they hasten from a
beginning that is alien towards what is their proper end. But only
'circular motion' is always uniform since it is always equidistant from
its beginning and its end, which is the centre, moving neither away 1317,30
from it nor towards it, but around it. For the 'beginning' and the 'end'

are not 'on' the circle, so that it may move from it and to it, 'but outside', or rather inside.[368] This is why it moves around it evenly, always being an equal distance away. But when what is moving naturally gets near its proper place, it moves faster either since it is moving towards something of the same kind, or because what im-
1317,35 pedes and is divided[369] is less when <the moving thing> approaches <its natural place>, and what is less is easier to divide, or because
1318,1 what is near its proper place is more similar to it and not as contrary to it as the place from which it started the motion, or because it starts its motion to its proper place when it has just changed from its contrary and has not yet wholly got rid of its contrary's qualities from which it was changing into these, but it becomes more pure as it
1318,5 proceeds. So if it is because of the change into what it has become that it undergoes this particular motion, when it changes to it wholly and <becomes> purely <it>, it will undergo the motion faster. For then it also possesses more completely the tendency in virtue of which it moves.

265b17-266a5 That locomotion in respect of place is the primary kind of motion [is testified by everyone who has made mention of motion. For they attribute its origins to things that cause this kind of motion. For separation (b20) and combination are motions in respect of place, and this is how Love and Strife produce motion: the latter separates and the former combines. Also Anaxagoras declares that mind, which is the first mover, separates. Likewise for everyone who says that there is no cause of this kind, but that motion takes place on account of the void. They (b25) too say that nature undergoes motion in respect of place (since motion that is due to the void is locomotion[370] in respect of place), but they think that no other kind of motion belongs to the primary entities, only to their compounds. For they declare that these increase, decrease and undergo alteration because the indivisible bodies are combined and separated. (b30) The same is true for all those who bring about generation and perishing through density and rarity, since it is by means of combination and separation that they produce these. Also, in addition to these people are those who make soul the cause of motion. They declare that that which moves itself is the origin of things that undergo motion, and animals and all living things (266a1) make themselves undergo motion in respect of place. And we say that the only thing that is in motion in the strict sense is that which is in motion in respect of place.[371] But if it is at rest in the same place, while increasing or decreasing or happening to undergo alteration, we say it is moved in some respect,] but not that it is moved (a5) without qualification.

Aristotle's practice after his demonstrations is to introduce the testi- 1318,10 monies of his predecessors as agreeing with his demonstrations, in order on the one hand to teach and compel his readers through his demonstrations, and on the other to make the belief more certain in his hearers through the testimony of the others; he does not employ the testimony of his predecessors[372] as a demonstration, as is the 1318,15 practice of more recent writers. This is what he does now, confirming that motion in place is the primary kind of motion also from the fact that everyone who had declared his opinion on these matters was brought to this view as if guided by nature itself, as he himself says elsewhere.[373] As evidence that the locomotion 'in respect of place' is posited as the primary kind of motion he introduces the fact that even the earlier natural philosophers who mentioned motion assigned as 1318,20 'the origins' and causes of motion the things that cause 'this kind of motion'. 'For separation and combination' 'are' 'motions in respect of place'. And 'Love and Strife', the efficient causes in Empedocles, cause motion by virtue of locomotion; the former is said to combine the elements, the latter to separate them. Empedocles says: 'At one time 1318,25 all coming together into one by Love and at another each being borne apart by the hatred of Strife.'[374] And again: 'For thus in its course it sometimes chanced to meet with <the other elements> in this way, but often otherwise.'[375] And in Anaxagoras, Mind, which arranges and moves the homoeomeries from the beginning, is said 'to separate[376]' 1318,30 them.[377]

This is the view of those who put forward an efficient cause of the generation of existing things. But <it is the view> also of those who do not mention an efficient cause but who do speak about motion, like Democritus and his followers. 'They too say that nature', i.e. the natural, primary, indivisible bodies, 'undergoes motion in respect of place' on account of the void. For these are what they called 'nature' and they said that these things, moving by virtue of the weight in 1318,35 them, move in respect of place through the void, which yields and does 1319,1 not resist.[378] For they said that they 'will wrestle around'.[379] And this is not only the primary motion but also the only kind of motion that they assign to the elements, whereas they assign the others to the things composed of the elements. For they declare that 'these increase, decrease and undergo alteration' and are generated and perish when the first bodies are joined with <one another> and separated off <from one another>. Now if they ascribe motion through the void to the first <bodies> and 'motion that is due to the void is locomotion in respect of place', it is clear that they considered this to be the primary kind of motion. For the motion of the primary <bodies> is primary.

It must be acknowledged that whereas many manuscripts have the
1319,10 text as follows: 'they declare that' 'because the indivisible bodies are combined and separated', Alexander writes 'combined and altered' and expounds accordingly: it is because of the different position and order of the atoms that they say that the compounds alter, since it is not the case <that they alter> because the atoms do – for he proclaims that the atoms do not have any other motion except motion in place.
1319,15 But 'altered' has been said instead of 'the indivisible bodies taking a different[380] position and order'.

After saying how those who posit a plurality of principles of things that are generated – both those who employ efficient causes and those who arrange the generation of things without this – reckoned locomo-
1319,20 tion to be the primary kind of motion, he next says how those too who say that the principle and element is one, like Thales, Anaximenes, Anaximander, Heraclitus, and their followers,[381] who deal with nature too, posit locomotion as the primary kind of motion. For they explain generation and perishing in terms of condensation and rarefaction, and condensation is a form of combination and rarefaction a form of separation. In fact, I think that both those who say <that there
1319,25 is> one <element> and called the affections that befall it condensation and rarefaction, and those who say that there are more <than one> and called their affections combination and separation, spoke appropriately. Both the former and the latter terms indicate motion in place.

The fourth group of witnesses he calls are 'those who make the soul the cause of motion', as Plato, after proving in the *Phaedrus* that the
1319,30 soul is a self-mover, says that it always moves itself and 'is the' source and 'origin' of motion for everything else that is moved.[382] These people too ascribe motion in place proximately to the soul. For even if according to them the soul is the cause of the other kinds of motion as a source and origin of every motion, still it does so via motion in place, as Aristotle too declares that it is by means of the first mover
1320,1 via the heavenly bodies' motion in place, that the other motions, which occur in the sublunary region, also take place.[383] Also for the Aristotelian self-mover, the animal that moves itself by virtue of its soul, the motion it causes in itself is motion in place. After saying that 'animals' move themselves, he adds 'and all living things',[384] not
1320,5 meaning by 'living thing' his generalized term 'animal' which holds of plants too (since plants do not move in place[385]), but after saying that 'animals' move themselves he adds 'and all living things', meaning that which has in itself a soul that causes motion and through which the animal moves itself.

In addition to this he also introduces testimony from the customary use of terms to confirm that locomotion is the primary kind of motion. For, he says, we say that 'the only thing that is in motion' without qualification 'and' 'in the strict sense is that which is in motion in respect of place', while what is at rest in respect of place but undergoing any of the other kinds of motion, such as increasing, decreasing, or altering, we say is not 'moved without qualification', but only 'in some respect'. For we say it is increasing or altering, but what is moving in place we say is moving without qualification. At any rate we say that the heavenly bodies too move without qualification because their motion too is in respect of place. Democritus and his followers said that motion in place is the only kind of motion, for they said that even things that are being altered are changing in place but this escapes notice because they do not change place as a whole but in parts.[386] The Stoics said that motion in place underlies every motion, existing either over great distances or over distances that are observable by reason.[387]

> **266a6-9** Therefore, we have stated that there has always been motion [and always will be for all of time. We have also stated what the origin of the eternal motion is, and also what kind of motion is primary and what kind of motion alone can be eternal.] And we have stated that the first mover is unmoved.

Then, coming to the end of his discussion, he first briefly reminds us of what has been proved, on the basis of which he demonstrates the rest. It was proved immediately at the beginning of the book that motion is eternal, being ungenerated and imperishable, and therefore never started to exist and does not have an end.[388] It was also proved 'what the origin' and proximate cause 'of the eternal motion is' – not something that is moved by anything else (for that way we must go to infinity), but what is moved by itself.[389] It was also proved 'what kind of motion is primary' according to all the significations of primary: locomotion is prior to the other species of motion,[390] and of locomotion circular motion is prior to rectilinear motion, and this 'alone can' 'be' one and continuous, and therefore 'eternal'.[391] It was also proved that the primary mover must be 'unmoved' – both the motioncausing element in self-movers and what is completely transcendent, that is not moved even incidentally, in the way in which the motion-causing element in self-movers was said to be moved.[392]

<CHAPTER 10>

> **266a10-23** [Let us now state] that this [the first mover] must be without parts [or magnitude, after first determining <the premises> that are prior to this <conclusion>. One of these is that

nothing finite can cause motion for an infinite time. There are three things: the mover, the moved, and, third, that in which <the motion takes place>, namely, the time. These are either all infinite, (a15) or all finite, or some of them (i.e. two or one) <are finite>. Let A be the mover, B the moved, and C an infinite time. Let D cause some part, E, of B to move. The time <of this motion> will not be equal to C, since what is larger <is moved> in more time. And so the time, F, is not infinite. Further, in this way by adding to D[393] I will exhaust A and (a20) <by adding> to E <I will exhaust> B. But I will not exhaust the time by continually subtracting an equal amount, because it is infinite. And so all of A will make the whole of B move in a finite time that is a part of C.] Therefore nothing can be moved with an infinite motion by a finite thing.

Having proved earlier that the primary mover is unmoved and eternal,[394] he now proposes to prove that it is 'without parts', and
1321,5 consequently that the motion it generates is not bodily or due to force. For this[395] is how motions generated by bodies occur that are related in such a way that the greater power causes motion over a greater length of time and makes the same thing move in a shorter time,[396] while the smaller power causes motion over a shorter period of time and <makes the same thing move> in a longer time.[397] This is how
1321,10 bodily motions take place. For a mule might move a mill for a whole day, and a beloved human only until midday, and the mule turns it more quickly, the human more slowly.

He proves that the primary mover is incorporeal and 'without parts' or magnitude by eliminating the opposite claim, that it possesses magnitude. For it must either possess magnitude or be without magnitude. That it does not possess magnitude he proves on the basis
1321,15 of a division which he puts as follows at the end of the book when he concludes the argument:[398] if it possesses magnitude, it must be either finite or infinite. But there cannot be an infinite magnitude, as was proved in the third book of the present treatise.[399] So, if it is finite in magnitude, it possesses either finite or infinite power; there is no other possibility.

1321,20 Now if what has finite power cannot cause motion in an infinite time (as he will prove in the first argument), and if nothing finite in magnitude possesses infinite power (as he proves in what follows), what causes eternal motion in an infinite time will clearly in no way possess magnitude or be divisible. Therefore it is not divisible into parts and is without magnitude.[400]

He says that for the supporting argument for this claim it is first
1321,25 necessary to prove in advance the prior lemmas from which the argument will be conducted. For premises are prior to deductions

because deductions depend on premises. He will first prove that no finite power can 'move'[401] anything finite over 'an infinite time', and second, 'that' there cannot 'be' 'an infinite' 'power' 'in a finite magnitude'.[402] When this is proved, it is inferred that what is moved for an 1321,30
infinite time is moved by no magnitude, if in fact no infinite magnitude exists, no finite magnitude can possess infinite power, and nothing can be moved for an infinite time by a finite power.

He first proves that nothing that is finite and possesses finite power can move anything finite in an infinite time. He takes for granted a principle that in fact he evidently assumes in the following 1321,35
argument, that something that possesses a greater power moves the same thing in less time than a smaller power does. The proof is as follows. There being three things – the mover, the moved, and, third, 1322,1
the time in which the moved is moved – these must be 'either all finite' 'or all infinite', or some are finite and others infinite (as a person would claim who held that something finite is moved by something finite in an infinite time, which is precisely his task to eliminate). Now he hypothesizes that both 'the' magnitude 'that causes motion',[403] which he calls 'A', and 'B', 'the moved', are finite, and that the 1322,5
time C, in which B is moved by A, is infinite. Then, taking some part, D, of the finite mover A, he hypothesizes that it causes 'some' 'part' of 'B', namely 'E', to move. But it is necessary to assume that D, the mover, and E, the moved, are not in the same proportion to each other as that in which A, the whole mover, is to B, the whole thing moved 1322,10
by it, but that D, the mover, exceeds E, the moved. It is necessary to supply this in thought if what follows is to be inferred consistently – that D be a third of A, for example, and E a tenth of B. In this way D will move E in less time than A moved B. And so, the time in which E is moved by D is not infinite. For B was moved by A in an infinite 1322,15
time, while having a bigger proportion to A than E does to D. But 'what is' proportionally 'larger' is moved 'in more time', and the larger power moves the smaller thing in less time.

Now let F, which is finite, be the part of C, the infinite time, in which E is moved by D. Then if I add another such thing to D, i.e. 1322,20
another third of A, and again another such, 'I will exhaust A', that is, I will take the whole of A. Now if D moved E in time F, which is a part of the infinite time C, and if something else of the same magnitude is added to the mover, clearly the power D when doubled will move E in less time – in half of F. And if an equal amount is added again 1322,25
to the moving power, it will be the whole power A, if D was a third of it. And it is clear that it will move E in still less time. If, then, we hold A, the mover, fixed, and add equal amounts to E, the moved – if we add one, A will move it in double the time it took to move E; if we add two, <it will move it> in three times <the time>. And this will be F, in which D, one third of A, moved E, which we have now tripled. For 1322,30

when both are tripled – both the mover (D) and the moved (E) – the same time remains. So when we reach the tenth stage in adding to E, we will have taken B, of which E was one tenth. So B will be moved
1322,35 by A in ten times the time in which E was moved by A. But it was posited that B is moved by A in C, which is infinite. Therefore the same thing will be moved by the same thing both in a finite and in an infinite time, which is impossible. For by adding to E until we take the whole of B, either we will exhaust the time C too, or we will not. But if we exhaust it, it is not infinite. For what is measured out is not
1323,1 infinite, and it will be measured out by F, which is finite, which was the first <period of> time that was taken.[404] But if we do not exhaust it and some time is left over after the subtraction of all the times equal to F that were in B, A will not have moved B in the whole of C but in
1323,5 a part of it equal to F taken as many times as the things in B were taken. This is what he means when he says 'by adding to D I will exhaust A' (since the mover is given as finite) 'and <by adding> to E <I will exhaust> B' (since it is given that what is moved is finite too). 'But I will not exhaust the time', since it is infinite.

But[405] it is possible to keep the mover (A) the same, and subtract
1323,10 a part of what is moved (B), and A will clearly move this in less than the infinite time; therefore, in a finite time. Now if we add together the part of B that is taken and that is moved in a finite time, as many times as correspond to the fraction it is of B,[406] the time[407] too will be as many times more than the time in which A moved the part that was subtracted. Therefore it will be finite. Therefore B will be moved
1323,15 by A in both an infinite and a finite time, which is absurd. This conclusion was drawn from the facts that B is finite, and that everything finite is exhausted by subtracting equals from it.

He does well to use the word 'I will exhaust' rather than 'I will measure out'. For if when the things being subtracted are equal to the original <part that was subtracted>, and the last one is equal, we
1323,20 have not only exhausted A, but have also measured it out by D. And this will occur if at the beginning D is subtracted in such a way that A can be divided by it into some number of equal parts. But if the last part of A that is left over is no longer equal to D but less than it, A is still said to be exhausted by this kind of subtraction, but it is not also said to be measured out by D. The same account holds for B and E
1323,25 too. No matter whether A and B are measured out by the parts that are taken from them, or are only exhausted, the same account holds. For the absurdity followed not from its being measured out but from its being exhausted.

After proving this, he continues, 'therefore nothing can be moved with an infinite motion by a finite thing', since this follows from the
1323,30 fact that a motion that takes place in an infinite time is infinite. He says that '*nothing*' 'is moved[408] an with infinite motion' 'by a finite

thing', even though he has now proved that *what is finite* is not moved for an infinite time by a finite thing, since he had already proved that there is no infinite magnitude at all.

> **266a23-b6** [It is evident that what is finite cannot cause motion for an infinite time.] But that (a25) in general there cannot [be an infinite power] in a finite magnitude [is clear from the following considerations. Suppose that the greater power is always that which does an equal amount in less time – heating, for example, or sweetening or throwing, or in general causing motion. Therefore, what is affected must also be affected to some extent by a finite magnitude with infinite power, (a30) and in fact more than by anything else, since the infinite power is greater. But then in fact there cannot be any time <in which the thing is affected>. For if A is the time in which the infinite force were to be heating or pushing, and a finite force <were to accomplish the same result> in <time> AB, then by continually taking in addition to this (266b1) a greater finite force, at some time I will reach a stage at which it has caused the movement in time A. For by continually adding to a finite amount I will exceed any determinate amount, and likewise by subtracting I will fall below <any determinate amount>. Therefore, the finite <power> will cause a motion in an equal time to the infinite <power>. (b5) But this is impossible.] Therefore, nothing finite 1323,35
> can possess infinite power.

After proposing to prove that what proximately causes eternal motion (i.e. the primary mover), is not only unmoved but is without parts or magnitude either, and having as a result proved that there is no 1324,1
infinite magnitude at all among existing things, he first proves that it cannot be a finite magnitude that possesses finite power, since such a thing cannot cause an eternal motion. It is left to prove that it cannot be a finite magnitude that possesses an infinite power either. He 1324,5
proves this point not for the primary mover alone, but generally 'that in general there cannot be an infinite power in a finite magnitude'.

He proves this assuming as evident the axiom he used previously, that what has greater and more power moves the same thing[409] in less time than a smaller power does. Consequently, what is moved by 1324,10
a finite power would also be moved by an infinite power (which is more) in less time. Assuming these claims as evident, he proves that anyone who says that there is infinite power in a finite magnitude will have to admit something contrary to what he has postulated. For what has been moved by a finite power will not be moved by an infinite one, as he will prove. 1324,15

It was posited that what is moved by a smaller power is also moved

by a greater one, and in less time. That what is moved by a finite power will not be moved by an infinite one he proves by reductio ad impossibile, where the impossible consequence is that the same thing is moved in an equal time both by an infinite and by a finite power,
1324,20 whereas this is impossible since it is agreed that the greater power moves the same thing in less time. He says, 'but then in fact there cannot be any time' in which the infinite power will move it – for just as an infinite power has no ratio to any finite one, neither does the shortest time (in which the infinite power will cause the motion) <have any ratio> to the time in which the finite <power causes the
1324,25 same motion>. If, then, every finite time has a ratio to every finite time, there will be no time in which the infinite power will move it. He skips over the absurd consequence of this (that it does not cause motion at all if <it does> not <do so> in time, and further, that that infinite power is not something that causes motion[410], since there is no time in which <it does so>) and he proceeds to prove it, hypothesizing the opposite, namely that it does cause motion in time.

For, he says, let 'A' be 'the time' 'in which' the finite thing moves
1324,30 or changes something by means of its infinite power. Then what has finite power will move the same thing 'in' a longer time, 'AB'. Now if we double the thing that has the finite power, it clearly changes the same thing in a time shorter than AB, and if it <again> increases similarly, again <it will change it> in less time, and as long as we continue to make it increase and double the magnitude, and the power
1324,35 in proportion, the time will become smaller. And by continually adding to the magnitude and subtracting from the time, we will at some point reach A, the time in which it was posited that this same thing is moved by the infinite power. 'Therefore' 'the finite' power 'will move[411]' the same thing 'in an equal time' 'to the infinite <power>'. For since it is increasing by double the amount, the power always
1325,1 remains finite since every double of a finite <amount> is finite. So if it is absurd that a finite power and an infinite power move the same thing in an equal time, and the absurdity was a consequence of the claim that an infinite power causes motion in some time, and since the claim from which something absurd and impossible follows is itself absurd and impossible too, therefore, there is no time in which
1325,5 an infinite power in a finite thing will cause motion. Therefore it will not cause motion. And so, there will not be an infinite motive power in a finite thing either. If there were, it would cause motion and what is moved would be moved in time.

Alexander remarks, 'we should not start as Aristotle did by hypothesizing that some particular thing that is moved <is moved> in
1325,10 time A by a mover with infinite power, and infer as he did that the same thing will be moved in an equal time by something with finite power too. For', he declares, 'if this assumption were made, it will be

vulnerable to the objection that it is not true. For no one will grant
that what has been moved by an infinite power can also be moved by
a finite one. Instead, we should assume the converse. We should 1325,15
hypothesize that what has been moved in some time by a finite power
is also moved by an infinite one in a much shorter time. For clearly,
what is moved by a smaller power will also be moved by a larger one.
In fact he signifies as much by saying "Therefore, what is affected
must also be affected to some extent by a finite magnitude with
infinite power, and in fact more than by anything else, since the 1325,20
infinite power is greater." Here he assumes that what is affected by
a finite power is also affected the same[412] by an infinite one because
the infinite one is larger. And what is affected by a smaller power
must also be affected by a larger one.' This is what Alexander wrote,
in his very words.

But perhaps it is well and necessary that Aristotle starts with what 1325,25
possesses infinite power, and there is no need to assume the converse.
For his present purpose is to prove that no finite magnitude possesses
infinite power, and he proves this by reason of the fact that an infinite
power (if indeed one were to exist) causes something or imparts
motion in no time, which he proclaims by saying 'but then in fact there
cannot be any time'. The absurdity that is a consequence of the claim 1325,30
that an infinite power does not cause or impart motion in time, he
omits as evident. For if everything that comes to be or is moved in
any way whatsoever, comes to be or is moved in time, then anything
that does not come to be or is not moved in time neither comes to be
nor is moved. Therefore, that infinite power neither causes anything
nor imparts motion, and does not exist at all. That it does not cause 1325,35
anything or impart motion in time he proves, as I said above, by
hypothesizing the opposite, that it does cause <something> or impart
motion in time. And in inferring an impossible consequence, that a
larger and a smaller power, or rather, an infinite and a finite one (for
this is more absurd) move the same magnitude in an equal time, he
reasonably presupposes in such a line of argument that an infinite 1325,40
power moves some finite magnitude in finite time A. For this was the 1326,1
opposite of the claim that 'but then in fact there cannot be any time'.

Then, assuming that the same thing is moved in a longer time by
a finite power, he assumes nothing absurd, as Alexander supposes.
For he does not assume without qualification that what is moved by
an infinite power is also moved by a finite one, but assuming only that 1326,5
a finite magnitude is moved by a finite power in a finite time, which
is evident, he proves that a finite thing that is moved in a finite time
(no matter what magnitude both it and the time are), can, when the
power continually increases and the time becomes smaller, be moved
by a finite power in a time equal to A, the time in which the infinite 1326,10
power moved it. And this is absurd. The absurdity followed from the

hypothesis that an infinite power imparts motion in a finite time no
matter how small. For it must either not impart motion in time, so
that it does not impart motion at all, or if it imparts motion it must
do so in some time, and in a time shorter than that in which the finite
1326,15 power does, so that it does so in a finite time. Thus, this consequence
follows for one who claims that an infinite power imparts motion. For
by continually adding to a finite power it is possible to exceed any
power that is taken; and by subtracting from the shortest time[413] it is
possible to make the time still shorter to the point where the power
is increased and the time becomes smaller down to A, which proves
1326,20 that the same thing that is moved in time A by a finite power that
has increased, is moved in an equal time by an infinite power. Thus,
Aristotle does not assume that what is moved by an infinite power is
also moved by a finite one. He assumes only what is evident, that
what is finite is moved in a finite time by a finite power. He then
1326,25 proves as a consequence for anyone who claims that an infinite power
moves something in a finite time, that a finite power imparts motion
in an equal time to <the time of the motion caused by> the infinite
power. So if what I say is true, the argument does not need the
converse assumption.

'But it is possible to prove in the following way too', says Alexander,
'that a finite magnitude cannot contain infinite power. If we subtract
1326,30 a part of the magnitude, it is clear that this will not possess infinite
power; therefore <it will possess> finite <power>. Thus by continually
subtracting from the magnitude parts equal to the one that has been
subtracted, each of these parts possessing finite power, we will divide
the whole into finite magnitudes and powers. For there will be no part
of the whole that possesses infinite power (for it is posited that the
1326,35 whole is what possesses infinite power), since it is also possible to
prove in the same way for the part that *it* cannot possess infinite
power: neither can any part of *it* possess infinite power.'[414]

That Grammarian[415] whom I mentioned at the beginning of my
discussion of this book[416] thought it would be a great achievement if
1327,1 he could get a large number of lay people to despise the heaven and
the entire world on the grounds that those things are just as perish-
able as they[417] themselves are. Clearly <he would get them> to despise
the very creator too, if the maker of the world (which is subject to
generation and perishing) were demonstrated to be not a creator or
the god, father and author of existence of all existing things during
1327,5 the entire period before he brought things into creation (if in fact at
that time no existing thing had come to be). On the basis of gigantic
thoughts like these that man even had the gall to write in opposition
to what Aristotle proves in the first book of the *De Caelo* about the
eternity of the heaven and the world,[418] though he did not understand
what is said there, as I attempted to demonstrate in my commentary

on that work.[419] Further, he objects to the proofs that motion and time
are eternal that are found at the beginning of this book, confronting 1327,10
him [Aristotle] at the outset, as can be learned in detail from what I
said against him there.[420] He thinks that even the present demonstra-
tion of Aristotle, through which he proves that no finite body pos-
sesses infinite power, contributes to this goal. For if both the body of
the heaven and that of the world are finite, they possess finite power, 1327,15
and he thinks that what possesses finite power has immediately been
demonstrated to be perishable. Moreover, at the end of the argument
he says, 'Therefore it remains that both the whole heaven and each
of its parts possess finite power, and so do all sublunary <elements>'.
Indeed it is surprising that he doesn't add that he does too, since the 1327,20
entire goal of their religious zeal is to prove that both the heaven and
its creator are no different from them.

But I think those who read his arguments should notice that here
too he obviously fails to pay close attention to what Aristotle says and
misses its entire point. He thinks that Aristotle believes that it is the
same thing to possess infinite motive power and to be able to continue 1327,25
in motion for an infinite time. In fact he writes the words that made
it clear to us that Aristotle holds that no finite body possesses a power
that lasts an infinite time. Whence he declares that what can cause
eternal motion is a power that has existence in a body. And he does
not notice that (a) it is one thing to possess an infinite power all at 1327,30
once together that causes motion eternally and (b) quite another to
be able to continue in motion eternally, and that (a) what causes
motion eternally has the attribute of possessing infinite power all at
once together in actuality, but that (b) being infinite in the sense of
being able to be moved ad infinitum is an attribute of what is moved
eternally, and this not in virtue of an active motive power that exists
all at once together, but in virtue of a passive power that belongs 1327,35
potentially.

At the end of his comments on this book even Alexander, whom
that man holds up as support, says that in the case of what is moved
ad infinitum we should not speak of a power in virtue of which it is
moved, except homonymously. Therefore, even though infinite motive
power cannot belong all at once together to a finite body, nothing
prevents a finite body from having the attribute of being moved ad 1327,40
infinitum, because in the case of what <is moved> ad infinitum, the
power applied at any moment[421] is finite. In fact there is no need for
it to have a rest, when one <part of the power> after another is being
applied successively,[422] both because of what causes motion eternally
and because what is always inherent potentially in what is moved[423]
does dwell in it. Thus here too, as before, in his ignorance of the 1328,1
difference between 'infinite all at once together' and 'ad infinitum'
(the latter being infinite and sufficient to qualify something as eter-

nal), and in his ignorance of the fact that the former is an attribute of
what causes motion eternally, while the latter <is an attribute of> what
is moved eternally, that man held that anyone who proves that a finite
body cannot contain an infinite motive power all at once together has
thereby proved that a finite body cannot be moved ad infinitum either.
1328,10 But in this he is ignorant of what Aristotle states in the third book
of the present treatise about 'ad infinitum', some of which I cited at
the beginning of my discussion of this book,[424] and some of which I
shall cite now too. 'For in general the infinite has this mode of
existence: one thing is always being taken after another, and what is
taken is always finite, but always different.'[425] This property belongs
1328,15 to things whose existence is in coming to be. Motion and time are of
this sort, and also the generation of humans, as Aristotle too says,
and the division of magnitudes.[426] For since motion is in what is
moved, not in the mover, as was proved in the discussions of motion,[427]
what is always causing motion primarily, being unmoved, must
possess infinite power because <it causes motion> always, and this
1328,20 power is infinite all at once together. For if something is being moved
(as opposed to causing motion or being unmoved) in a way that
involves an infinity that comes to be, it will require something else
as its mover. And a finite body that is always being moved, is in
motion ad infinitum without possessing the infinite power all at once
together in the way that the mover <must possess it>. For what is
always imparting motion primarily is incorporeal and self-consti-
1328,25 tuted, and is the bestower[428] of its infinite power to itself, while what
is always being moved, being a finite body and deriving from the
primary mover its properties of always existing and always being
moved, cannot receive the motion all at once together, since in that
case it would no longer need to be moved by anything else, nor would
it be moving at all. For motion does not exist all at once together but
has its existence in coming to be. For if what is moved received it all
1328,30 at once together, it would have stopped and would not be moving. I
believe this is sufficient on the topic of the mover's property of being
infinite all at once together,[429] which is why it cannot be a finite body,
and also about how what is moved proximately by it changes ad
infinitum.

But that man, in his ignorance of the difference between these two
things, thought that what Aristotle proves here, that no finite body
1328,35 possesses infinite motive power, contributes to his claim that the
heaven has been demonstrated to be perishable. And yet, even if he
fails to notice the difference between causing motion and being
moved, and between what is actually infinite and what is ad infini-
tum, he should at least have realized that Aristotle would not have
contradicted himself so blatantly in this single book, proving at the
beginning that there must always be motion and that what is moved

is eternal,[430] and next that the body that undergoes circular motion 1328,40
is what is capable of undergoing a single, continuous eternal mo- 1329,1
tion,[431] and now saying that the heaven is perishable since it pos-
sesses a finite body, if this is how he understands not possessing
infinite power. And again, in the first book of the *De Caelo*, which
follows the present book, although Aristotle clearly proves that the
heaven and the entire world are eternal,[432] that man even thinks that 1329,5
he has contradicted these demonstrations. But what kind of chame-
leon could have performed the kinds of transformations this man has
supposed Aristotle underwent? In fact, he does not even notice this
because he does not even think that the arguments for the eternity
of the heaven, or the present argument, which proves that a finite
body possesses finite power, need interpreters of Aristotle's treat- 1329,10
ment. For it was obvious that in this discussion Aristotle denies that
infinite motive power is found all at once together in cases where the
body that is subject to motion is finite.

But since he is ignorant of the whole aim of the Aristotelian
demonstration and thinks that anyone who proves that a finite body
does not possess infinite motive power will have proved immediately 1329,15
that the heaven is perishable because it is finite, and since he sets
himself to prove both in this way and by means of arguments of his
own that a finite body possesses finite power, and infers, as if it
followed of necessity, that the heaven is perishable – come then, let
us run through these follies of his as well.

The first argument he puts contains this inference: the heavenly 1329,20
bodies are composed of matter and form; things composed of matter
need matter for their existence; things that need anything are not
self-sufficient; things that are not self-sufficient do not possess infi-
nite power. And he infers from this that the heavenly bodies by virtue
of their own nature do not possess infinite power and consequently
are perishable. However, it is clear that if the heavenly bodies needed
perishable matter for their existence, it would be reasonable for them 1329,25
to be perishable. But if in fact a person who holds that their matter
or substrate (whatever it is) is imperishable is not refuted, how will
that which needs an imperishable substrate be thought to be perish-
able? Further, the fact that it needs something and is not self-suffi-
cient in this respect does not necessitate that it is also perishable. For
what is not self-constituted but has its existence in something else 1329,30
might not possess infinite power; still, there is no immediate neces-
sity for it to be perishable in such a way that it will ever perish. For
as we have said, what is actually infinite all at once together is one
thing and what is ad infinitum is another.

'Second, if the essence of matter consists in being suitable to receive
all the forms, and it does not possess this power in vain, and the same 1329,35
matter cannot receive several forms at once, then by virtue of its own

definition matter cannot permanently harbor any single form. But if
this is so, then nothing that is composed of matter and form will be
imperishable, on account of its material nature.' He says that he
proved in the fourth book of *Against Aristotle* that one and the same
1330,1 prime matter underlies both the heavenly and the sublunary <ele-
ments>. What he says there received the appropriate examination
when I attempted to expound the first book of the *De Caelo*.[433] Here I
think I should remark only that if the matter is the same and is
1330,5 related similarly to all the forms, heavenly bodies and sublunary
<elements> must change into one another – and what could be more
impossible than that? There is no point in spending time on this. For
in fact, after conceiving this absurdity this man says, 'Even if someone
grants that the same matter does not underlie both the heavenly
bodies and the sublunary <elements>, still, since the matter of all the
heavenly bodies, which they considered a prime substrate, is one and
1330,10 the same, the so-called fifth body, and since in fact there are different
forms of this matter (viz., that of the sun, that of the moon, that of
each of the other stars, and also of each of the spheres), it is clear that
the matter of the heavenly bodies too is suitable to receive each of the
heavenly forms, even if on account of some stronger and transcendent
1330,15 cause it did not receive them. If, therefore, the matter of the heavenly
bodies, to the extent that it can be receptive of all those forms,[434] will
receive different forms at different times, it follows that on account
of the potentiality of their own matter none of the heavenly bodies
will be imperishable.'

The first thing worth remarking in this is that he thinks the prime
1330,20 matter of the heaven is the fifth body. But if (as he says earlier) the
matter of all heavenly and sublunary <elements> is one and the same,
how does it make sense that the fifth body is the matter of sublunary
<elements>? Next, if because of some stronger and transcendent
cause the matter of the heavenly bodies does not receive any other
form than what it now possesses, there is no need on account of its
matter for the heaven to be perishable; nor is there if, as that man
1330,25 says, contradicting himself, 'this too will receive different forms at
different times'. In this way too there is no need for the entire heaven
to be perishable, but if <it is perishable> at all, <it can> only <be> its
parts <that are perishable>, with the contraries changing into one
another as happens with sublunary <elements>. So, if there is a
single matter, all that follows from his arguments is this, that things
that are up evidently come to be down, and things that are down come
to be up, as drunken people suppose, with the result that the sun is
1330,30 here and a human is in heaven, and heavenly bodies grow up among
the things here. But what could be more absurd[435] than this? Alter-
natively, if the matter of the heaven, being single, can receive all the
celestial forms, either it has this ability in vain or at some time those

things too change into one another. Yet not even so will the result come out that he is so eager for – to prove the world perishable. For there is no need for even the sublunary <elements> ever to fail to be 1330,35
simply because their parts change into one another in virtue of their underlying matter. For where the perishing of one thing is the generation of another, how can such a composite whole perish? How much better and more befitting the creator god to say that – since he is always in the same condition and unvarying, and is unchanging in his substance, his potentiality, and his actuality – (a) what is proxi- 1331,1
mately created by him is eternal, lest he sometimes be a world-maker and sometimes be left idle, barren and isolated, while (b) the sublunary realm, whose generation is due to his eternal and eternally moving creations, (i) possesses parts that come to be and perish, so 1331,5
that even the furthest part of the world may have generation thanks to the goodness of the creator, but (ii) since he is always, he contrived that this too be eternal by making its generation perpetual, with the perishing of one thing being the generation of another.

In his third argument, thinking that he is following Plato, he tries to demonstrate that even if they do not perish, the heavenly bodies do not by their own nature possess infinite power but are held 1331,10
together by a power mightier than their own nature. '(a) If the heavenly bodies are composite', he says, 'and (b) if there is an account of the undoing[436] of any composite, and (c) if when there is an account of something's undoing there is also an account of its perishing (for the dissolution of a composite into its elements is its perishing),[437] and (d) if when there is an account of something's perishing that thing does not have infinite power, therefore, (e) the heavenly bodies by their own nature do not possess infinite power.' We will know shortly 1331,15
how Plato declares that though there is an account of undoing that the creator knows, yet he does not undo what was well fitted together by him.

He[438] continues: 'Even those who say that the heaven is not made of the four elements, but of the fifth body, even they declare it to be a compound of substrate (the fifth body) and form (solar or lunar). And if someone subtracts the forms of all things, clearly he will leave 1331,20
only their three-dimensional extension, in respect of which none of the heavenly bodies will differ from any other, or from the bodies found among us either, and so since there is an account of the composition of the heavenly bodies too, there will also be an account of their undoing and perishing. Consequently they do not possess infinite power even if they never perish, being held together as Plato says,[439] by a mightier bond than their own nature, the will of god.' In 1331,25
fact, it is clear from this argument too that because of his vain belief he has even forgotten his original purpose. For it was his task to demonstrate that the heaven is subject to generation and perishing.

But now he concedes that although through its own nature it possesses finite power and is perishable, still, through the will of god it
1331,30 remains unable to be undone and imperishable. In fact it is clear that the will of god would not grant this good to what is not suitable to receive it. But even if the heaven, being subject to motion and finite, does not possess infinite motive power all at once together, nevertheless it has a natural tendency to undergo motion ad infinitum.

Further, he does not do well to hypothesize that the fifth body is
1331,35 composite in such a way that there is an account of its undoing. For where there is an account of a thing's undoing, that thing has a substrate that does not harbour forms, and it is composed of opposite forms that change into one another. However, the heavenly bodies possess the form of circular motion, and it does not have a contrary, since circular motion has no contrary, as Aristotle proves in the first
1332,1 book of the *De Caelo*.[440] I think I have demonstrated that this man who has spoken against these demonstrations has understood nothing that Aristotle says.

He next attempts to prove that the total amounts of the elements have finite power, and the heavenly bodies likewise. This is not at all
1332,5 absurd, since in fact being finite they do not possess infinite power all at once together. But it is very absurd to think that what is perishable is the same as what does not possess infinite power all at once together. He says it was proved in the *Physics* that it is not true that any chance form, either composite or simple, exists in any chance magnitude. A human could not become the size of a finger or a
1332,10 mosquito a cubit long.[441] Nor does water exist in any chance magnitude, but when it is carved up so finely that the magnitude that results from the process of cutting is smaller than <a magnitude> that is of a nature to receive the form of water, the water is destroyed. This is what happens when we make water disappear by touching it with our fingertip, spreading out the moisture and rubbing it and
1332,15 dividing up the form of the water into very small bits. 'From this it will be proved', he says, 'that the total amounts of the elements do not participate in infinite power either. For we see that the smaller the magnitudes of the elements become, the faster they perish, and the larger they become the slower <they do so>. Let it be posited then that a ladleful of water can last one year. Then each amount of water
1332,20 of the same magnitude will last the same time. Yet in fact, since the mass of water as a whole is finite, clearly the entire mass will be measured out by the ladle, and the whole amount of water will be divided into a finite number of ladlefuls. But each of the parts has
1332,25 finite power, so what is composed of them all will have finite power too. The same holds for the other <elements> too. Thus, up to now it has been proven through these arguments', he says, 'that each of the

four elements in its total amount has finite power.' This is what he says in pretty much these very words.

Now we should first notice what need he has of the assumption that not any chance form occurs in any chance magnitude. It is not needed by the hypothesis about the ladleful of water, that if the water is divided 1332,30
into ladlefuls each of which lasts a year, the whole will last a year too; for the time of the divided ladlefuls[442] is not added together.[443]

So much for the unsoundness of the hypothesis. But we should realize that when he hypothesizes that the entire amount of water perishes in a finite time, he does not take into account the other elements that change into water during that time. In fact, if parts of 1332,35
the other elements change into water just as parts of water change into other elements when they perish, and this is perpetual and eternal because the perishing of one is the generation of another, and if the cause of perishing and generation of sublunary elements is chiefly the eternal motion along the ecliptic, how can the total 1332,40
amounts of the elements perish?

Thinking that he has proved this soundly, that man adds, 'we will 1333,1
prove by the same argument that all the heavenly bodies have finite power too' – although in fact he does not even advance the same argument. He says that none of the heavenly forms is of a nature to be established in any chance magnitude, and neither is the entire world – <a principle> that[444] he did not employ previously. 'Now if every body', he declares, 'is divisible ad infinitum, and the things in 1333,5
the heavens[445] too are bodies, they too are consequently divisible ad infinitum in terms of their own definition insofar as they have extension, even if they are not divided – just as matter too by its own definition is formless, even if it is never without form. Further, what has extension in three dimensions is in terms of its own definition without qualities even if it is never without quality. And the cause of 1333,10
the fact that it is not divided or is difficult to divide, or that it is not divisible by us (as is the case with adamant) is the form that supervenes upon it from without. So if, being bodies, the things in the heavens are divisible ad infinitum in virtue of being magnitudes, if someone were to divide them in virtue of their definition as we separate forms from matter in virtue of *its* definition, it is clear that the process of cutting will reach some magnitude so small that none of those forms will exist in it. Consequently, the forms will perish as 1333,15
soon as such a division is made. So if, in that they are bodies that admit division because of the proper definition of their nature, they admit perishing – our thought being the thing that brings into actuality what belongs to them potentially, and if nothing for which there is an account of its perishing has infinite power by nature, therefore none of the heavenly bodies will have infinite power by its own nature. For if something has infinite power by nature it 1333,20

should not be suspected that there is an account of its perishing or that it perishes by mere thought, as is the case with things that are in every way simple and separate from any relation to bodies. For we cannot even imagine what the natural definition[446] will be in virtue of which they might be thought of as perishing. For this is what is
1333,25 now being investigated: what is the natural definition of each thing, and not what additional attribute some things acquire from a transcendent cause.

'So even if it were granted that it is because they are held together by the will of god that the heavenly bodies do not perish, still in no way does it follow that they will not, by their own nature, also admit an account of their perishing. But if there is an account of their perishing, they do not have infinite power. What would keep this account of their perishing which is comprehended by our thought from ever coming into
1333,30 actuality?[447] For the fact that they undergo none of the things that lead to perishing is no demonstration that they are imperishable by nature too, as we proved in the fourth book of *Against Aristotle*.'

I have cited this at such great length both in order to display the
1333,35 man's state to his readers and to make it still more evident that this man thinks that to possess finite power and to be perishable are the same thing, and that he has got no idea of what has its existence in being ad infinitum. Further, he clearly does not know how an entire body is said to be divisible ad infinitum. This point[448] is directed against those who hold that bodies are composed of indivisibles. In fact, he [Aristotle] proves that every body has extension and is
1333,40 divisible into parts, and that it too has parts of this sort,[449] and is not
1334,1 composed of indivisibles. Now what is divisible ad infinitum is not said to be so on the grounds that it is being completely divided and undoing its own continuity (for then it would no longer be divisible ad infinitum), but on the grounds that everything that is taken is divided into parts that are not pulled apart from one another but are separate and divided in extension[450] – although <they are> not <separate and divided> as things that are being cut and pulled apart
1334,5 from one another <are>; for even if they are never pulled apart, the bodies are still partitioned in extension and are divided into parts. For while what is per se incorporeal is without spatial extension and fits onto itself as a whole to a whole and is not divided into parts, an extended body is dispersed in its parts, different parts in different places, and is divided in this way. And this divisibility ad infinitum
1334,10 of bodies is an attribute generally of all bodies qua bodies. Even if my right side were not by nature capable of being cut off from my left side, all the same they are divided from one another and distinct and partitioned in actuality. Bodies that can be affected are also of a nature to be cut into bodies that are themselves cuttable, but those that cannot be affected are not cut either in virtue of their definition

or in the thoughts of sensible people, but they too are divided and 1334,15
partitioned in actuality in their corporeal extension even though they remain continuous.

But when that man heard that every body is divisible, or rather when he misheard it, in his own thought he carved up the heavenly bodies into smallest parts. This too is mad and deranged, although less so than to postulate that this kind of thought of his could bring 1334,20
things to actuality. For he says, 'They also admit perishing – our thought being the thing that brings into actuality what belongs to them potentially.' Now if the divisibility ad infinitum of bodies does not necessitate that the heavenly bodies ever be carved up into very tiny parts, neither is there any necessity for them to lose their form through the will of god or by their own nature. For in fact, the will of 1334,25
god does not bestow benefits upon any chance things, but only to those things that are suitable for them. And it is worth remarking that he thinks that whatever he may imagine in this way can come to be. For, he says, 'we cannot even imagine' the perishing of things that are unrelated to bodies and that are imperishable by nature. And he might say by contraposition (of which too he is ignorant) that things 1334,30
whose perishing we can imagine are not naturally imperishable – so much power did he assign to his own imagination! And what is surprising if since he is able to imagine that the heaven is perishable he immediately contends that in fact it perishes, since he suffered this same madness in discussing god too? But failing to understand the meaning of Plato's account in the voice of the creator of the 1334,35
undoing of the heavenly bodies,[451] he alters it into an account of their perishing, and as a result, imagining that the heaven perishes he speaks arrogantly against the heaven itself, saying 'What would keep this account of their perishing which is comprehended by our thought from ever coming into actuality?' He is not embarrassed that it has not yet suffered anything like what perishable things do. 1334,40

But all that he has written about this, as he says, in the fourth 1335,1
book of *Against Aristotle*, I have knocked down as best I could in expounding the first book of the *De Caelo*.[452] But for now I must object that for things that come to be and perish, the time from their prime to their end is equal to or greater than that from their beginning to 1335,5
their prime. That man thinks that the heaven came to be over six thousand years ago and he is certainly pleased to suppose that it is now in its last days.[453] How is it, then, that it has given us no indication that it is past its prime and heading towards its end? In fact, even if nothing else, we should certainly notice at least that it is moving slower if it is reaching the extremity of old age. But as things 1335,10
are, it is not making the days or nights or hours any longer now. This is clear when we compare activities that take place in those periods of time, such as farming, travelling and sailing, with those that took

place in the past. For a day's distance travelling is the same now, and
oxen plough the same amount – or even less – in a day, and water
1335,15 clocks that are made by the same methods take in and let out the
same amount of water each hour as before.

We must proceed to the fourth and last, as he says, of the argu-
ments, in which he wants to prove that the body of the heavens does
not possess infinite power. In fact, this is true even according to
1335,20 Aristotle, who denies to every body the property of possessing infinite
power all at once together, although he does attribute the properties
of being moved and existing ad infinitum to the body that is in circular
motion. But since that man believes he is eliminating its eternity
when he eliminates its infinite power, he thinks he can immediately
replace it with perishability. So he says that if the heavenly body
possesses infinite power, each of its parts must possess either infinite
1335,25 or finite power. 'But it does not possess infinite power', he says, 'since
what has infinite power is self-sufficient. But each of the parts of the
heaven needs the other parts too for the existence of the whole as well,
if indeed parts are relational entities. For they are parts of the whole
and they share their own powers with one another, as is most clearly
shown by the moon and by their combined activity on the things
1335,30 here.[454] Now if the parts are not self-sufficient', he says, 'they do not
possess infinite power either.'

And he thinks he proves this point in a different way too. For if
each of the parts possesses infinite power, 'the whole', he says, 'will
have power either equal to each of the parts or greater. But if it has
greater power, there will be something greater than the infinite,
which is impossible, and besides it will follow that the same part is
1335,35 both infinite and finite: infinite by hypothesis, and finite in that it is
exceeded by something. But it is also impossible for a whole to have
equal power to a part. For it is evidently seen everywhere that the
whole is more powerful than each of its parts. For even if the quality
of cold is the same in a part of the water as in the whole, still the
power that maintains their existence is not the same; the larger parts
1335,40 the bodies have, the longer they last. Therefore if the power of each
1336,1 part is not infinite, it is finite, so that the whole is finite too. For what
is composed of parts that are finite both in number and in power
cannot possess infinite power. Therefore it remains', he says, 'that the
whole heaven and each of its parts have finite power. And so do all
1336,5 the sublunary <elements>.'

The same reply can be given briefly to this too, that the infinite
power to cause motion all at once together, which Aristotle states is
an attribute of the primary mover which causes motion eternally, can
belong neither to the whole that is subject to motion nor to its parts.
Nor do they possess all at once together infinite power of being moved
and coming to be (for both motion and coming to be have their

existence in coming to be),[455] but they do possess it in virtue of being 1336,10
ad infinitum. And in this respect infinity is equally an attribute of
both whole and parts, if indeed the whole cannot exist without the
parts or the parts without the whole. For in fact with things that last
for some time, such as animals, we see that the existence of the parts
must be coextensive with that of the whole, although when they are
taken in respect of their own being and not as parts, some are actually 1336,15
longer lasting than the whole. For example, bones are longer lasting
than the integral state of the animals: when it is broken up the bones
persist. So it is not true that the extension of existence in the whole
is composed <of the extension of existence> of each of the parts and
of the power <of existence> in it. It is not true either that when added
together the parts endow the whole with a power of existence that is
as many times greater <than the power of existence in the parts> as 1336,20
the whole lasts in respect of existence, especially where the whole is
always composed of the same parts. For since the total amount of
water and of each of the elements is located in different parts at
different times, because some come to be and others perish, it is not
at all surprising if it is longer lasting than its parts in respect of
existence. On the other hand, although the total amount of the heaven
is always made up of the same parts, not even that man would say 1336,25
that it is longer lasting than its parts. So existing ad infinitum holds
equally of both the whole and the parts, and does not differ in respect
of extension in time (if indeed, as he too agrees, parts and whole are
relational entities and must coexist[456]) but in being more complete
and comprehensive in respect of its existence and perhaps in respect
of its time too. For the time of more complete things is more complete, 1336,30
not in that it extends for more years, but in that it includes in itself
the times of the more part-like things just as the existence of the parts
is also included in the existence of the whole. But enough of this man
who by now, I think, has received a sufficient examination on the
present points.

It is worth remarking that here too Aristotle puts forward the same 1336,35
thoughts as his teacher in different terms.[457] Plato too presents the
difference between the separate cause of the heavens and the heavens
themselves as due to their being. The cause is something that has 1337,1
real being, being always in the same condition and unvarying, and
not undergoing motion or change in its substance, its potentiality, or
its actuality, while he declares the heaven to be corporeal, and
consequently not self-constituted or possessed of real being.[458] For
what has real being is prior in nature to what comes to be; what is 1337,5
self-constituted <is prior in nature> to what has its being from
something else; and what is unchangeable <is prior in nature> to that
which changes in any way. And clearly, what has real being and is
self-constituted and unchangeable is the proximate cause of existence

of what is coming to be always and not <only> at some time, and of what undergoes change in respect of its activity, though not in respect of its substance – like things that undergo motion only in respect of
1337,10 place. For the procession[459] is continuous and undergoes gradual descent of level and does not descend immediately to what is altogether extended.[460]

But since what is generated proximately by that which is self-constituted and has real being, is not itself self-constituted but is created by what is self-constituted (this is why it is in a process of coming to be and is not), it does not receive its existence from the cause all at once together (for in that case it would no longer be something that
1337,15 was in a process of coming to be, but something that is). Moreover, it does not stop receiving it, both on account of its own nature (because after what always is, what is always in a process of coming to be has its subsistence prior to what is sometimes coming to be) and because what is always creating is engaging in activity towards what is being created, in the same respect and unvaryingly. And so it receives existence that is finite, but it is always receiving it.[461] This is why the transfer by what has real being and the reception by what is proxi-
1337,20 mately generated by it are both inexhaustible. And this is what is infinite in this way, i.e. ad infinitum.

This is what Plato signified in a few words after he introduced the creator in the *Timaeus* addressing the heavenly gods and saying, 'O gods of gods, the works of which I am creator and father cannot be undone – as long as I am willing.[462] All that is bound can be undone, but only an evil thing would be willing to undo what is well fitted
1337,25 together and in good condition. Therefore, since in fact you have come to be, you are not altogether immortal, but you will in no way be undone nor will you meet with death as your portion, since you have obtained the benefit of my will – a still greater and more authoritative bond than those with which you were bound together when you came to be.'[463] He says this, or rather he fills with the creator's thoughts the substances in the world that are able to understand the pre-existing
1337,30 arrangement of the universe[464] in the father. These substances exist, unitary and intellective, next in order,[465] and so already[466] do, in the case of each god, the souls that depend on these intelligences in virtue of the unitary fitting together of the whole incorporeal <world>. To these he says, 'gods of gods'. His word is creative production and he creates the gods of the gods – making the incorporeal gods the gods of the corporeal gods[467] because although it is more evident[468] that the
1337,35 revolving images of the heavenly gods should be called gods – receiving that appellation on account of their swift motion,[469] the gods of
1338,1 these gods are those that projected their[470] originative causes incorporeally and intellectually. For every god is an origin of divinity for something.

And he declares that he is the creator of the perceptible gods and the father of the works in them.[471] For the corporeal universe is created and brought to completion through a process of work. He calls the corporeal 'all that is bound' since it is visible and tangible, and therefore composed of fire, earth and the intermediate elements, and bound by the bond of proportion. For this, he says, is the best bond for the composition of bodies.[472] Everything that is bound can somehow be undone because the separation of the simples is the undoing of the compounds, and this separation pre-exists only in the creator and does so in virtue of the causative effect of his embracing[473] the simple elements in his mind. This is why he says that they can be undone only by the creator himself. For if the elements of the heavenly bodies in the intellective beings are unified through the unification that is accomplished only by intellect and are separated through a kind of purification that is brought about intellectually and in no other way, he reasonably declares that the account of undoing, that is, separation, pre-exists in him, but that things that are well fitted together are not undone; they were fitted together according to the harmony of the model of the world in[474] the creator. So he says, 'since in fact you have come to be', i.e. since you have projected the special corporeal nature that is composite and not self-constituted, and that did not remain in what has real being, but that has its existence in coming to be, 'you are not altogether immortal', for you will not have the kind of immortality that is all at once together (this is what 'you are not' means), but you will have immortality that is in a process of coming to be and is under construction. And you will not 'be undone' even if you have a composite nature, because you are constituted next in order[475] in virtue of the unification supplied by the intellective goodness.[476] For since this is how they were by nature, they received their unification as something adventitious.[477]

But the first composition cannot be undone even if it is a coming together of simples, since it was brought about next in order by the agency of the intellective unification together with simples that had not previously been separate. And thus they are not altogether immortal but they do not obtain death as their portion since they receive immortality that is in a process of coming to be ad infinitum. And this is how from the difference in their substances Plato brought to light the proximate cause of the heavenly bodies and the special nature of the heavenly bodies themselves.

Aristotle, on the other hand, more like a natural scientist,[478] begins with motion, and first proves that motion must be eternal,[479] that what is moved <must be eternal> even more so, since motion is in what is moved;[480] moreover, that what is moved must certainly be moved by something,[481] and that the primary mover, which causes eternal motion, must possess infinite power and be without parts,[482]

whereas the thing that is moved primarily by it in a single, continuous
motion is the body that is in circular motion.[483] Further, this is the
only moving thing that in its own right undergoes motion that is
eternal, single, and continuous,[484] and in addition it is always simul-
taneously at the end of its rotation and at the beginning. For every
1338,40 part of a circular motion is both the end of the prior rotation and the
beginning of the future <rotation>.[485] But when everything we take
1339,1 is both beginning and end, such a thing must be inexhaustible in its
own nature, in the same way that according to Plato what is moved
proximately by that which is always in the same condition and
unvarying, must by its own nature be eternal. But what is moved
1339,5 eternally is finite and not self-moved, but gets its motion and, accord-
ing to Plato, its existence, from something else,[486] and consequently
is subject to generation and is moved by something else. Therefore,
it cannot receive all at once together the power of causing motion or
of being moved. For motion has its existence in coming to be, like a
day, and is not all at once together. Therefore, if it undergoes eternal
1339,10 and infinite motion but not all at once together, it will clearly possess
eternity and infinitude as it comes to be, always undergoing motion
by virtue of its own nature and through the unmoved activity of the
mover, but not at the same time having the property of <being>
always, because it is moved by something else and is finite. And this
is what being moved ad infinitum is, any motion that is taken being
always at an end, and one being followed by another inexhaustibly.

1339,15 So just as Plato called the generation ad infinitum of the generable
able to be undone and not altogether immortal, but also unable to be
undone and not meeting with death as its portion,[487] so Aristotle
declares his opinion that motion ad infinitum is both finite in power
and eternal, since limit belongs to the heavenly body by its own nature
since it is corporeal, moved by something else, and finite, while
1339,20 infinitude <belongs to it> through its own nature proximately by the
agency of what is unmoved and always in the same condition – since
it exists and undergoes circular motion precisely because of the
motionlessness of what creates[488] and moves it. And so, according to
both philosophers the heaven is eternal and always moving both on
account of its own nature and on account of the cause that creates
and moves it.

1339,25 I suppose everyone who listens to Aristotle's arguments attentively
would investigate whether the primary mover causes motion tempo-
rally or atemporally. If temporally, (a) when he discusses that which
has infinite power, what does Aristotle himself mean by saying 'there
cannot be any' 'time' in which it causes movement? And (b) if the
1339,30 primary mover is unmoved and unsusceptible to any change in its
substance, in its potentiality, or in its actuality, how will time, which
is the measure of motion, measure the unmoved actualities of the

primary mover? But if it does not cause motion temporally, how is the primary thing that is moved by it moved in time? To these questions the response is that everything that causes motion and is itself moved, like things that cause motion corporeally by pushing, pulling, 1339,35
and throwing, causes motion temporally, not simply because it causes motion, but because it is moved. For time is the measure of motion, but the motion is not in the mover, but in the moved. But since the primary mover, being unmoved in both substance and actuality, is established as <being> beyond every motion, it both possesses its actualities atemporally as being beyond time and creates motion that 1339,40
stems from itself in what is moved by it. For if what is moved is not 1340,1
the first entity because it is moved by something, it is by the agency of what is unmoved that motion must exist in what is moved, and it is by the agency of what is atemporal that time must exist in what is moved temporally. For it is not the causing of motion but being moved that is measured by time, since motion is in the moved. But just as the motion of what is moved primarily is infinite but does not possess 1340,5
its infinitude all at once together, but as coming to be ad infinitum, so also the time that measures the primary motion is infinite in that it proceeds ad infinitum. We must next return to the continuous text of Aristotle from which we digressed.

266b6-20 Neither, then, can there be a finite <power> in an infinite <magnitude>. [Although there can be[489] more power in a smaller magnitude, all the more <can there be> more <power> in a larger <magnitude>. Let AB be <an> infinite <magnitude>. Then BC possesses a certain power which (b10) caused D to move in a certain time, call it EF. Then if I take twice the power of BC, it will cause the motion in half the time of EF (for let this be the proportion), and so it does so in FG. Therefore by continually taking things in this way I will never go completely through AB, but I will be taking a time that is increasingly smaller than the given time. Therefore the power <of AB> (b15) will be infinite, for it exceeds every finite power, if indeed[490] the time of every finite power must be finite (for if a certain amount of power causes motion in a certain time, a greater power will cause it in a time that is smaller, but determinate, in inverse proportion) and the whole power <of AB> is infinite; just like a number (b20) or a magnitude] that 1340,10
exceeds every determinate <amount>.

In the preceding passage he proved that by necessity no finite magnitude possesses infinite power, in order to prove in consequence that the primary mover, which possesses infinite power, is not a finite magnitude. But if there is no infinite magnitude among existing

1340,15 things at all, the primary mover has clearly been demonstrated to be without parts or magnitude, and this is why, as I said, he proved that by necessity no finite magnitude possesses infinite power. He now proposes to prove that an infinite magnitude does not possess finite power either – a point that is not necessary for his present purposes (since no infinite magnitude exists in actuality at all). But since it is
1340,20 the converse and is akin to the point previously demonstrated, and since it has theoretical interest in its own right, he proposes to demonstrate it too. He proves it, sketching the demonstration in a few words to begin with, as follows. He appears to object to his own proposal that 'in an infinite' magnitude there is not a 'finite' power,
1340,25 by saying, 'although there can be more power in a smaller magnitude'. We see this with the poisons of wild beasts and the seeds of both animals and plants. From a fig-seed comes a fig tree that is so large because of the great power that is in so small a body, and elephants are generated from tiny sperm, as are the numerous offspring of prolific animals. Now if this is true, what is to prevent the apparent
1340,30 converse of this from being true too, that also the largest body of all, the infinite, contains finite power? The prefixed adversative adverb 'although' shows that he is bringing this as an objection. Alexander says, 'it would be more useful for his present point to say "although there can be less power in a larger magnitude", since this does a better
1340,35 job of indicating that even an infinite magnitude can contain a finite power. But he said "although there can be more power in a smaller magnitude", indicating that he is assuming the former claim along with the latter.' Thus Alexander. But while to a person who says that
1341,1 no infinite magnitude contains a finite power someone who objects 'although there can be less power in a larger' says the same thing without any explanation, someone who says 'although there can be more <power> in a smaller' introduces the opposite point beginning with the opposite term. For 'less in a larger' is opposite to 'more in a
1341,5 smaller'. And perhaps it is even more evident that 'there can be more power in a smaller magnitude', as he actually puts it. For in fact it is disbelieved that so small an amount of poison has so much power of destruction, but it is even less evident that certain larger magnitudes have less power. So he proves what is less evident from what is more
1341,10 so. Such is the objection. He refutes it by saying 'all the more <can there be> more <power> in a larger <magnitude>'. For if there is more power in a smaller magnitude, in a larger magnitude of the same kind the power will be even more than in a smaller one. So if the magnitude is extended ad infinitum, its power too will be extended ad infinitum regardless of whether the original magnitude in the proportion with
1341,15 the power was smaller or larger. For as magnitude progresses ad infinitum, it draws the power along with it up to infinity, since the magnitude does not increase without the power <increasing too>.

After demonstrating in this way (by refuting the objection) that an
infinite magnitude will not possess finite power, he demonstrates the
same result through an illustration, as is his usual practice, since he 1341,20
in fact desires to geometrize nature. He assumes an 'infinite' magni-
tude 'AB', which by hypothesis possesses finite power, and from it
subtracts BC, which clearly possesses finite power too, since the
infinite line AB had finite power as well. He says, let the finite
magnitude BC with its finite power move a finite magnitude D 'in' 1341,25
'time' 'EF' (which is clearly determinate). For it has been proved that
no finite magnitude moves anything finite in an infinite time. So if
we subtract from the infinite AB AC, twice BC – a double magnitude
which possesses double the power – it will move D in half the time it
took its half, BC, to move it. But if in turn we subtract from the infinite
magnitude AB something that is double the magnitude CA, which 1341,30
possesses double the power found in AC (which moves D in the time
FG), the power taken will in turn move D in half of the time FG. So
if we keep doing this, AB will never cease providing us with an infinite
number of magnitudes for subtraction, each having double the power
of the one that was previously subtracted from it, nor will the time 1341,35
taken ever cease to have halves. Therefore, if the power does not give
out either, it too will be 'infinite' just like the magnitude. For an
infinite power is one that exceeds 'every' 'finite' 'power'. In general,
the infinite is that of which there is always something left over outside
of everything that is taken from it, and that exceeds what has been 1341,40
taken, as in fact is observed <to happen> with the power possessed
by an infinite magnitude.

Then, after proving on the basis of the increase ad infinitum of the 1342,1
magnitude being subtracted from the infinite that the power of an
infinite magnitude is infinite since it exceeds 'every' 'finite' 'power',
he proves the same thing on the basis of the decrease of time ad
infinitum, which he employs for this purpose, saying 'I will be taking
a time that is increasingly smaller than the given time'. Just as a 1342,5
magnitude is indefinite and infinite when it is always possible to take
more than that which is added and which is definite, so also time in
which it is always possible to take less than what is added, is
indefinite and becoming smaller ad infinitum. Therefore, if it has
been proved that a finite power causes the motion in a finite time,
while a time that is finite is determinate, it follows that also a finite 1342,10
power in an infinite magnitude, no matter how large it is (as long as
it is finite) will cause the motion in a determinate time. For if a power
of such and such a magnitude causes the motion in some determinate
time, if a power is taken that is larger than this by any amount
whatsoever, it too 'will cause it in a time that is smaller, but deter-
minate'. And indicating the ratio of the decrease in time, he says 'in
inverse proportion'. For the greater one power is than another, the 1342,15

less the time in which the greater power causes motion than the time
in which the smaller power does. Therefore he says in 'inverse
proportion', because the smaller time is to the larger as the larger
power is to the smaller. So if the power is always greater by the same
1342,20 degree as the time is less, and the time in which the power of the
infinite magnitude causes motion is less than any determinate
amount, the power that causes motion in this time exceeds any
determinate power. But in fact, the power that exceeds any deter-
minate power is infinite, just as an amount or magnitude that
exceeds any determinate <amount or magnitude> is infinite.
1342,25 Therefore, the power in an infinite magnitude is infinite, not finite.
This is how he infers that the power in it is infinite, on the basis
both of infinite magnitude and of the diminution ad infinitum of
the time in which it causes motion.

Now it can <be thought that there are> either one or two demon-
strations for these conclusions, but rather two, since in fact each
conclusion is drawn separately – the one being based on the magni-
1342,30 tude, when he says 'therefore the power will be infinite, for it exceeds
every finite power', the other being based on the time, when he says
'the whole power is infinite' since it exceeds 'every determinate
<amount>', 'just like a number or a magnitude'. After saying in his
illustration, 'then if I take twice the power of BC, it will cause the
1342,35 motion in half the time of EF', he continues 'for let this be the
proportion'. And so, the double magnitudes that are subtracted also
have double the power, and the magnitudes and powers are in the
same ratio. If this is hypothesized, then if half the magnitude has
made D move in time EF, double the magnitude will make D move
'in half' of the time 'EF', which he calls GF. And even if the things
1343,1 being subtracted from the infinite do not have a ratio of double to half,
so that double the magnitude straightforwardly has double the power,
the proof will not be impeded at all. For if they are not related in that
way, we will not look to the magnitude and in virtue of *that* subtract
double of what was already taken away, but we will look to the power
1343,5 in the magnitude that was subtracted first. And so, what is subtracted
from the infinite magnitude is a magnitude that possesses double the
power of magnitude BC, whatever the size of the magnitude that is
subtracted and whatever ratio it has to magnitude BC. For if the
subtraction takes place in this way, D will each time be moved by the
magnitude that is subtracted next, in half the time of that in which
it was moved by the preceding magnitude, because what is subtracted
1343,10 is double the power. For as he says, it has been proved and agreed
that the larger power moves the same thing in less time 'in inverse
proportion'.

266b20-7 This result can be proved in the following way too. [For we shall take a certain power that is in a finite magnitude, that is the same in kind as that in the infinite magnitude, and that will measure out the finite power in the infinite magnitude. (b25)] From these[491] considerations it is clear [that there cannot be an infinite power in a finite magnitude or a finite <power>[492] in an infinite <magnitude>.]

He also offers another demonstration of the same point, that there
cannot be a 'finite' power 'in an infinite' magnitude, in which he says 1343,15
some things clearly while leaving other things for us to fill in. 'For we
shall take' a kindred 'power that is in a finite' body, i.e. one that is of
the same kind[493] as the finite power that is 'in the infinite magnitude':
if the power in the infinite magnitude is heaviness <we will take> the
same power in the finite body too, and likewise if it is lightness. He 1343,20
takes a power of the same kind because he wants the power in the
infinite magnitude to be measured out by that in the finite one. For
powers that are of different kinds do not measure one another out.
Therefore, being of the same kind, the finite power in the finite
magnitude 'will measure out the finite' <power> 'in the infinite', in
such a way, however, as either to produce an even result or not to do
so, so that something is left over after the last application. But if this 1343,25
is so, then either the infinite magnitude will also be measured out by
the magnitude containing the power that measured out the power in
the infinite magnitude, and in this way the magnitude will no longer
be infinite, if in fact it is measured out – or the whole <magnitude>
will not be measured out when the power in it is being measured out,
and thus what is left of the infinite magnitude will not possess any 1343,30
power. So the finite power will not be in an infinite magnitude, but
in the finite magnitude that is measured out along with its power.
'But', Alexander declares, 'it would be more intelligible if we were to
subtract something from the infinite magnitude itself and use it and
its power to measure out the power in the infinite magnitude and the 1343,35
infinite magnitude itself. That way we will no longer need to add that
we should take a kindred power. For the power being subtracted from
the power in the infinite magnitude is of the same kind as the whole.'
But perhaps it was because he makes the previous demonstration
depend on subtraction from the infinite that Aristotle brings this one 1344,1
forward in a different way. He then draws the conclusions that have
just been demonstrated, that there cannot be 'an infinite power in a
finite magnitude' or a 'finite <power> in an infinite <magnitude>'.

266b27-267a12 It is a good idea [first] to deal with [a puzzle] that arises in connection with things that undergo locomotion.

[If everything that is moved is moved by something (except for things that move themselves), how are some things, such as things that are thrown, moved continuously (b30) if the mover is not in contact <with them>? If what causes the motion also at the same time moves something else, such as the air, which causes motion by being moved, it is similarly impossible that *this* is moved when the first <mover> is not in contact or causing it to move. Rather, they <must> all at the same time both[494] be in motion and have stopped (267a1) when the first mover stops, even if, like a magnet, it makes the thing that it moved such as[495] to cause motion. So on the one hand we must say this, that the first mover makes the air such,[496] or <makes> the water or something else that is of a nature[497] to cause motion and to be moved, able (a5) to cause motion. But it does not stop causing motion and being moved at the same time, but rather, when the mover stops making it move <it stops> being moved at the same time, but it is still a mover. This is in fact why it causes motion in something else that is next to it. And the same account holds for *this*. The motion begins to stop when the power of causing motion that occurs in the successive objects is less each time, and it finally stops when (a10) the prior object no longer makes <the next one> a mover, but only makes it be moved. And these must stop at the same time – the one as the mover and the other
1344,5 as what is moved,] and the motion[498] as well.

He proved what he had just proposed – that the primary mover which causes eternal motion is not only unmoved and eternal, but also without parts or magnitude – from the fact that if it had magnitude it would need to be either infinite or finite: if it has an infinite
1344,10 magnitude, it does not exist at all in actuality, but on the other hand it cannot be a finite magnitude because what causes eternal motion possesses infinite power, and a finite magnitude cannot contain infinite power. The entire argument about the primary mover depends on the axiom that 'everything that is moved is moved by something'. (If this were not assumed and if it is not necessary for
1344,15 anything that undergoes an infinite motion to be moved by something, what has been demonstrated about the primary mover will then have no place.[499] For if some things are moved without anything moving them, what is undergoing an eternal motion need not be moved by anything.) Therefore before drawing the consequence appropriate to what has been proved, he wants to put forward an objection that is brought against the axiom that everything that
1344,20 undergoes motion is moved by something, to set out the unsatisfactory refutation that was advanced, and to advance the appropriate one himself. The objection goes as follows: 'If' 'everything that is

moved is moved by something', 'how' do certain things that do not
contain a mover in themselves but that are moved from without,
remain 'continuously' in motion even when the mover is no longer
causing them to move, as happens with 'things that are thrown'? They
remain moving continuously for a long time even though what threw
them is no longer moving them. For they will then seem to be being 1344,25
moved by nothing and to eliminate <the principle> that 'everything
that is moved is moved by something'. He says 'if the mover is not in
contact', supposing that it is no longer in contact or remaining in a
condition of contact throughout the entire time of the motion. Then
he mentions the solution with which some people[500] attempt to solve
this puzzle. They say that the thrower 'moves' 'something else' – 'the 1344,30
air' – along with what is thrown, and that 'it moves'[501] what is thrown
'by being moved' by the thrower. He declares that this is 'similarly'
'impossible' as <in the case of> 'the first <mover>'. The argument
merely substitutes the air for what is thrown; the same argument
holds for the air too. If this is not moved by itself either (this being a
property of things with souls), but by the thrower, the air should have 1344,35
stopped being moved as soon as the thrower stopped moving it, and
'all' things that are moved should 'at the same time both be in motion 1345,1
and have stopped when the first mover stops' causing motion. So what
is thrown and the air must 'at the same time' 'both be in motion and
have stopped' being moved if indeed what causes motion and what is
moved are relative to one another: they are being moved as long as
the thrower causes motion and they stop when the mover stops. For 1345,5
if the air remains in motion, it will follow that '*this* is moved' when
there is 'not' anything that 'is in contact with it or causing it to move',
so that the air will be moved but will not be moved by anything. In
this way again it will not be the case that everything that is moved
is moved by something.

The text is transmitted variously: either 'but everything <must>
at the same time have been in motion and have stopped' or 'but they 1345,10
<must> all at the same time both be in motion and have stopped'.[502]
Perhaps the latter is better, since to have been moved is the same as
to have stopped. After saying that 'they' must 'all at the same time'
'be in motion and have stopped when the first mover stops', he
continues, 'even if, like a magnet, it makes the thing that it moved
such as to cause motion', i.e. even if it does not touch what is moved, 1345,15
but moves it via something in between, as a magnet moves the second
iron ring without being in contact with it, when the magnet attracts
the one next to it and *that* one attracts the one after *it*. For when that
happens the magnet makes the iron ring that it moved move the next
one, because it not only moves the one that is next to it but also
imparts motive power to it, so that *it* causes motion as well as being 1345,20
moved. And therefore, he says, if something causes motion in this way

'like a magnet', it must simultaneously 'both be in motion', when the primary mover moves it, 'and have stopped' its motion, 'when the' primary 'mover stops' causing motion. Taking the magnet as his starting point, he uses the same approach to solve the puzzle. For as with the magnet, where the air is not responsible <for the motion of
1345,25 the second iron ring> nor does what is moved remain <moving> when the mover has stopped, but the motive power given to the next ring[503] makes what is moved[504] cause motion as long as it is active, this is the account we must give in the other cases too, that 'the' primary 'mover' 'makes' what is proximately moved by it 'able to cause motion' – <it makes> 'the air such[505] or the water'. For these are above all the media through which things that are moved are moved, and they too
1345,30 are able to move them. He adds 'or something else' perhaps because of the magnet and iron or because of amber and straw. Also the bone of the hierax fish is said to attract gold, and this too may impart the same power to what it attracts proximately. Therefore, whether it is by pushing that it moves the things that are moved, as happens with things moved through air or water, or whether it is by pulling, as with
1345,35 the cases I have discussed, the first mover, e.g. the thrower, moves not only the missile but also the air or water through which the throw takes place, and not only does it move these things but it also imparts to them the power of causing motion, so that they cause motion as
1346,1 well as being moved, since they possess a natural capacity to cause motion and be moved simultaneously. Now this is what he declares we must say. For the sequel of the passage that says 'so on the one hand we must say this' is as follows. 'On the one hand' 'we must', then, if in fact we solve the puzzle,[506] 'say this', that 'it' will 'not' stop 'causing
1346,5 motion and being moved at the same time'. For what has been moved, he declares, and has inherited the power of being moved by possessing the appropriate natural capacity, does not at the same time stop being moved and causing motion, as can be seen with the magnet. For the iron ring that is attracted by it when in contact with it is moved no longer, but continues to move what comes next; for it is moved as long
1346,10 as the mover causes motion, but when the mover stops, what is moved ceases to be moved too. So too for air. It does not follow what is thrown, being moved along with it, but it causes the <air> after it to move at the same time and also makes it cause motion. The same account holds for this too, but not in the same way. The power imparted by the first mover becomes continually smaller as it is passed on, and
1346,15 'finally' perishes when the last thing that is moved is 'no longer' able to cause motion. This is why what is thrown then falls, since it no longer has anything by means of which it will be moved. This is also how things that are heated by fire not only possess the property of being heated by the fire but also receive from it the power of heating. And after proceeding to become smaller down to a certain point it

finally stops when, he declares, the penultimate one (he calls this 1346,20
'prior') no longer has the strength to make the one after it cause
motion 'but' 'only' to make it 'be moved'. For if this no longer causes
motion, what follows it will no longer be moved. Therefore, 'the'
'mover' (the penultimate one) and 'what' 'is moved' (the last one) stop
'at the same time'. For when the one before it stops causing motion,
the last one also stops being moved, and when what is moved stops,
the motion stops too, since motion is in what is moved. And it is clear 1346,25
that it stops at the same time in the case of the last ones, since he
said earlier that 'it does not stop causing motion and being moved at
the same time', saying this for each of the things that are moved prior
to the last one; for the last thing that is moved does not cause motion.

Alexander does a good job of investigating in this connection why
the same puzzle does not remain even after this is said. For the air 1346,30
that took its power of causing motion from the thing that threw the
missile causes motion either being itself moved or not. But if it is not
moved, it will not cause motion, since things that cause motion
corporeally cause motion being themselves moved. But if it causes
motion being itself moved and ceases to be moved, then (because
mover and moved are relational entities) as soon as the thrower
ceases to cause motion, it too will cease to cause motion.[507] From this 1346,35
it follows that as soon as the thrower stops causing motion, what is
thrown stops being carried in its motion, which clearly does not occur.
After investigating these matters he says, 'perhaps by saying that the
air takes some power of causing motion from the original mover, he 1347,1
means that it possesses a power of its own that it took from what
caused it to move. But if it cannot cause motion without itself being
moved, it also <received the power> of being moved <from it>. And
so it is from the thrower that it has taken the origin and start of being
moved as well as of causing motion. But it has got from it the kind of 1347,5
power that makes it able to cause motion being moved *by itself*,
becoming in a way a self-mover for a little while, because its nature
is such that through its susceptibility it receives this power, which
<is imparted> by what moves it. Therefore it ceases from the motion
that comes from the thrower, stopping <from *it*> when the thrower
does, but it does not immediately cease from the motion it undergoes
that stems from itself, because in no other way could it move what 1347,10
comes after it; it took its own power to do this from the thing that
moved it and threw it in the first place. This is why the thing that is
thrown does not stop moving when the thrower does. The things that
throw what is thrown are naturally such[508] as to be able both to be
moved and to cause motion *by themselves*, since they have taken their
origin and start from what threw them. Consequently each of them, 1347,15
moving by its own power the next thing in succession, provides the
cause of the fact that what is thrown is moved for the entire period

in which it is moving. For just as what is changing from water, for
example, and has become fire heats, heated water heats too, and the
cause of heating for both of them is what is making them hot. But
1347,20 although the fire has its cause from what made it, it still heats as long
as it is fire, but the water does not heat as long as it is water, but only
as long as it is hot, because the former has become such by nature,
but the other is not so by nature, but has it as an attribute. This is
also the case for things that are moved by something else when the
thing that moved them is no longer present, and this category in-
1347,25 cludes those that are by nature moved and thrown. For the cause of
motion for both is what made them such. But things that are moved
by nature by having come to be such, are equally subject to the same
motion as long as they are, while in the case of things that are thrown,
the air <is moved> because it has been affected by the force of the
thrower while remaining in its own nature, because it is susceptible
to this. It will undergo this motion as long as it preserves the affection
1347,30 with respect to which it is moved, but not always. For such movement
does not belong to it by nature. But', he declares, 'Aristotle speaks
clearly about this in the third book of *De Caelo*.[509] He declares that
the intermediate bodies (water and air) have both attributes. Since
air participates in both lightness and heaviness, it moves in which-
ever direction the thrower gives it the origin and start of its motion,
1347,35 and <continues to move in that direction> as long as it preserves the
power that comes from the thing that provided it with the start. And
it [the air] moves and carries what is thrown as if <by a power> within
itself.' I have copied all this practically verbatim from the passage of
1348,1 Alexander because of the cogency of the puzzle and because of the
originality of the solution, whose main point he presents in these few
words, 'And so it is from the mover that it has taken the origin and
start of being moved as well as of causing motion. But it has got from
it the kind of power that makes it able to cause motion being moved
1348,5 *by itself*, becoming in a way a self-mover for a little while.'

From what Aristotle says next we might be puzzled about what he
means by saying, 'but it does not stop causing motion and being moved
at the same time, but rather, when the mover stops making it move
<it stops> being moved', 'but it is still a mover'. For if it causes motion
1348,10 being moved by itself, and, since it causes motion corporeally, it
cannot fail to be moved too, how is it still causing motion when it has
stopped being moved? It may be possible to solve this through Alex-
ander's remarks. 'For the air', he says, 'is moved both by the thrower
and by itself. It stops being moved by the thrower immediately when
1348,15 the mover stops, but as regards the motion that comes from itself, it
does not stop immediately after what threw it stops.' But we might
be puzzled from what Aristotle says earlier about self-movers.[510] If in
order to solve the puzzle we posit that the air is moved by itself, it

follows that – since a self-mover cannot be moved as a whole by itself
as a whole, as has been previously proved, nor can it be moved as a
whole by a part or as a part by the whole – if the air proved to be a 1348,20
self-mover, some part of it must cause motion and some part must be
moved, as in the case of animals where the soul causes motion and
the body is moved. So what is the mover in the air, and what is the
moved? If someone were to say that its nature causes the motion and
its body is moved, he needs to be reminded of what Aristotle says in
this book about the four elements – that 'none of these moves itself 1348,25
but they contain a principle of motion – not of causing motion or of
producing it, but of undergoing it'.[511] Now perhaps Alexander did well
to say not that the air comes to be a self-mover without qualification,
but that it is 'in a way a self-mover', through the agency of the first
mover, on the grounds that the motion imparted to it by the mover is
active even after the mover stops, but not on the grounds that the air 1348,30
or water possesses one thing that causes motion and another that is
moved, as happens with self-movers in the strict sense. For in these
cases the mover, i.e. the soul, is unmoved and causes motion per se
even if it is moved incidentally, whereas the air that is moved
proximately by the thrower is moved as a whole and it is by being
moved that it moves the air next to it, employing a kind of push which 1348,35
the first mover employed on it. But it causes motion not by means of
something unmoved, the way the soul of the thrower moves the body,
but <it does so> the way the body of the thrower moves the air by its
own motion – this is how the air moves the air next to it, since the
motion remained in it, and how that air moves the consecutive air,
until by being transmitted the motion is gradually exhausted. That 1349,1
the motion remains in it for a time even when the mover has stopped
(when the power that caused the motion is significant and the thing
that is moved is suitable) is also shown by tops, which rotate for a
long time after the spin that the mover gives it, and bronze disks that 1349,5
are struck and resound for a long time because they remain in motion
and keep moving the air. For after taking from the mover the origin
of their motion, because of their disposition towards being moved they
remain moving even when the mover withdraws. And this is what
Alexander calls 'becoming in a way self-movers for a little while', since
the thing that is moved remains in motion and by being moved moves 1349,10
what comes next.

But what does Aristotle mean by saying, 'it does not stop causing
motion and being moved at the same time, but rather, when the mover
stops' '<it stops> being moved at the same time, but it is still a mover'?
Since what causes motion and what is moved are relational entities
and thus simultaneous, when what causes motion withdraws from
its relation to the other one, what is moved stops; however it causes 1349,15
motion by being in motion itself too, not because the mover is still

moving it, but because the motion imparted by it remains. This is why it does not cause motion being unmoved, but being moved. Not, however, because the other one is still moving it or because it turns out to be unqualifiedly self-moved by its own agency, but because the
1349,20 motion imparted by the other one remains in it and moves the next thing, though it has stopped being moved qua being related to a mover. How, then, is what is moved still moved by something, if what moves it stops and *it* keeps moving? Perhaps it is because when what causes motion imparted the motion to what is moved, even if what causes motion stops, the thing that is moved is said to be moved by
1349,25 that which had imparted the persisting motion to it. For the account does not make evident that the mover must certainly be present, but that the motion must be imparted by it to what is moved. But if we say that the one who throws the missile imparts a persisting motion to the air, why don't we say that it imparted this to the missile, so that we no longer need the air, and are not compelled to say that it is moved and also causes motion? Perhaps it is because the intermediate
1349,30 elements (water and air), as Aristotle himself says,[512] are suitable for both upward and downward motion, and consequently for lateral motion as well. Therefore, since they receive more persisting motion we declare that they contribute to the persistence of the lateral or upward motion of earthy things, which are not of a nature to move in
1349,35 those directions, since a certain persisting motion is certainly imparted to them too by the thing that moves them in any direction whatever. I have said this since I want the two arguments to fare well – both the one that says that everything that is moved is moved by something and the one that says that anything that causes motion corporeally causes motion being moved, and also because I want
1350,1 Aristotle's claim that it does not simultaneously stop being moved and causing motion to appear consistent with the previous points, and moreover because I think we should accept Alexander's helpful point on these matters, that the thrower makes the adjacent air 'become in a way a self-mover', and that that <air> does the same to
1350,5 the air next to it, because by imparting persisting motion the mover in one way makes it cause motion and in another makes it be moved, as animals possess one thing that causes motion and another that is moved, and what causes motion is unmoved. For these <parts of the air>, by being in motion, push the adjacent parts. However, if anyone can address these claims more plausibly, I shall consider him not an enemy but a friend.

1350,10 **267a12-20** This is the kind [of motion] that occurs in things that can [be in motion at one time and at rest at another, and it is not continuous, but appears to be, since the things that are in motion are either consecutive or in contact. For (a15) what causes motion is not one thing but things that are next to one another. This is why this kind of motion, which some call mutual replacement, occurs in air and water. But it is impossible to solve the puzzles except in the way we have stated. Mutual replacement makes all things be moved and cause motion at the same time, so that they also stop at the same time. But as things are, it is obvious (a20) that there is some single thing that is moved continuously.] By what, then? Not the same thing.

He has now refuted the objection against the argument that every-
thing that is moved is moved by something. The objection was based
on things that are thrown and that remain in motion even after the
thrower has stopped, and he refuted it by having the first mover make 1350,15
what it moved able to move what comes after it, and so on (since the
same argument holds for all). He now adds that this kind of trans-
mission of motion does not occur with eternal things but is an
attribute of 'things that can be in motion at one time and at rest at
another'. A little below he adds, 'This is why this kind of motion occurs
in air and water'. And he shows that this motion is 'not' 'continuous', 1350,20
both because it belongs to things that are sometimes in motion and
sometimes at rest (such motion is interrupted by the periods of rest)
and because what moves the thing that is thrown is not one thing but
more than one thing, and these are 'either' 'consecutive' 'or' in contact.
Things are consecutive when there is nothing of the same kind in
between,[513] and are in contact when there is nothing at all in between 1350,25
but they are next to one another.[514] So if motion caused by one mover
is continuous, and they are not one, the throw is not continuous but
only apparently so. After saying that the motion is 'not continuous'
but occurring in succession, he continues, 'this is why this kind of
motion', i.e. motion from throwing, 'occurs in air and water' – because
when these things are moved by something, they are of a nature to
cause motion in the consecutive things and likewise for the things 1350,30
next to *them*. This kind of motion, he declares, 'some' call 'mutual
replacement'. Mutual replacement occurs when one body is expelled
from its place by another and they exchange places, and the one that
expelled the other stands in the place of what was expelled, and the
one that was expelled expels what is next to it and that one expels
the next one, when they are several, until the last one comes to be in 1350,35
the place of the one that was first to expel <something else>. But, he
declares, even if mutual replacement takes place in the case of objects
that are thrown, still this is not the cause of the motion of what is
thrown. For 'mutual replacement' 'makes all things be moved and
cause motion at the same time', i.e. be expelled and expel. But if all 1351,1
things are moved at the same time, they 'also' stop <at the same

time>. And the thing that first expelled <something else> does not
occupy the place of the first thing that is expelled unless the latter
occupies the place of the next one, and that occupies the place of the
one next to it, and the last one occupies the place of the one that was
1351,5 first to move, and the motion of them all takes place at the same time,
and so does the settling into each other's places. But the occupying of
one another's places, which is what mutual replacement is, contrib-
utes nothing to the motion of what is thrown unless we hypothesize
(as he himself says) that things that cause motion also make the very
things that are moved able to cause motion. For in this way what
occupies the place of what is thrown will both move what is thrown
1351,10 and make it move the consecutive things. And so 'it is impossible' 'to
solve' 'the puzzles' about thrown objects 'except' by hypothesizing that
what is moved takes motive power from the mover. For indeed when
mutual replacement occurs motion takes place, but not *because of*
mutual replacement. For just as mutual replacement occurs when
1351,15 someone walks but mutual replacement is not the cause of the
walking, so in the case of things that are thrown something else and
not mutual replacement is the cause of the motion. For mutual
replacement is merely an exchange of places and locations. That it is
not the cause of motion he proves from the fact that things that
undergo mutual replacement 'all' 'at the same time' both 'are moved'
'and cause motion', 'and cease'[515] from their motion, as was said a little
1351,20 above. But what is thrown is obviously a single thing that undergoes
continuous motion. 'By what, then' will this be moved? Not, to be sure,
by the things that are undergoing mutual replacement. For all these
are moved and exchange places simultaneously, and when one stops
moving they all do. For what is thrown is pushed forward by the
thrower, and the air in front of it takes its place simultaneously,
1351,25 moving around to where it moved from. What, then, is the cause of
its progression and locomotion after this? For the air that took its
place is not larger than the place it occupied, so as to force and push
it ahead because it[516] is constricted.

Since Alexander declares that it is Plato's opinion that things that
are thrown move by mutual replacement, we must know that he
1351,30 [Plato] too thinks that mutual replacement occurs when things move,
and that he wants bodies to exchange places with one another so that
no void is left in the universe – but that he thinks the cause of motion
is not mutual replacement but non-uniformity and inequality.[517] But
when Aristotle was saying that everything that is moved is moved by
1351,35 something[518] and was supplying his discussion with so many demon-
strations, and was putting forward objections to it and refutations of
the objections, Alexander should have said at that point that this
account, according to which what is moved is certainly moved by a
mover, is Plato's. But it is perhaps not a bad idea to offer a few

comments that show that this view *is* Plato's and what he thinks the cause of motion is and how he supposes the exchanges of place occur.
For he says in the *Timaeus*, 'There will never be motion in a state of 1351,40
uniformity, for it is difficult, or rather impossible, for there to be 1352,1
something moved without a mover or for there to be a mover without something moved, but when these are absent there is no motion.'[519] And so, according to Plato too, what is moved is moved by some mover and movers move something that is moved. 'It is impossible', he says,
'for these ever to be uniform. Thus, let us posit that stability always 1352,5
occurs in a state of uniformity and motion in a state of non-uniformity. And the cause of non-uniformity, moreover, is inequality.'[520] Bodies that are equal and similar would not move one another, but (as he adds a little below), 'when small bodies are put among large ones and the smaller among the larger and break them up, while the larger
cause the smaller to combine, they all change positions, up and down, 1352,10
towards their own proper places. For as they each of them change their magnitude, they also change the location of their places. In this way, then, and by this means the generation of non-uniformity is always preserved, and perpetually causes these things' incessant motion both now and in the future.'[521] This is how he says that
exchange of places occurs: 'But when in turn the fire is expelled from 1352,15
there, since it does not go out into a void, the nearby air is pushed and simultaneously pushes the moist <air>, which is easily moved, into the places previously occupied by the fire and mixes it with itself.'[522] However, those who want to examine this topic directly can learn about it more accurately elsewhere.[523]

267a21-b6 Since there must be continuous motion among existing things, [and since this motion is single, and a single motion must be <the motion> of a certain magnitude (for what is without magnitude is not moved), and in fact of one thing moved by one thing (otherwise it will not be continuous, but divided, with one <motion> next to another), then (a25) if the mover is one thing, it causes motion either being itself in motion or being unmoved. If, then, <it causes motion> being itself in motion, it will follow necessarily both that it itself changes, and that at the same time (267b1) it is moved by something, so that <the series of movers> will come to a stop and will arrive at a point where motion is being caused by something that is unmoved. But[524] this need not change together with <what it moves>, but it will always be able to cause motion (for causing motion in this way is effortless) and this motion alone, or above all, is uniform, since the mover (b5) undergoes no change. And the thing that is moved must not undergo any change in relation to it either,] in order for the motion to be similar.

After coming to the end of the issues he put forward, he then reviews what was said and briefly recalls the demonstrations. For since it has been proved that there must always be motion and it did not begin, not having been before, and will not stop so as no longer to be, but is equally without beginning or end in time;[525] and that continuous motion is primary,[526] and a continuous motion is 'single' (for a motion that is not single is not continuous[527]), and a single motion is <the motion> of a single 'magnitude' that is being moved ('for what is without magnitude' 'is not moved' per se) and where the mover is single (for unless both what is moved is one and the mover is one, the motion will not be 'continuous' and one, 'but divided, with one <motion> next to another')[528] – now since this has been proved, that in a single eternal motion the mover is one and what is moved is one, he first reviews the attributes of the mover and then he adds the attributes of what is moved. For 'if' 'the' 'mover' 'is one thing', 'it causes motion either being itself in motion or being unmoved'. But 'if' <it causes motion> 'being itself in motion' clearly it must be a body, for it has been proved that everything that is moved per se is a body. Therefore this 'will follow' 'necessarily' for what is in motion – both that 'it is moved' 'by something' and that 'it itself' is moved. And as to what moves it (supposing that it is moved, since it is moved by something), the mover too must either cause motion being moved and being moved by something, or else be unmoved. But to avoid a regress ad infinitum, we will eventually arrive 'at a point where' 'motion is being caused' 'by something that is unmoved'. For if there were no origin, the things that come after it would not be either. But it has been proved that what is self-moved is prior to things that are moved movers,[529] and in self-movers what causes motion is unmoved per se, but moved incidentally. But what causes motion in the strict sense 'need not change together with' what is moved, even incidentally.[530] This is why he adds, 'this' 'need not change together with <what it moves>', distinguishing it from the element in a self-mover that causes motion, which must incidentally change along with what is moved by it. What causes eternal motion (which is the subject of the discussion) should 'cause motion'[531] without effort, for nothing that expends effort can last for long;[532] and what does not change along with what is moved and is not moved in any way but is established in the same state of everlasting perfection, is the furthest removed from effort. The single continuous motion should be uniform too, since if it is non-uniform it will be neither single nor continuous.[533] But the motion caused by what is unmoving is uniform, since what is unmoved always stands in the same relation to what is moved by it, since it changes in no respect. The words 'alone, or above all' can have been said with philosophical caution. Someone might say that what has been said so far concerns the mover alone, since Aristotle directs the majority of his discussion at <the question of> uniformity, and it is mainly because of the mover that the motion is uniform. But he also requires the thing moved to be the same in order for the motion to be the only motion that is strictly speaking uniform.

And indeed after bringing his review of the attributes of the mover this far, he adds about what is moved that it must always preserve the same relation to the mover, 'in order for the motion' to remain 'similar' and truly uniform. For the only motion that is strictly and entirely uniform is that in which both the mover and the moved stand in the same relation to one another unchangeably and forever. But it is clear that the body of the heavens is of this sort, since it is in circular motion and is moved proximately by the unmoved moving cause. And so the eternity of the heaven, and consequently the eternity of the entire world are demonstrated together with the eternity of motion and the unchangeable sameness of the primary mover, even if he is going to demonstrate the eternity of these things again in *De Caelo*.[534]

> **267b6-9** But[535] it must be either at <the> middle or on[536] <the> circle, [since these are the principles <of the circle>. But the thing[537] that is closest to the mover is moved most quickly, and the motion of the circle is of that sort.] Therefore the mover is there.

After describing the primary mover and the thing that is in eternal motion, and stating how they are related to one another, since it was apparently customary to investigate where the mover of the heaven is located (at any rate the Pythagoreans seemed to say that it is in the middle[538]), he thinks it appropriate to add this topic too to his discussion of the mover. He says that 'it must be' at the principle of what is moved, which is the most basic part.

A sphere has two principles: the middle and the circumference. Therefore he says the mover 'must be' 'at' one of these. But since 'the thing that is closest to the mover is moved' 'most quickly', 'and the motion of the circle' (i.e. of the circumference) is quick, 'the mover' should be 'there'. For the middle, being unmoved, does not seem appropriately related to the mover. Eudemus says it is on the celestial equator.[539] For this moves most rapidly, and the mover seems to begin <causing motion> where it will cause motion most rapidly and easily.

Alexander poses puzzles about both the external circumference and the great circle: if the moving cause is on one of these,[540] how will it not be moved incidentally? After saying that if it is not at the poles of the circumference, it must be outside, he poses the additional

puzzles that there are two of them[541] and they are unmoved, but Aristotle and Eudemus say that what is nearer the mover is moved fastest. In solving the puzzle he says that if it were to be in any part of the outermost circumference, it would be moved incidentally because the motion of the sphere is located in its parts. But if it is on the entire circumference (for in that way it will be at the fastest
1354,20 moving part), it will no longer be moved incidentally, because the entire circumference does not move or change place but always remains in the same place. Therefore he accepts that it is on the entire circumference of the outer sphere. 'For in this way', he says, 'it will be nearest to what is moved fastest, and it will not be moved incidentally, and will always have the same relationship to what is moved
1354,25 by it. This is why it will always cause a similar motion.'

Alexander posed these puzzles and solved them in this way, although he had previously done well to say that the mover should not be understood as occupying some place (since it had been proved to be without parts), or as if there is a form in which it is, but <that it should be understood> as an incorporeal substance in existence in its own right. 'For', he says, 'even if the body that undergoes circular
1354,30 motion is ensouled and is moved in virtue of the soul in it, still it requires something else to provide the origin of its motion; for with all ensouled things something external turns out to be the cause and origin of their motion in place that is due to the soul, since the motion in place of ensouled beings comes to be because they aim at something.'

1354,35 Alexander wrote this accepting that Aristotle speaks of the primary mover only as a final cause. But if, as many arguments show, he thinks it an efficient cause too, then he too thinks, like Plato, that the soul is the cause of motion for what changes place, and that the unmoved mind is the cause of its always being moved around the same things and in respect of the same things and in the same place, according to a single account and a single order.

1355,1 Now if the mover is without parts and is in every way separate from bodies, all by itself, perfect and independent of the entire corporeal world, it is good to say that it is both everywhere and nowhere. If the Pythagoreans say it is established at the centre,
1355,5 Aristotle declares it to be in the fixed sphere;[542] they hold that the centre is more appropriate than the other parts of the universe with a view to participation in the goodness of the creator that holds things together and establishes them, while Aristotle on the other hand thinks that it is the fixed sphere that primarily enjoys the creative motion. This is why the Pythagoreans called the centre the place of
1355,10 Hestia and the tower of Zeus,[543] while Aristotle says the motion of the fixed sphere is the measure of the other motions, since it is first and

is included <in their motions>, and through its swiftness contains an image of the partlessness of the mover.[544]

Therefore we should not be afraid of making the primary mover incidentally undergo motion by saying that it is in the fixed heaven. For it is not strictly speaking in the heaven, but the heaven is in it, since what is in something is surrounded by that in which it is. But *it* surrounds the whole world by virtue of its infinite power.

But perhaps not even the assistance that Alexander thought of is sufficient to guarantee that it is not moved even incidentally, since it[545] still stems from the thought that the primary mover is in the heaven. 'If it is not in some part', he declares, 'but in the entire circumference, since the entire circumference does not move or change place, but remains in the same place, it will not be moved even incidentally. Therefore, if the entire circumference is not moved and the primary mover is said to be in what is moved fastest, clearly it is not in the entire circumference. Next, if it is in the whole circumference as something appointed to be there, and the whole circumference is moved in respect of its parts, clearly the primary mover too will be moved incidentally along with the parts. But if it is present in the whole unrelatedly and transcendently, what is to keep it – since it is also present in all the parts at once unrelatedly, transcendently, indivisibly and immovably – from moving along with them incidentally?' And Eudemus, setting out the problem soundly, says, 'If the primary mover is without parts and does not touch what is moved, how is it related to it?'[546] How can what is without parts and does not touch what is moved but is unrelated to it and transcendent, be moved along with what is moved, so as to be moved incidentally? Alexander states that Eudemus says that the primary mover is in the celestial equator since this is moved most rapidly. But I found the following text in the Eudemian work: 'On a sphere the place around the poles moves most rapidly.'[547] And it is worth remarking that Alexander thinks that Aristotle says that the primary mover is at the circumference of the sphere. But perhaps when he says 'on' the[548] 'circle' he means in the whole heaven.

267b9-17 There is a puzzle as to whether (b10) something that is moved can [cause motion continuously, but not the way something that pushes repeatedly <causes motion continuously>, in the sense that <pushing> consecutively is <pushing> continuously. For either the same thing[549] must[550] be pushing or pulling or doing both, or something else <must do it>, one thing after another following in succession, as was said above <to happen> in the case of things that are thrown, if the air and the water,[551] being divisible, cause motion by one <part> always

> being moved <after another>. (b15) But either way the motion cannot be single, but only one after another. Therefore the only continuous <motion> is that which the unmoved <mover> causes. For since it is always similar,] it will be similarly related to the thing that is moved too, and continuously as well.

After proving that the only motion that is continuous is motion caused
by what is unmoved, he confirms this by eliminating the opposite
1356,5 claim, that 'something that is moved' can 'cause motion continuously'.
Everything that causes motion being moved is corporeal and causes
motion corporeally, and things that cause motion corporeally move
by pushing or pulling or in both ways (like people who move mills
with their hands); and what pushes or pulls or does both 'either' 'must'
be 'the same', 'or' 'one thing after another' following in succession
must cause one or another of these kinds of motion. This[552] was shown
1356,10 a little above <to happen> 'in the case of things that are thrown',
where not only is one thing moved after another, but also the cause
of motion for what is thrown turned out to be the fact that they cause
motion successively. (Since 'the air and the water, being divisible', in
the sense of being easy to divide and easily moved, 'cause motion'. But
how does the air cause motion? 'By one <part> always' 'being moved'
and then another[553] and thus causing motion.) So if this is how
something is moved by a thing that is moved, the motion is not single
1356,15 and continuous, but consecutive. But if it does not cause motion
through a succession <of movers>, but the mover is one, as with things
that keep pushing or pulling, this motion will not be continuous
either. For even if the mover in these cases is one and not several (as
happens with things that are thrown), still for each push or pull there
is a beginning, since the thing that pulls or pushes rests, because
1356,20 causing motion in this way requires effort. Therefore, if everything
that causes motion being moved, causes motion by pushing or pulling
or both, and if it is completely impossible for there to be one continu-
ous uniform motion no matter whether the subsequent <mover> is
the same or <there are different movers> in succession, it is clear that
'the only continuous <motion>' is 'that which the unmoved <mover>
causes', since everything that is moved is moved either by something
moved or by something unmoved. He adds the cause of the fact that
1356,25 this <kind of motion> is continuous, saying, 'for since it is always
similar', if it is unmoved and unchangeable in its substance, potenti-
ality, and actuality, 'it will be similarly related to the thing that is
moved too', since it always has the same relation to it, so as to move
it similarly and uniformly. And after saying, 'similarly related to the
thing that is moved too', he adds, 'and continuously as well', indicating
1356,30 that the similar relation of what is unmoved to the moved is continu-

ous too and is not interrupted and does not stop, so that the motion that the unmoved causes proximately is eternal too.

Alexander does well to pose the puzzle why we should not also say that the motion that the spheres of the planets undergo as they are
carried along with the sphere of the fixed stars is continuous, eternal 1357,1
and uniform, although they are made to undergo this motion by the sphere of the fixed stars, which is moved. He first gives in to the puzzle and says that Aristotle was riddling when he spoke in this way. But it is a puzzle whether something that is moved can cause motion
continuously and not like something that causes motion by pushing. 1357,5
After posing this puzzle he shows the cases in which it is not possible, leaving aside the puzzle now under consideration. Then he starts again and goes on to say, 'Perhaps their motion is not caused by the sphere of the fixed stars since it does not cause this motion in virtue of a motive power that is in it. But since it is moved by the mover, their movement follows it and is itself caused by what causes *this*
motion.' That the motion from east to west of the entire heaven, taken 1357,10
as a single thing, is one, is clear, I think, from the fact that it takes place about the same poles, and also from the fact that not only are there eight spheres but also the entire heaven is one – and a motion is one when it is the motion of one thing caused by one thing. Aristotle himself indicates this too when he says that the primary mover is not
in the sphere of the fixed stars but in the entire body that is in circular 1357,15
motion: 'The thing that is closest to the mover is moved most quickly' he declares, 'and the motion of the circle is of that sort. Therefore the mover is there.'[554] I think that by 'circle' he means everything that revolves in a circle.

Eudemus does not pose the same puzzle that Aristotle does, whether something that is moved can cause motion continuously. Instead he poses the puzzle whether what is unmoved can cause motion. 'What causes motion in respect of place', he declares, 'seems
to cause motion by pushing or pulling. And if not only in these ways, 1357,20
in any case it touches it either directly or through one or more intermediaries. But what is without parts cannot touch anything; it has neither beginning or end, whereas things that touch have their ends together. How then can what is without parts cause motion?' He solves the puzzle, saying, 'Some things cause motion being moved and
others cause motion being at rest. Things that are moved cause 1357,25
motion by touching, while those that cause motion without being moved do so otherwise, though not all in the same way: the primary mover does not cause motion in the same way as the earth causes the ball thrown upward to move towards it. For the primary mover does not cause it to move in circumstances where there is a prior movement, since that way it would no longer be the *primary* mover. On the

other hand, the earth will never cause motion primarily, being at rest.'[555]

1357,30 **267b17-26** Now that these points are determined, [it is evident that the unmoved first mover cannot have any magnitude. For if it has magnitude, it must be (b20) either finite or infinite. Now it was proved earlier in the *Physics* that there cannot be an infinite magnitude, and it has been proved just now that a finite <magnitude> cannot possess infinite power and that nothing can be moved by a finite <magnitude> for an infinite time. But the primary mover causes an eternal (b25) motion and does so for an infinite time. So it is evident that it is indivisible, without parts, and has no magnitude.]

After reviewing the account of the primary mover a little above, and recalling that on the basis of the proof that there is continuous eternal motion among existing things, it was proved that the primary mover is one and unmoved, he goes on to say where in the thing that is moved we should think the mover is.[556] And after again establishing its
1357,35 immobility by proving that nothing that is moved can cause a continuous motion,[557] he adds a review of the third conclusion that was
1358,1 drawn about the primary mover – that it is without magnitude or parts. For three things were proved of it: that it is one, unmoved, and without parts. So after briefly recalling the two proofs, he now recalls the third, which proved 'that the' primary 'mover' which causes
1358,5 eternal motion and which is 'unmoved', 'cannot' 'have' 'any' 'magnitude'. For the magnitude will be 'either' 'infinite' or 'finite'. Now if 'there cannot be an infinite magnitude', as 'was proved' in the third book of this treatise[558] (he is accustomed to call the first five books of the *Physics* by the special term *On Natural Principles*, and the last three *On Motion*[559]), and further he proved that it cannot be
1358,10 a finite body either, since a finite body must possess power that is either infinite or finite. Therefore if 'a finite' body does not possess 'infinite' 'power' (as 'has been proved') and if a finite power cannot cause motion 'for' the 'infinite time' (as 'has been proved' as well), 'but the' 'primary mover' which causes the 'eternal' 'motion' was
1358,15 proved to be causing motion over 'an infinite time' (for the primary motion had to be imparted by the first mover) – since these results have been proved, 'it is evident that' the primary mover 'is indivisible, without parts, and has no magnitude'.

At this point Alexander did well to make exactly the remark one should when the proposal is being demonstrated – that in the case of
1358,20 something that is moving ad infinitum, we should speak of a power only homonymously. 'For in the case of a mover or a maker', he says, 'a power is spoken of like a kind of strength that must grow weary in

the case of things that cause motion in this way, whereas in the case of what is moved, it should be spoken of simply as a suitability to be affected. For in the case of lifting, since what is lifted can be lifted as long as it exists, it does not immediately follow that the lifter is always 1358,25
lifting it, for the lifter lifts by means of some power and strength that grows weary since it is finite and in a finite thing'. In fact, the Grammarian should certainly have noticed these words of Alexander and should not have thought that both Aristotle and Alexander hold that what does not in the strict sense possess infinite power can be moved ad infinitum. For what is moved ad infinitum possesses a 1358,30
suitability for this by its own nature, and this ability to be moved is in it, although since it gets its motion through the agency of something else it possesses it as something that comes to be, not all at once together. For motion has its existence in coming to be. After investigating here again how (since it is finite) the sphere of the fixed stars will not possess an infinite power of causing motion, even if it is always causing the spheres of the planets to move, Alexander solves 1358,35
the problem under investigation by saying the same as he said shortly before, that in a way even the planets are moved by that unmoved thing since they move together with the sphere of the fixed stars. In virtue of its corporeal nature the fixed sphere is subject to motion but does not cause motion. The Grammarian should have noticed that Alexander did not here investigate how the finite fixed sphere is eternal, but how being finite it possesses an infinite power of causing 1359,1
motion, since he knew that nothing prevents what is finite from being moved and from existing ad infinitum, since it possesses motion and existence that are coming to be – not from some particular time, but ad infinitum – but being finite it does not possess its motive power all at once together.

In this way the truly marvellous Aristotle brings his instruction 1359,5
about the principles of nature to culmination in theology,[560] which is above nature, and proves that the entire corporeal structure of nature is dependent on the incorporeal intellective goodness that is above nature and unrelated – here too following Plato. But it was from the very existence of the body of the world that Plato discovered the intellective god who is the creator of the world. He distinguishes what 1359,10
has real being from what comes to be, defining the former as what is always in the same state and condition, in that it has obtained its existence all at once together, without intermission and indivisibly for eternity, and defining what comes to be as that which has its existence in coming to be, in that it is changing and being moved.[561] And he posits that every corporeal structure is subject to generation because its being is also extended in respect to the extension of its 1359,15
existence, which is located in time and undergoes change, and because it has its existence in coming to be and consequently is depend-

ent on a cause, since it cannot be self-constituted. 'For it is impossible', he says, 'for anything to obtain coming to be without a cause.'[562] And what has real being is the cause of what comes to be, or else we will go ad infinitum putting one thing that comes to be before another.
1359,20 For the same reason the unchangeable <is the cause of> what undergoes change. Therefore by going backward to the unchangeable cause from the existence of the world that is subject to change, he discovered that the creator of the world is an intellective god that has real being and is always in the same state and condition, established for eternity.

Aristotle too proceeds from motion and change and from the
1359,25 subsistence of bodies, which is finite and has extension, to the unmoved, unchangeable, unintermittent cause. For he proves that there must be eternal motion among existing things, and consequently that what is moved <is eternal too>, since motion is in what is moved. He also proves that everything that is moved is moved by something, and that the primary mover must be an unmoved and unchangeable cause
1359,30 of eternal motion in the things that are proximately moved <by it>. That coming to be in Plato and motion in Aristotle signify change is easy to learn from the fact that Plato contrasts what comes to be, as changing, with what is in the same state and condition, while when Aristotle says that everything that is moved is moved by something,
1359,35 he is talking not only about things that are moved in the strict sense, but also about things that are generated and perish, and in general of things that undergo change. Moreover, he frequently calls the unmovable unchangeable. For it is beyond not only motion in the strict sense, but generation and perishing as well. But this wonderful man seems to me clearly to refuse to apply the term 'generation' to
1359,40 eternal things, because the imagination easily suggests a temporal origin for things that are said to be generated. And this is precisely
1360,1 the mistake many have made in their inability to comprehend eternal creations in their thoughts. They have supplied a temporal origin for what exists and is said to be generated through a cause, and they think people learn more easily when a temporal beginning, middle
1360,5 and end of creation are hypothesized. And in fact this is how most of the wise engage in cosmogony, aiming at making it easy for their audience to learn. They say that things come to be present first, second, and third. And they may think it pardonable if the theologians too do not refuse to reveal the births of the gods in order to make it easy for their audience.[563] But Aristotle, apparently perceiving that
1360,10 these are the words of people who have gone mad and who conceive of a temporal origin as well, does not undertake to engage in cosmogony, and evidently refuses to say 'subject to generation' in the case of eternal things, but employs the term 'motion', which signifies the same thing but does not demand a temporal origin. He makes it clear in the third book of this treatise that he does not refuse to speak of

coming to be in the case of things that have their existence ad infinitum, saying concerning what goes ad infinitum, 'Since existence 1360,15
has many meanings, the infinite exists in the way a day or an athletic contest exists – by virtue of one thing always coming to be after another'.[564] Now for both men the starting point of the demonstration is thus the same, leading from the changing to the unchangeable. And in what immediately follows one of them says that what is moved is moved by something, while the other says that what comes to be gets its generation from a cause. And the one proves that the primary 1360,20
mover is unmoved, unchangeable and without parts, while the other proves that the cause of what comes to be is what has real being; but this describes what is without parts and existing together as a whole and always in the same state and condition, and this is what being wholly unchangeable signifies.

Some think that Aristotle says the primary mover – which he hymns as mind,[565] eternity[566] and god[567] – is only a final cause and not 1360,25
also an efficient cause of the world and in particular of the heaven, since it is eternal and consequently ungenerated. They think this because they hear him often saying that it causes motion as the object of love, and often celebrating it as a final cause.[568] It is a good idea, then, to prove that here too he is consistent with his teacher in calling god not only a final cause but also an efficient cause both of the entire 1360,30
world and of the heaven. From what he says in the *Timaeus* ('Let us then state the cause by which the creator established coming to be and this universe: he was good.'[569]), Plato clearly calls god the final and efficient cause of the world. And when he says, 'He proceeded to construct the universe, establishing intelligence in the soul and soul 1360,35
in the body, in order that the work he accomplished might by nature be as beautiful and good as possible',[570] throughout practically the entire dialogue he hymns the creator as looking to the good. And in the creator's speech to the heavenly beings he clearly shows that he himself proximately causes the existence of the things in the heavens, 1361,1
while the sublunary things are <caused> by the heavenly beings. For the first creator says to the heavenly beings, 'O gods of gods, the works of which I am creator and father,'[571] and further on he declares, 'Three mortal kinds are left. If these do not come to be, the heaven will be incomplete,'[572] now calling the world heaven, like Aristotle. But, he 1361,5
declares, these things too must come to be 'if it is going to be sufficiently complete. But if these things were to come to be and participate in life through my doing, they would be made equal to the gods; therefore, that they may be mortal and this universe may be truly all, turn yourselves naturally to the creation of living things.'[573] The expression, 'if these things were to come to be through my doing' reveals that things that arise from a cause that is in the same state

1361,10 and condition (or, as Aristotle would say, by the agency of what is unmoved) must be eternal.

No one disputes that Aristotle calls god or the primary mover a final cause; but <to show> that it is also an efficient cause I think it is sufficient that in his distinction <of the kinds> of causes in the second book of the *Physics*, he calls the efficient cause the source from
1361,15 which motion has its origin: 'Also, the source from which the change or rest has its first origin, as the person who deliberated is a cause, and the father <is the cause> of the child, and in general the maker <is the cause> of what is made.'[574] What could be clearer than this to say with a view towards showing that the primary mover is an efficient cause? Also in the first book of *De Caelo* he clearly says that
1361,20 neither god nor nature does anything in vain,[575] after saying in the same book that eternity 'has taken its name from always being, and is immortal and divine, and from it depend existence and life for the others – in some cases more exactly, in others more obscurely.'[576] And it is clear that just as all things are made good through the final cause, so they exist and live through the creative <cause>. Also in the first
1361,25 book of *De Generatione*, while investigating the causes of perpetual generation he shows that the primary mover is an efficient cause too, writing as follows. 'One cause being the source from which, we say, the motion has its origin' (clearly speaking in this way of the efficient cause), 'and another cause being the matter, this latter is the kind of cause to be discussed. Regarding the other, we stated earlier, in the
1361,30 account *On Motion*, that there is something unmoved for all time and something always being moved.'[577] Therefore, he too declares that there are two efficient causes: the unmoved one is the cause of all things, and the heavenly bodies are the cause of the sublunary ones. In expounding these words Alexander says 'At any rate the first mover is the efficient cause of the motion of the divine body, which is ungenerated.' Further, in the first book[578] of the treatise *Metaphysics*,
1361,35 praising Anaxagoras and Hermotimus before him for not only assign-
1362,1 ing material causes of the universe but also recognizing mind as an efficient and final cause, he writes as follows: 'Someone, declaring that in nature as in animals mind is the cause both of the world and of all its order, appeared like a sober person in comparison with those who had earlier spoken at random.'[579] Now after saying that
1362,5 Anaxagoras and before him Hermotimus touched on these discussions,[580] he continues, 'Now those who thought like this posited the cause as both the principle of well being for things that are and as the source from which their motion arises.'[581] Therefore he praises the men who posit mind as both a final and an efficient cause, as shortly above he was praising Anaxagoras for calling mind a principle of
1362,10 motion and so preserving it impassive and uncontaminated.[582]

Alexander and some other Peripatetics hold that Aristotle believes

in a final and motive cause of the heaven, but not an efficient cause
– as indeed the passage of Alexander cited shortly above revealed,
which says, 'The primary mover is the efficient <cause> of the motion 1362,15
of the divine body, which is ungenerated.' Come, then, let us prove
that he considers mind to be also the efficient cause of the heaven. It
is sufficient that he defines this particular efficient cause as the
source from which motion has its origin, and that he calls mind, or
the unmoved cause, that from which the heavenly motion proxi-
mately has its origin. For it is through the motion of the heaven that
the unmoved is the origin of the motion of sublunary things too. 1362,20
However, in the second book of the *Physics* he calls luck and chance
incidental causes that supervene upon things that are efficient
<causes> per se, viz., mind and nature ('For', he says, speaking of luck
and chance, 'as to the type of the cause, both of them are among the
sources from which motion has its origin.'[583]), and he adds this: 'Since 1362,25
chance and luck are causes of whatever things might have mind or
nature as causes, when something comes to be the cause of those
things incidentally; and since nothing incidental is prior to what is
per se, <it is clear that the incidental cause is not prior to what is the
cause per se either>;[584] therefore, chance and luck are posterior to
mind and nature, so that no matter how true it is to say that chance
is the cause of this heaven, mind and nature must be a prior cause of 1362,30
many other things and of this universe in particular.'[585] An argumen-
tative person might perhaps find an escape from this by saying that
Aristotle does not prove in these words that mind and nature are
causes of the heaven, but only that anyone who says that chance and 1363,1
luck are efficient causes of the heaven will be forced to admit mind
and nature as prior causes. But he should pay attention to the fact
that what is moved by something else must also get its subsistence
from something else, if in fact existence is superior to movement. But
since according to Aristotle the power that every finite body has is 1363,5
finite – that is to say, the power that causes motion and is constitutive
of existence – it is necessary, then, that just as it has its eternal
motion from the unmoved cause, so also it receives its eternal corpo-
real existence from the incorporeal <cause>. My teacher Ammonius
has written an entire book[586] that provides many proofs of the fact
that Aristotle considers god to be also the efficient cause of the entire 1363,10
world, and I have here taken over some points sufficiently for my
present purposes. His more complete instruction on this topic can be
found in that book.

If someone inquires why in the world Aristotle does not say that
god is an efficient as evidently as <he said that he is> a final cause,
I will now again state the account I gave earlier about what is subject
to generation.[587] For since what works as an efficient cause produces 1363,15
something that is generated, and what is generated seems to bring

with it a temporal origin of its generation, this is why he refuses to speak of eternal bodies as coming to be and to identify their cause frequently and evidently as efficient. And perhaps if someone were to say that the expressions 'what is generated' and 'what works as an efficient cause' apply strictly to things that are subject to generation
1363,20 and perishing, bringing with them a particular <reference to> time, he will use other terms in discussing eternal things. Notice that Aristotle does not refuse to call motion eternal even though motion has its existence in coming to be. But he does not choose to call generation eternal as applying to the same thing, because what is generated seems to be generated not having existed previously, and tends to perish afterwards.

1363,25 I have done my best to articulate the present book part by part, but it may not be a bad idea to add a brief guide to the main points, indicating the exact order of the topics and assisting my readers' memory. At the beginning[588] he proposes to investigate whether
1363,30 motion always exists and neither has been generated without existing previously nor will perish so as afterwards no longer to exist. He first proves this from the practically unanimous agreement of the natural philosophers, rejecting Anaxagoras and Empedocles who are the only ones who appear to say that motion does not always exist.[589] Second, he proves from the definition of motion that it is not possible to take
1363,35 a first or a last motion; there is always one before the first and one after the last, since motion always exists.[590] A third proof is based on what is earlier and later in time,[591] and a fourth on the instant, which is always an intermediate in time between past and future.[592]

After remarking next that it seems contrived to say that previously
1364,1 there was an infinite period when motion did not exist, but that it did exist later (as <it seems contrived to say> 'that is how it is by nature'),[593] he then sets out three arguments that raise objections against the view that motion always exists. The first attempts to demonstrate that there is no such thing as continuous eternal motion
1364,5 from the fact that every change takes place from contrary to contrary and from the definition of contraries.[594] The second seems to prove that motion is generated without existing previously from the fact that things that have no soul are moved later without being moved earlier.[595] The third is based on things with souls, claiming that these too are moved later although they were not moved earlier, but the cause that makes them move is not evident as it is with things that
1364,10 have no soul, and consequently in these cases especially motion seems to come to be without existing previously.[596] He then refutes the objections. The first <he refutes> on the grounds that although changes from contraries to contraries are finite, nothing prevents there being some single motion that is continuous and eternal while

the rest depend on the existence of other motions earlier than the
motions that are apparent.[597] Then, referring the argument to a
principle, he makes a division of the relations things have to motion 1364,15
and rest,[598] and proves first that it is not true that all things are
always at rest – both because it is evident and also from various
considerations[599] next, that it is not true that all things are always
being moved either,[600] and third, that it is not true that some things
are always at rest and the rest always in motion[601] or that all things
are sometimes at rest and sometimes in motion.[602]

Then, after reviewing the division and briefly refuting the afore-
mentioned cases,[603] he confirms the remaining one, which says that 1364,20
some things are always at rest, some are always in motion, and others
are sometimes at rest and sometimes in motion.[604] In this he makes
a division between things that cause motion and things that are
moved,[605] with a view to the immediately following demonstration
that everything that is moved is moved by something.[606] From this he
proves in turn that some existing things are always in motion, that
others are always at rest, and others are sometimes in motion and 1364,25
sometimes at rest.[607] He also proves that self-movers are prior to
things that both cause motion and are moved,[608] and among things
that cause motion what is unmoved is prior without qualification.[609]
In proving that everything that is moved is moved by something other
than itself, he proves that the simple natural bodies do not move
themselves but are moved by something else – they are moved forcibly
by what throws, pushes, or pulls them, and they are moved naturally 1364,30
as a result of what changes it[610] from potentiality into actuality.[611] He
also proves that the person who removes an empediment causes it to
move, but incidentally.[612] Then, concluding the argument, he proves
briefly that everything that is moved is moved by something –
whether what moves it is itself or something else.[613] But if this is so,
then we must either go ad infinitum saying that one thing is moved
by another, or arrive at something moved by itself.[614] 1364,35

Then he takes another approach to show that what is self-moved
is prior to things that both cause motion and are moved.[615] After
discovering in this way that the primary mover causes motion as
something that is moved, he next discovers that it is unmoved,
proving that not every mover that is moved moves itself either
incidentally or per se. First he proves that <it does> not <move itself> 1364,40
incidentally.[616] In the meanwhile he says that since there is some-
thing that is only moved and something that simultaneously causes 1365,1
motion and is moved, it is reasonable that there be something that
only causes motion and is not moved.[617] Then he proves that neither
is it necessary per se that everything that causes motion do so being
moved by something.[618] He then proves generally that the first thing
that is moved will be moved either by something unmoved or by

1365,5 something that is self-moved.[619] And he adds that what is self-moved is more an origin of motion than what is moved by something else.[620] At that point he investigates how the self-mover causes itself to move,[621] and proves that it causes motion through a part of itself, the soul, which is unmoved, and it is moved through another part, the body.[622] And it does not cause motion and undergo motion as a whole, nor does it cause a part to move by means of the whole nor the whole
1365,10 by means of a part.[623] To this he adds that no part of what is self-moved is self-moved.[624] And he then concludes that the primary mover is unmoved, causing motion either proximately or through something that is self-moved as an intermediary, in which what causes motion again is unmoved.[625]

He next proves that the primary mover is unmoved and eternal, as being beyond the motion and change that have to do with generation and perishing;[626] also, that it is one.[627] He proves both results
1365,15 hypothesizing for now that there is a continuous eternal motion.[628] Since this motion is primary, it is reasonable and necessary that it is the first mover that causes it. (He will prove later that the motion of the heavenly bodies is of this sort.[629]) From the other principles of motion he proves that the primary mover is unmoved even inciden-
1365,20 tally and is eternal.[630] He next adds consistently that what is proximately moved by the unmoved eternal cause must be eternal too.[631] And then he proves what he hypothesized, that there is a single continuous motion,[632] assuming that the primary motion should be of that sort,[633] and after demonstrating in many ways that locomotion is prior to the other kinds of motion and change.[634] Also, <he proves> that among locomotions only circular motion can be continuous, not
1365,25 motion that turns back on a finite straight line or combined motion, since what turns back comes to a stop in between.[635]

After mentioning the points that divide and terminate straight lines, he now takes the analogous instants in time and does well to remark that the instant that divides the time is, in relation to time,
1365,30 the end of one part and the beginning of another, but in relation to the thing that is in time, it will be added to the later part.[636] <He remarks> also that time is not composed of or divided into things without parts,[637] as he also proved earlier than this book.[638] He also proves through a more dialectical and general approach that opposite motions must be interrupted by a stopping,[639] and he extends the
1365,35 demonstration more generally to cover every change,[640] proving on this basis that things that move in a straight line do not undergo a single continuous motion. He then proves the same result generally and in a way appropriate to the present point.[641] He then proves that only circular motion can be continuous, since it is not composed of contrary or opposite motions as rectilinear motion is, and is not
1365,40 interrupted by rest like rectilinear motion and all other motions and

changes.[642] Therefore, if locomotion has been proved prior to the other 1366,1
kinds of motion, and circular motion is the primary kind of locomotion
since it alone can be continuous and one, it is clear that of all motions
circular motion will be primary.[643] This is why it is also the measure
of the others.[644] It likewise follows from the fact that it is the measure
that it is primary. For the measure is simpler than what it measures 1366,5
and is prior, as the unit is <simpler than and prior to> number and
the finger or foot to a larger magnitude;[645] and it is reasonable that
the primary motion is the only one among the rest that is both uniform
and a measure.[646]

Then, as is his practice, he draws on the opinion of the other
natural philosophers and on the customary use of terms to confirm
that locomotion is prior to the other kinds of motion.[647] After review- 1366,10
ing the main points of what had been demonstrated previously,[648] he
adds that the primary mover is without parts or magnitude,[649] since
there cannot be an infinite magnitude and since a finite magnitude
cannot contain infinite power, as he proved.[650] He also adds the
converse of this, that an infinite magnitude cannot contain finite
power either, even if we hypothesize that there is an infinite magni- 1366,15
tude.[651] But since the entire discussion depends on <the principle>
that everything that is moved is moved by something, he makes the
discussion more credible by introducing and refuting the objection
from <the behaviour of> thrown objects.[652] And after all this, he
concludes in summary fashion his account which proves that the first
mover is without parts and indivisible,[653] as a result of which the
philosopher proved that the entire structure of nature depends on a 1366,20
cause that is above nature, and that the study of nature depends on
first philosophy, thus placing this fairest end on the treatise concern-
ing the principles of nature.

Notes

1. *to prôtôs kinoun*. I translate this phrase throughout as 'the primary mover'. 'First mover' translates *to prôton kinoun*.

2. *kinêsis*. Without great enthusiasm I employ the standard translation of this term, 'motion', although for Aristotle and Simplicius the term *kinêsis* covers not only motions in place (*phora*, which I translate 'locomotion') but also changes in quality (*alloiôsis*, 'alteration') and in size (*auxêsis*, 'growth', and *phthisis*, 'decline'). But it does not cover all changes. In particular it does not cover *genesis* ('generation', 'coming to be') or *phthora* ('perishing', 'destruction'). Again, following tradition, I use 'change' as a translation of *metabolê*, which covers all changes: all kinds of *kinêsis*, and generation and perishing as well.

3. *Phys.* 8.5, 257a33-258b9.

4. *koinon*, i.e. common to all kinds of motion.

5. *Phys.* 8.1. But there it was proved that there is always motion, not that any single motion is eternal.

6. *Phys.* 3.3, 202a13-14.

7. *Phys.* 7.1, 241b34-242a49.

8. *Phys.* 8.4.

9. Incidental motion is mentioned at *Phys.* 8.5, 257b20-1, b32-4. Something is moved incidentally if it is not moved per se, but is located in something that is moved. See *Phys.* 5.1, 224a21-30. As the next two sentences and 1251,22-3 show, Aristotle and Simplicius regard souls as things that are moved only incidentally.

10. This narrow view of the soul is appropriate to the discussion of *Phys.* 8, which is based strictly on considerations of motion.

11. This is a general reference to *DA*.

12. cf. *Phys.* 8.6, 259a6-20.

13. Simplicius has *kath' hauto* for Aristotle's *haplôs*.

14. *kai* before *kata sumbebêkos* could be part of the quotation. It is translated 'and incidentally' in the lemma.

15. Simplicius has *de* where Ross has *dê* (b16).

16. Simplicius has *bouloito* where the Aristotle MSS have *bouletai* (b16).

17. Simplicius reads *tôn men* where Ross has *tôndi* (b28).

18. Simplicius has *kinountôn*, which Ross omits (b29).

19. i.e. the particular moved things.

20. Simplicius reads *sunekhôs* where the MSS of Aristotle read *sunekhous* (b29).

21. i.e. having the property of causing the continuous and eternal generation and perishing of things in the world.

22. Simplicius has *aïdion ex anankês*, where the Aristotle MSS (which Ross follows), agreeing with Themistius, have *aïdion kai ex anankês* (b31).

23. Simplicius has *arkhai*, which Ross omits.

24. Simplicius along with many Aristotle MSS has *kinousôn*, where Ross, following one Aristotle MS, prints *kinountôn* (a1).

25. The objection is stated in the parentheses that follow.

26. i.e. an unmoved origin of motion.

27. i.e. that they sometimes are and sometimes are not.

28. Since things without parts are indivisible and since (by *Phys.* 6.4, 234b10-20) things that are indivisible are not subject to change, it follows that things without parts are not subject to generation and destruction. If the motion-causing elements in self-movers are indivisible, it follows further that they are principles of motion (see n. 26 above).

29. This argument corresponds to nothing in Aristotle's text.

30. *Phys.* 6.10, 240b8-241a26. That which is without parts is defined at 240b12-13 as 'that which is indivisible in respect of quantity'. Examples of things without parts are instants and points.

31. Nothing in the following argument corresponds to the text of the *Physics*.

32. Simplicius not unreasonably relies on the converse of the principle on which the previous argument depended (see n. 28). Given that something exists at one time and does not exist at another, then if it is a body (and consequently has size and thus parts) the transitions between not being and being, and between being and not being which it undergoes are instances of generation and destruction.

33. i.e. this is the view Aristotle is refuting.

34. i.e. the collective cause of the eternity and continuity of the process.

35. The reference may be to *Phys.* 8.6, 259a15-20.

36. This premise is not in the *Physics*.

37. viz., not existing simultaneously, but coming to be ad infinitum.

38. A good deal of charity must be exercised in order to make sense of this claim. Perhaps Simplicius just means that for each sublunary thing, the immediate cause of its coming to be is another sublunary thing that comes to be before it. Alternatively, he may have in mind a single genealogical chart where in each case the parent belongs to the generation previous to the child.

39. Eternal: *Phys.* 8.1. At 8.6, 259a16-17 Aristotle argues that eternity implies continuity. At 1255,34-1256,30 Simplicius implies that Aristotle has not yet proved that motion is continuous.

40. viz., non-eternal movers.

41. Simplicius adds *tês* before *kinêseôs*.

42. viz., self-movers.

43. *Metaph.* 12.10, 1076a4 (cf. Homer, *Il.* 2.204); but this reason is not given here.

44. This reason is not given by Aristotle.

45. An almost exact quotation of *Cael.* 2.5, 288a2-3, also of *PA* 4.10, 687a16-17.

46. For Anaxagoras, see *Phys.* 1.4, 187a26-b7. Aristotle is our only evidence that Anaxagoras posited an infinite number of basic substances (not elements, properly speaking). The view also appears at *Metaph.* 1.3, 984a13. The nearest Anaxagoras himself comes to asserting this doctrine in the surviving fragments is in DK 59B4, which mentions 'seeds of all things, having all kinds of shapes and colours and flavours'. For Democritus, see *Phys.* 1.2, 184b20-1. Aristotle (and his commentator Simplicius) are the best sources for the view that Democritus and his predecessor Leucippus believed in an infinite number of different kinds of atoms – where the kind of an atom is determined by its size and shape. See Aristotle, *Metaph.* 1.4, 985b4-19 (DK 67A6), *GC* 1.1, 315b6-15 (DK 67A97), *Cael.* 3.4, 303a11-15 (DK 67A15), Simplicius, *in Phys.* 28,4-26 (DK 67A8, 68A38).

47. *Phys.* 1.4, 188a17-18, 1.6, 189a15-16. Aristotle frequently refers to Empedocles' doctrine of four elements (fire, air, water, and earth), which was the predecessor of his own physical theory. See Empedocles, DK 31B17, 31B6, 31B21, 31B26.

48. Simplicius has *to gar* where the Aristotle MSS have *kai gar to*, which Ross prints (a16).

49. Simplicius calls this argument hypothetical (as opposed to categorical) because it contains premises that are conditionals.

50. Simplicius does not bother to discuss the possibility that consecutive movement could contingently, even if not necessarily, contain no gaps.

51. viz., the succession of movers.

52. The closest Aristotle has come to this claim is at *Phys.* 8.5, 256b14-15.

53. The closest Aristotle has come to this claim is at *Phys.* 5.1, 224a34-b7.

54. Aristotle asserts (rather than proves) this claim at *Phys.* 8.6, 258b31-2.

55. *Metaph.* 12.8, especially 1073b26-38.

56. *Phys.* 8.1.

57. *Phys.* 8.1, 251a23-b10, b28-252a5.

58. The argument is a syllogism in Camestres.

59. *Phys.* 8.1, 251b28-252a5.

60. Simplicius presumably means *Phys.* 8.2, 252b28-253a2.

61. *Phys.* 8.7, 261a28-30.

62. *Phys.* 8.8.

63. Simplicius agrees with Philoponus and all the Aristotle MSS in reading *tôn kinountôn*, which Ross omits.

64. Ross, who follows the majority of the Aristotle MSS reading *engignesthai* (b4), wrongly reports that Simplicius has *ginesthai* (agreeing with MS E of the *Physics*).

65. Simplicius has *heautou* where Aristotle has *hautou*.

66. *Phys.* 8.5, 257a33-258b9.

67. *Phys.* 8.6, 259b6-15.

68. *Phys.* 8.6, 258b16-20.

69. cf. n. 63.

70. Since elsewhere Simplicius follows Aristotle in holding that Heraclitus believed in universal flux (1313,8-10), his hesitation on this matter in the present passage is surprising. In fact, there is no fragment of Heraclitus that states 'all things are in motion'.

71. *Phys.* 7.1, 241b34-242a49.

72. *Phys.* 8.5.

73. *Phys.* 8.5, 257a33-258a27.

74. *Phys.* 8.6, 259b1-3, cf. 8.4, 255a5-10.

75. *Phys.* 8.4, 255b30-1.

76. *Phys.* 8.2, 252b17-28, 253a7-20.

77. *Phys.* 8.2, 253a14-15.

78. For this doctrine, see *Somn.* 3.

79. i.e. the things that affect the self-movers.

80. viz., the body that undergoes circular motion.

81. i.e. it too has an origin of motion and its origin of motion is external to it.

82. This basic tenet of Aristotle's cosmology is frequently mentioned by Simplicius. Aristotle treats it at length at *Cael.* 4.2-5.

83. *Cael.* 1.8, 277a28-9.

84. *Phys.* 8.1.

85. i.e. as a whole, it stays in the same place; its movement involves its various segments coming to occupy one another's places.

86. i.e. the mover.

87. Plato, *Phdr*. 245d4-6 (approximate quotation).

88. viz., the origin.

89. viz., by the first mover.

90. i.e. the first mover.

91. Simplicius' lemma reports *kineisthai*, but Simplicius' commentary gives *to kineisthai*, in agreement with most of the Aristotelian MSS, which Ross follows.

92. Simplicius omits *de*, which is found in all the Aristotle MSS and is accepted by Ross.

93. Aristotle says that it holds of 'some origins of the things in the heaven that undergo a plurality of locomotions'.

94. viz., rotate.

95. This statement is a drastic simplification on the planetary theories of Eudoxus and Callipus, a version of which Aristotle adopted in *Metaph*. 12.8.

96. Cf. above, 1260,25.

97. i.e. the planetary spheres.

98. Reading *kineisthai* instead of *hêgeisthai* (1262,5).

99. Simplicius has *esti ti aei toiouto to kinoun*, where Ross, following the Aristotle MSS, has *estin ti aei toiouton, kinoun* (b32).

100. Simplicius, followed by Ross, omits *tên autên* (a4) which is found in some Aristotle MSS.

101. Simplicius omits *aei* (a4).

102. Simplicius (along with some Aristotle MSS) omits *hupo kinoumenôn men* (a6), which Ross reads.

103. Simplicius reads *ê* where the Aristotle MSS have *de* (a6).

104. *Phys*. 8.1.

105. Eudemus, fr. 121, p. 50,13-20 (Wehrli).

106. The proof of these claims is the business of *Phys*. 8.6.

107. The sphere of the fixed stars.

108. viz., things in the sublunary realm.

109. sc. but not both.

110. Simplicius gets this wrong. As his words stand, they describe the sphere of the fixed stars, but this does not 'come to be in opposite places'. Aristotle correctly says, 'what is moved (a) by the unmoved or (b) by what is already moved' – i.e. (a) the sphere of the fixed stars and (b) the planetary spheres, including that of the sun.

111. *GC* 2.10, 336a17-18.

112. i.e. sublunary things.

113. i.e. the planetary spheres.

114. sc. in the sublunary realm.

115. sc. in the planetary spheres.

116. viz., the sphere of the fixed stars.

117. viz., of the planetary spheres.

118. viz., the sphere of the fixed stars.

119. viz., the sublunary motions.

120. i.e. other sublunary motions.

121. viz., the souls of the planetary spheres.

122. viz., the motion of the celestial spheres.

123. viz., in the sublunary realm.

124. Plato, *Phdr*. 246b6-7.

125. i.e. the souls'.

126. i.e. in the heaven.

127. i.e. the souls'.

128. i.e. on earth.

129. *Phys.* 8.6, 260a14-17, but Aristotle does not talk there in terms of the movements of the heavens.

130. *Phys.* 8.3, 254a19-22.

131. Simplicius leaves open the question how many such movers there are. There is one unmoved mover for each motion, but it is a separate question whether all motions are ultimately due to the same unmoved mover. In *Phys.*, Aristotle gives reasons to think that there in only one unmoved mover (8.10, 267a21-b9), but in *Metaph.* he argues that there are many (12.8, 1074a14-16).

132. *Phys.* 8.6, 258b10-259a20, 259b32-260a19.

133. *Phys.* 8.7, 260a26-8.

134. *Phys.* 8.7, 260b16-261a26.

135. *Phys.* 8.8.

136. *Phys.* 8.7, 260a20-1.

137. *Phys.* 8.7, 261a28-b22.

138. *An. Post* 2.1, 89b29.

139. Ross reports that Simplicius has *gar* where Ross (following most of the Aristotelian MSS), reads *d'* (a26). However, Simplicius' lemma has *de*.

140. *prôtos* normally has the superlative meaning, 'first' or 'primary', in contrast to the comparative *proteros*, 'prior'. Not infrequently, however (see index *s.v.*) it has the comparative meaning, a usage that goes back to Homer. Occasionally, as in the present passage, it must be translated first one way then the other.

141. *Phys.* 8.7, 260a29-b5.

142. *Cat.* 14, 15a30.

143. *dunamei*, rendered 'potentially' in the translation of the Aristotle passage.

144. *Phys.* 8.9.

145. For Democritus, see Aristotle, *GC* 1.1, 315b6-15 (DK 67A97), 1.8, 325a23-6 (DK 67A7), Plutarch, *Adv. Col.* 8, 1110F-1111A (DK 68A57), Simplicius, *in Cael.* 242,21-6 (DK 67A14) and Aristotle, *On Democritus* ap. Simplicium *in Cael.* 295,11-24 (DK 68A37). For Anaxagoras, see DK 59B17. For Empedocles, see DK 31B8, B9, B11, B15.

146. Simplicius follows Aristotle here in the incorrect belief that Anaximander posited a single originative substance that is intermediate between fire and air. See Kirk, Raven and Schofield, *The Presocratic Philosophers*, 2nd ed., Cambridge, 1983, pp. 111-13.

147. i.e. generations and perishings of substances.

148. Plato, *Laws* 10, 893e1-4 (approximate quotation).

149. Plato, *Laws* 10, 893e6-7.

150. Not a quotation, but a reference to Plato, *Laws* 10, 893e7-894a6, 894b10-11.

151. Simplicius has *kai tôi khronôi*, where Ross, following all the Aristotle MSS, has *kai to tôi khronôi* (b19).

152. viz., what is complete.

153. Closing the quote here (1268,6); Diels neglected to do so.

154. *Cat.* 12, 14a35-b7.

155. Adding *mê* in accordance with *Cat.* 12, 14a30.

156. Simplicius calls this book delta.

157. From here to 1269,5 we have a close paraphrase with abundant quotations of *Metaph.* 5.11, 1018b10-1019a11.

158. Aristotle has 'some rule' (*tina logon* instead of Simplicius' *ton logon*). I take it that the point depends on the fact that the rank order of these things is implicit in the definition of the term: *parastatês*, 'he who stands next (sc. to the first one)', *tritostatês*, 'he who stands third', *paranêtê*, 'that which is next to the *nêtê*'.

159. The lowest string on the standard ancient lyre was called the *nêtê*, the second-lowest string the *paranêtê*. The lowest string on a lyre had the highest pitch. See M.L. West, *Ancient Greek Music*, Oxford, 1992, ch. 8.

160. *Metaph.* 5.11, 1019a2-4.

161. *Phys.* 8.1, 251a8-b28.

162. *Phys.* 8.5.

163. *Phys.* 8.8.

164. *Phys.* 8.7, 261a27-b26.

165. Placing the comma after *phoran* at 1270,3 (with MS A).

166. Aristotle does not prove this claim. He argues at length for the different claim that one kind of locomotion (namely, circular motion) is the only kind of motion that can be infinite (*Phys.* 8.8).

167. I punctuate differently from Diels, who makes this a separate sentence.

168. *Phys.* 2.2, 194b13.

169. *Phys.* 8.6, 259b32-260a1.

170. *Cael.* 1.12.

171. In Simplicius' lemma, the text reads *holôs te ei phainetai*, where all the Aristotle MSS, followed by Ross, omit *ei* (which I do not translate).

172. *Metaph.* 5.11, 1019a2-4.

173. This passage marshals two separate pieces of evidence for Aristotle's view. (1) Unlike lower forms of life such as plants, higher forms of life (animals) can move in place. (2) Animals acquire the ability to move in place after their capacities to undergo other kinds of change are first actualized. The word *teleios* and associated words (*atelês*, etc.) are put to work in describing both of them. Plants are 'less perfect' (*atelestera*) than animals, and mature animals have reached their 'state of completion' (*teleiotês*). To translate these words consistently here would sound contrived. Cf. also *telos* (1271,36), translated there as 'end'.

174. i.e. a final cause.

175. The mature adult specimen is what reproduces.

176. This terminology and way of looking at syllogisms is Stoic, not Aristotelian. See Sextus Empiricus, *Adv. Math.* 8.227. The form of the argument is 'if *a* then *b*; a; therefore *b*'.

177. *Proterou.* These are the same as the ways of being primary mentioned above. See also n. 140 above.

178. *Phys.* 8.5, 257a28-31.

179. This claim was not proved previously. Aristotle simply asserts that 'most of all it is clear' (261a23).

180. Plato, *Laws* 10, 893e1-4 (approximate quotation).

181. Plato, *Laws* 10, 893e6-894a1 (approximate quotation). Simplicius omits *mê* before *menousês*. Plato has 'if it does not persist'.

182. This is not a quotation, but a reference to Plato, *Laws* 10, 893e7-894a6, 894b10-11.

183. Simplicius omits *to* before *mê on horoi* (a34).

184. i.e. in the way that the theorem currently under discussion is universal and useful.

185. Reading *monêi* at 1273,31 with the MSS, which Diels misprints as *monê*.

186. Simplicius is thinking of 261b3-14, but in fact Aristotle prepares the way for this distinction at 261a33, where he speaks of *hai kinêseis kai metabolai*.

187. *Phys.* 5.1, 225a20-b1.

188. Following Diels' suggestion, ad loc.

189. *megethos* tr. here as largeness, usually (including just below) as magnitude.

190. This argument relies heavily on the definition of a motion in terms of its endpoints. When something that is becoming black has come to the end of the motion, it is black. If something is becoming white and becoming black simultaneously, then at the endpoint of the motion it is both white and black. In order to avoid this contradiction, Aristotle says that nothing can be becoming white and becoming black simultaneously; since the state of being white is different from the state of being black, the change to being white is different from the change to being black and they cannot be simultaneous, hence nothing can simultanenously be becoming white and becoming black. In the case considered, it first becomes white and then becomes black, and while it is becoming white it is not yet becoming black.

191. *ei de pausamenon tou leukainesthai leukon genoito* (1274,36). The point is not that it achieves a state of being white either *while* the process of becoming white is still going on or *after* the process has ceased, but at the instant at which it ceases. Cf. *Phys.* 8.8, 263b9-26.

192. Punctuating with a full stop after *anthrôpon* (1275,11).

193. i.e. things that are coming to be and perishing.

194. Aristotle discusses contradictory and contrary opposites at *Int.* 7, 17b16 9, 18a33 without making precisely this claim. But as *Int.* 12, 21b37-22a1 shows, he is committed to this claim. Cf. also *Int.* 14, 24b9.

195. *Phys.* 5.6, 229b23-31; but there Aristotle does not call it 'contrary' but 'opposite as a privation'.

196. Reading *metabolêi* for Diels' *metabolês.*

197. *Phys.* 5.6, 229b23-31.

198. *summetros.* Aristotle uses this word in discussing the virtues (*EN* 2.3, 1104a18). For the doctrine that virtues are opposed to vices of excess and deficiency, see especially *EN* 2.8, 1108b11-19.

199. Diels does not mark the end of this quotation. I conjecture that it continues to the end of the paragraph (1277,14).

200. i.e. if A is the change from X to Y and B is the change from Y to X, where X and Y are contraries, nothing prevents A and B from turning out to be both of them A.

201. i.e. because the changes are opposites of one another.

202. Simplicius takes *touto esti* with 'that in which', whereas Ross takes it with the following clause and punctuates accordingly.

203. Simplicius has *megethos ê eidos*, where Ross follows some Aristotle MSS in reading *eidos ê megethos* (a4-5).

204. *Phys.* 8.7, 261a31-b26.

205. If A and B are limits, the claim is that in moving from A to B and back to A, the thing that undergoes motion 'turns back' at B, and (in consequence of *Phys.* 8.7, 261b1-3) pauses at B in between the motion from A to B and the motion from B to A.

206. Diels notes that MS a completes the lacuna as follows: 'the combined motion is impossible too'. But Diels believes that the missing passage is more extensive. Perhaps the following is more appropriate: 'to be continuous, the combined motion cannot be continuous either'.

207. *Phys.* 3.5, 204b4-206a8, 3.6, 206a14-b27, 207a22, 3.8, 208a5-22.

208. *Phys.* 1.5, 188a24-5, 3.2, 205b32-3.

209. *Phys.* 8.7, 261b1-15.

210. *Phys.* 5.4, 227b20-6, 228a20-2.

211. *eidos*, the same word used just above to refer to changes in substance; here (as in Aristotle's text, 262a2, a5) it is used generally to refer to what was just called 'place, affection, magnitude, and form'.

212. Simplicius is thinking of the four elements.

213. *kath' hous.* The masculine pronoun refers to *topôn* ('places') (1279,5). I have changed Diels' punctuation to bring this out.

214. *Phys.* 8.8, 261b32-6.

215. *Phys.* 8.8, 261b36-262a5.

216. Simplicius has *eis ta plagia* where Ross, following all the Aristotle MSS, has *eis to plagion.*

217. Aristotle is thinking of contraries as physically opposing one another.

218. *tekmêriôdeis.* Simplicius seems to be using this word in a looser way than when he speaks of 'proof by necessary signs' (*tekmêriôdês apodeixis*) as the method by which the physicist grasps the principles of physics. See D. Morrison, 'Philoponus and Simplicius on Tekmeriodic Proof', pp. 1-22 in *Method and Order in Renaissance Philosophy of Nature*, ed. E. Kessler, Aldershot, 1997.

219. See *Cat.* 10 on the types of opposition.

220. 'A sign (*sêmeion*) is a proposition stating that if one thing, A, is, was, or comes to be, then another thing, B, is, was, or will be' (*An. Pr.* 2.27, 70a6). Aristotle also calls a 'sign' the thing or fact or event (*pragma*) A whose existence or occurrence permits one to infer the existence or occurrence of B. One crucial distinction Aristotle makes is between signs that hold only for the most part and signs that hold always and necessarily. Aristotle calls the former signs *sêmeia*, and reserves the term *tekmêrion* for necessary signs' (Morrison, op. cit., p. 3).

221.

222. Where Aristotle talks of *motions* stopping one another, Simplicius talks of *things in motion* stopping one another *when they meet.*

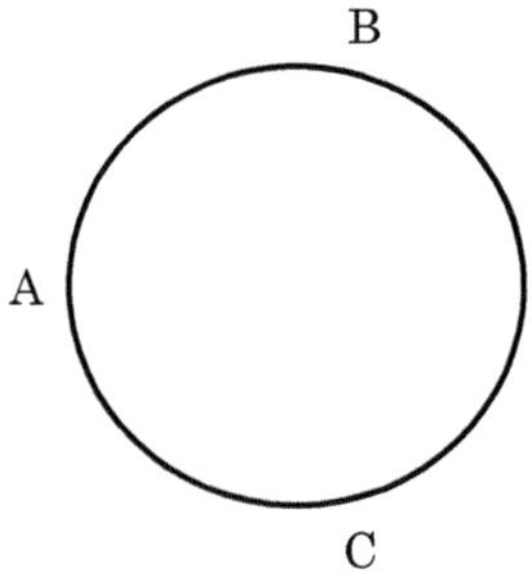

223. Simplicius omits *kai allôi* (b2).

224. Simplicius and the manuscripts of Aristotle have *de*, which Ross emends to *dê* (b5).

225. That is to say, by stopping at a point, what is undergoing motion actualizes that point, which was previously only potentially a point. See below, 1281,38-1282,15.

226. *idion.* Clearly this is not a property in the technical sense of *Top.* 1.5, 102a18-30, in which a property of something is a non-essential attribute which is an attribute of that thing alone.

227. This would actually be the case in the kind of back-and-forth motion under consideration.

228.

229. Simplicius' point is that the fact that the middle is numerically one does not prevent it from being more than one in definition; see 262a21.

230. *stasis*, the word translated below as 'stopping' and 'stationary state'. 'Stops there' translates *stasis* at 1282,3.

231. *Phys.* 8.8, 262a32-b1.

232. *Phys.* 4.11, 220a5, a9-10, 4.13, 222a10-20.

233. *Phys.* 6.1, 231b9-10.

234. *Phys.* 4.10, 218a18-19.

235. Most of what follows is not a direct quotation.

236. Reading *oukh hama ara* (262b15) with some Aristotle MSS and Simplicius (1285,31), where Ross prints the conjecture (wrongly reported to be supported by Simplicius) *ou gar hama*.

237. Following Diels' suggestion for filling the lacuna.

238. i.e. the endpoint.

239. i.e. they are ends potentially.

240. i.e. Alexander's preferred reading.

241. Simplicius takes 'potentially' as going with 'reach' and interprets 'potentially reach' as meaning 'potentially be at'.

242. I take it that Simplicius' comment on Alexander's text stops here, and punctuate accordingly.

243. i.e. the end point.

244. Simplicius seems not to read *kai axiountas* (263a5), which Ross omits, but which all the Aristotle MSS contain.

245. Ross reports that Simplicius has *peperasmenôi khronôi apeira* where the Aristotle MSS have *peperasmenôi apeira*, but I see no clear evidence that Ross is correct.

246. *Phys.* 6.9, 239b9-240a18.

247. This is an expanded version of the argument which Aristotle sketches in a line and a half (263a5-6). Note that in Aristotle's version the argument does not depend on the impossibility of traversing an infinite number of things *in a finite time*.

248. *ouden*, translated 'not at all' in the lemma.

249. cf. *Phys.* 3.6, 206a21-5.

250. i.e. one and continuous.

251. i.e. with the addition of 'potentially'. See Diels ad loc.

252. This curious claim is based on the view (which Aristotle does not hold) that bees are generated spontaneously in the decaying flesh of bulls. This belief is found most famously in Vergil, *Georgics* 4.281-314, 550-6; also in Sextus Empiricus, *PH* 1,41 and Nicander, *Theriaca* 741, which is quoted almost verbatim by Olympiodorus (*in Meteor.* 278,9), Philoponus (*in Phys.* 179,6) and Simplicius (*in Phys.* 239,19).

253. Magnitude: see references in n. 207 above; plurality: *Phys.* 3.6, 207b1-15.

254. *Phys.* 3.6, 206a19-21 (approximate quotation).

255. *Phys.* 3.6, 206a27-9 (approximate quotation).

256. *Phys.* 3.6, 206a21-3 (approximate quotation).

257. *Phys.* 3.6, 206a25-7 (approximate quotation). Aristotle actually says that the nature of the infinite is not the same in these three cases.

258. *Phys.* 3.6, 206a30-3 (approximate quotation).

259. Simplicius has *an* where most Aristotle MSS, followed by Ross, have *ean* (b9).

260. Simplicius apparently read *tou husterou*, which is absent from the Aristotle MSS, but which Ross adds (b21).

261. Simplicius omits *ou* (b21) along with some Aristotle MSS.

262. Ross reports that Simplicius has *ei* (b21), which is missing from some Aristotle MSS (which Ross follows but 1295,26 does not prove this claim).

263. Ross adds *to* (b22), which is not found in Simplicius or the Aristotle MSS.

264. i.e. the instant at which the division occurs.

265. It is said to be the first instant at which the thing is black, not the last instant at which it is white.

266. viz., such as it is in the later state.

267. Following Diels' suggestion to read *khronon*.

268. i.e. the instant that divides the earlier and later states.

269. On the former interpretation, 'the thing' is an affection; on the alternative interpretation, 'the thing' is the subject that changes.

270. *pathos* ('affection') stands for the outcome or result of all the kinds of motion here, and to that extent Simplicius can say that Aristotle uses the word in connection with all kinds of motion.

271. In this paragraph, 'thing' sometimes refers to D, sometimes to D in time A or D in time B, so that there are two kinds of 'things' involved, sometimes to what D is in time A or in time B (i.e. white or black), and sometimes to the affections themselves (i.e. whiteness, non-whiteness).

272. Simplicius slips here. C belongs to both A and B, which are time intervals.

273. viz., C.

274. viz., A's.

275. Reading *genomenon* for *ginomenon*.

276. This sceptical argument against the possibility of anything perishing is given by Sextus Empiricus (*Adv. Math.* 10.346 with context at 10.344-50). Also, *Adv. Math.* 9.269 and *PH* 3.111. Sextus attributes a similar argument to Diodorus Cronus (*Adv. Math.* 10.347).

277. *en tôi prôtôi perati*. The word 'first' is not wanted; it is the last instant of the process of dying. RRKS suggests that Simplicius may be thinking of it as the first instant our thoughts reach as they go backwards.

278. *Phys.* 6.1-2.

279. *Phys.* 8.6, 258b16-20, cf. 6.4, 234b10-20. Also, *Phys.* 7.3, 247b1-13.

280. An anonymous reader has suggested emending the text by adding *kai estin*, 'and is', as in Aristotle's text.

281. Simplicius is saying that Aristotle should have said 'already is' instead of 'has come to be'.

282. A was previously composed of indivisibles; now it is itself an indivisible.

283. Simplicius means that Aristotle should have said 'at another indivisible time … and is'.

284. In terms of the example Aristotle has in mind, we should continue to understand 'white' as the predicate.

285. Simplicius is here describing the refutation already given of the view that time is composed of indivisibles.

286. i.e. common to both views of the nature of time.

287. Approximate quotation. The divergence from the original requires the addition of the word 'point'. The original wording is quoted below, 1298,34-5.

288. i.e. a beginning or an end.

289. I adopt the anonymous reader's suggestion to emend A to D.

290. Ending the parenthesis after *ekeinôi* (1299,7) instead of *egeneto* (1299,8), with equal MS authority.

291. i.e. being white.

292. i.e. such as to change truth value indeterminately. See K. Hülzer, *Die Fragmente zur Dialektik der Stoiker*, Stuttgart, Bad Cannstadt, 1987, vol. 3, p. 1317.

293. cf. Diogenes Laertius 7.80 for the Stoic terminology: *proslêpsis* ('additional premise'), *sunêmmenon* ('conditional'). The form of this argument is what the Stoics called the first indemonstrable (ibid.).

294. *Phys.* 8.8, 263b9-26.

295. Ross's app. crit. wrongly reports that Simplicius omits *logois* (a7); see 1301,16.

296. *Phys.* 8.8, 262a14-b7.

297. At *An. Post.* 1.2, 71b17-25, Aristotle requires that a scientific demonstration (*apodeixis*) have principles (*arkhai*) that are 'proper' or 'appropriate' (*oikeiai*) principles of the fact that is demonstrated.

298. *logikôs*. This word is notoriously difficult to render into English. 'Dialectical' and 'general' are frequently given as translations, but neither works well here, since in the following sentence these deductions are called both *katholikôteros* ('more general', 1301,22) and *dialektikos* ('dialectical', 1301,23). 'Logical' is inappropriate, since the contrast is being drawn between two kinds of deductions ('proper' and '*logikos*') which are equally 'logical'.

299. Here as elsewhere I translate *katô* as 'downwards, down, beneath' and *to katô* as 'the bottom'. In this passage Simplicius is occasionally careless, as now, when he uses a directional word where he should be speaking of the terminus of motion.

300. Below, 1303,36-1304,9.

301. viz., they are contraries.

302. *Cael.* 2.2.

303. If locomotion were the only kind of motion under discussion, it would be appropriate to say 'somewhere'. But since the discussion applies to all kinds of motions, including alteration from the condition of having one quality to that of having another, I have given a translation that can cover all motions, which are changes from one 'thing' to another, be it a place, a quality, or a size.

304. Placing the comma after *eleusetai*, not after *kinoumenon* in 1303,3, with equal MS authority.

305. Since points are indivisible, Simplicius does not mean that the moving thing divides the halfway point. This will be a shorthand way of saying that it divides the continuum at the halfway point, thus actualizing the point.

306. If the text is not corrupt, it seems to mean that in a case where something moves up to the top and down again, the downward motion is a motion in its own right, not part of a single continuous motion. The following clause describes what we have seen would have to be the case if it *were* part of a simple motion (cf. 264a18-19).

307. Punctuating with a comma after *kineisthai* (1303,25), not a full stop.

308. Simplicius has *estai*, where the Aristotle MSS, followed by Ross, have *estin* (a26).

309. *Phys.* 1.7, especially 190b17-33 and 1.8, 191b15-16, cf. 1.5, 188a27-b26.

310. In Simplicius the verb *kineitai* (1304,11) has a plural subject, whereas in Aristotle (264a23) it has a singular subject.

311. From Aristotle's *hapan to kinoumenon* Simplicius quotes *to kinoumenon*.

312. i.e. motions that differ in that they are different parts of the same motion.

313. 'Contrary' as opposed to 'the privation of'.

314. *Cat.* 10, 12a27-34.

315. The per se attributes of a thing are (1) those that occur in the thing's essence (*An. Post.* 1.4, 73a34-b3), and (2) what a science proves by demonstration to hold of that thing (*An. Post.* 1.10, 76b6-10) rather than of any more general subject (*An. Post.* 1.4, 73b25-74a3). They follow from the 'proper' principles of the science (see above, n. 297).

316. *Phys.* 4.14, 222b30-1, 6.4, 235a13-14.

317. *Phys.* 6.5, 235b6-27.

318. *Phys.* 6.6, 237a10.

319. i.e. what is not white has ceased to be not white.

320. As the text stands, this must mean that the non-whiteness of the non-white thing has perished, but the expression is unusual.

321. *phantasia*. For Aristotle's views on the 'imagination', see *DA* 3.3, 427b27-429a9. *DA* 3.11, 434a5-11 claims that imagination can be affected by reasoning.

322. *Phys.* 5.3, 226b34-227a6, 6.1, 231a23.

323. *Phys.* 5.3, 227a10-12, 6.1, 231a22.

324. Punctuating differently from Diels.

325. If we consider the motions from black to white and from white to black as a single, continuous motion, this motion will have one end.

326. Closing quote here (Diels neglects to close the quotation).

327. Simplicius (1307,32, in a paraphrase, not a quotation) agrees with some Aristotle MSS in giving *ep' eutheias*, where Ross, following other Aristotle MSS has *kat' eutheian* (b14).

328. i.e. at the ends of.

329. i.e. the points *on the circle* that are on a diameter.

330. In the supposed case the contrary motions are reverse motions on the diameter.

331. i.e. the points.

332. Above, 1302,29-31.

333. In his proof of *Elements* 1.4, Euclid asserts that it is impossible that two straight lines enclose a space (which is equivalent to the claim that there is only one straight line from one point to another). Proclus claims that Euclid did not place this among his postulates or axioms because Euclid intended his first postulate ('To draw a straight line from any point to any point') to imply that it is one straight line, not two, that would always join the two points (Proclus, *in primum Euclidis librum commentarius*, 239,15-20 (Friedlein), translated and discussed by T.L. Heath, *The Thirteen Books of Euclid's Elements*, Cambridge, 1925, vol. 1, p. 195.)

334. Omitting the comma after *anakamptonta* in 1308,28.

335. Simplicius' lemma gives *heautou* (b18), which accords with the *hautou* of most Aristotle MSS (which Ross prints). In the commentary, Simplicius quotes the text as *apo tou autou* ('from the same <point>', 1309,13-14).

336. Simplicius' lemma gives *heauto* (b19), which accords with the *hauto* which Ross prints. In the commentary, Simplicius quotes the text as *eis auto* ('to it', 1309,14-15).

337. Simplicius omits *aph' autou*, which the Aristotle MSS have and which Ross prints (b19).

338. *Phys.* 8.8, 262a14-b7.

339. Simplicius implies that in pausing at A, X 'comes to be' at A both when it stops there and when it starts from there – an interpretation which is hard to understand and which makes no use of the fact that the subject of discussion here is *repeated* motion. I suppose that Aristotle had in mind the situation where X moves back and forth repeatedly between A and B, pausing (and therefore coming to be) at each endpoint *each time*.

340. Strictly speaking, of course, Simplicius holds that a thing that moves continuously is never *at* any of the points on its path.

341. Placing comma after *palin* (1310,5) (with MS A) instead of *pollakis* (1310,6) (with Diels).

342. Strictly speaking, rectilinear motion from A to B is not 'the same' as motion from B to A.

343. The feminine gender of *mias* (1310,22) is surprising, since Simplicius has been using *epi* with the neuter to refer to the substrate (*hupokeimenon*, 1310,18) over which the motion takes place.

344. Up to now Simplicius has interpreted 'the intermediates' as the endpoints of changes; from now on he takes them to be the range between the endpoints.

345. That is, on the assumption that the intermediates are actual states, if there is an infinite number of them, a Zenonian argument would apply.

346. Heraclitus is meant, together with those like Cratylus who took up Heraclitus' doctrines (Aristotle, *Metaph.* 1.6, 987a32-3). The doctrine that all things are always in motion would apply also to the fifth-century atomists (Aristotle, *Cael.* 3.2, 300b8-10) and presumably to Anaxagoras as well (DK 59B12 with 59B1 'when all things were together [i.e. at the beginning, before there was motion] nothing was manifest on account of smallness'). But in the *Phys.* text the people meant here also are said to have believed that things are in flux and that generation and perishing are a matter of alteration, and Aristotle (along with Simplicius) associates the doctrine of flux exclusively with Heraclitus and his followers, whereas Simplicius holds that Anaxagoras and the atomists along with Empedocles believed that generation and destruction are a matter of combination and separation (above, 1266,33-6).

347. Plato, *Tim.* 28a2-4.

348. The famous statement 'You could not step twice into the same river' is probably a misinterpretation of Heraclitus' thought, which he expressed as 'Upon those who step into the same rivers, different and again different waters flow' (DK 22B12, see Kirk, Raven and Schofield, op. cit., pp. 142-4). Even if it is a misinterpretation, it is a very old misinterpretation. It is quoted by Plato (*Crat.* 402a) and was known to Cratylus himself, according to Aristotle's citation at *Metaph.* 4.5, 1014a14.

349. *EN* 1.6, 1098a18-19.

350. Ross's app. crit. reports that Simplicius omits this last clause ('for ... destroyed' (a26-7). Indeed, it is not found in Simplicius' lemmas, but it is not clear to me that this shows that it was not in the text that Simplicius had before him.

351. *Phys.* 8.7.

352. *Phys.* 8.8, 261b28-9.

353. No actual infinity: *Phys.* 3.5, 204a20-206a8, 3.6, 206a14-b27, 207a22, 3.8, 208a5-22; cannot be got completely through: cf. *Phys.* 3.6, 207a7-8.

354. i.e. over all of it, rather than on a finite part of it.

355. Nothing in Aristotle corresponds to this sentence. Aristotle just says it will be incomplete and perishable (265a21-2).

356. *Phys.* 8.7, 260b16-19.

357. Simplicius here begins discussing 265a24-7.

358. *êremei*, the word Simplicius quotes here, has a singular subject in Simplicius, but a plural subject in the Aristotelian passage.

359. See *Phys.* 8.8, 262a21-5 for this doctrine.

360. *Phys.* 4.14, 223b12-224a2.

361. Alexander read *phora* where the manuscripts and modern texts have *periphora*.

362. i.e. the name of what circular motion has in common with other relevantly similar kinds of motion.

363. Closing the quotation from Alexander after *autês* (1317,6), where Diels has it go to the end of the paragraph.

364. Simplicius omits the redundant *pleion* (or *pleiô*) of the Aristotle MSS (b14), which Ross deletes.

365. i.e. on the circle, that is, the *path* of the circular motion.

366. Cf. above, 1316,9-23.

367. In the passage being commented on, Aristotle does not consider the two cases mentioned in this sentence.

368. Aristotle says that the centre is 'outside' (*ektos*) the circle (265b16), in the sense of 'off' the circumference. Simplicius adds 'or rather inside' (*êtoi entos*) to disambiguate.

369. Simplicius is thinking of the 'resistance of the medium'. That through which the object moves impedes its motion; the object moves by 'dividing' or cleaving its way through the medium. The closer the object gets to its natural place, the less medium there is to impede it and the less for it to cleave.

370. Simplicius omits *kai hôs*, which is found in all the Aristotle MSS and is printed in Ross (a26).

371. Ross follows Simplicius and one Aristotle MS in reading *kata topon* (a2), where one other Aristotle MS has *tên kata topon* and the rest have *tên kata topon kinêsin*.

372. Accepting Diels' emendation.

373. This expression does not occur in Aristotle. Simplicius will have in mind similar phrases, such as are found at *Phys.* 1.5, 188b29-30, *Metaph.* 1.3, 984a18-19, b9, and *PA* 1.1, 642a19.

374. Empedocles, DK 31B17 lines 7-8, translation as in R. McKirahan, *Philosophy Before Socrates*, Indianapolis, 1994.

375. Empedocles DK 31B53, translation from McKirahan, op. cit. (n. 374).

376. Simplicius quotes *diakrinein* from Aristotle (265b22), but in a syntactically different context that requires a different translation.

377. 'Homoeomeries' is Aristotle's word for Anaxagoras' basic substances. 'Homoeomery' is a concept which is important in Aristotle's physical system, but which does not fit well into Anaxagoras' system, in which there are no uniform substances, but 'there is a portion of everything in everything' (DK 59B11).

378. This passage is our only source for this usage by the atomists of the term '*phusis*'. For motion of atoms in the void, see *GC* 1.1, 315b6-15 (DK 69A97), 1.8, 324b35-325a36 (DK 67A7), Simplicius *in Cael.* 295,1-22 (DK 68A37). Whether atoms have weight is controversial. Simplicius here and elsewhere (*in Cael.* 569,5-9 (DK 68A61)), following Aristotle (*GC* 1.8, 326a9-10, *Cael.* 4.2, 309a1-14 (both DK 68A60)) holds that they do possess weight, but other ancient sources deny this: Aetius 1.3.18 and 1.12.6 (both DK 68A47). A reasonable compromise is to hold that the atomists did ascribe weight (in some sense) to the atoms, but not as one of their basic properties.

379. I translate the manuscript text (*peripalaisesthai*), which Diels prints but regards as unsound. He suggests *peripalassesthai*, 'are hurled about' (Democritus DK 68B168), but see J.B. McDiarmid, 'Phantoms in Democritean Terminology: *peripalexis* and *peripalassesthai*', *Hermes* 86, 1958, 294-5.

380. Note the close connection between the words translated 'altered' (*alloioumenon*) and 'different' (*alloian*).

381. Simplicius names the Presocratics whom Aristotle describes as 'all those who bring about generation and perishing through density and rarity' (265b30-2). In fact, this description applies certainly only to Anaximenes among the thinkers Simplicius names. See *in Phys.* 24,26-31 (DK 13A5), ps.Plut. *Strom.* 3 (DK 13A6), Hippolytus, *Ref.* 1.7.3 (DK 13A7).

382. Plato, *Phdr.* 245c7-9.

383. Simplicius refers to *Phys.* 8.6, 258b32-259a6 and 259b32-260a10, although those passages do not mention the heavenly or the sublunary bodies.

384. *empsukhon*, literally, 'having a soul in it, ensouled'.

385. This aspect of Aristotle's doctrine of the soul is presented in *DA* 2.2-3. Plants possess only the 'nutritive soul', whose capacities are limited to nutrition, decay and growth, and do not include the capacities of sensation and local movement, which are characteristic of animals.

386. Alteration is a matter of atoms changing arrangement and position (*GC* 1.2, 315b6-15 (DK 67A97)), which itself is a matter of motion in place.

387. Simplicius is referring to the motion of the *pneuma*, which for the Stoics is basic to the maintenance and structure of the universe. 'The movement and properties of individual bodies are a consequence of the dispositions of a single all-pervading dynamic substance' (A.A. Long, *Hellenistic Philosophy*, London, 1974, p. 157).

388. *Phys.* 8.1.

389. *Phys.* 8.5, 256a13-21; but this passage does not discuss eternal motion in particular.

390. *Phys.* 8.7.

391. *Phys.* 8.8.

392. *Phys.* 8.5, 256b4-258a20, 8.6, 258b11-259a6.

393. Simplicius and several Aristotle MSS have *tôi* D where Ross prints *têi* D (a19).

394. *Phys.* 8.5, 258b4-9, 8.6, 258b10-259a6.

395. i.e. corporeally and forcibly.

396. i.e. it takes a shorter time to make it move. This use of 'in a time' is prevalent throughout this section of the commentary.

397. As the following example shows, Aristotle means that the greater power can cause the object to move for a longer time than the smaller power (the human gets too tired to move the mill sooner than the mule does), and also while it is moving the object, the greater power makes it move faster than the smaller one does.

398. *Phys.* 8.10, 267b19-26.

399. *Phys.* 3.5, 204a34-206a8 proves that there is no infinite body.

400. Adding a full stop after *amegethes* (1321,24), as in MS A.

401. The word *kinein*, which Simplicius quotes from Aristotle (266a13), is transitive in Simplicius but intransitive in Aristotle.

402. Adding a full stop after *einai* (1321,29), as in MS A.

403. *to kinoun megethos*. The participle *kinoun*, here translated 'that causes motion', is translated 'mover' in the lemma at 266a13, when it is used substantively.

404. Actually, the first period of time that was taken was equal to one-third of F.

405. Simplicius now gives a simpler argument for the same point.

406. For example, if the part taken is one-third of B, we are to 'add together' three such parts.

407. That is, the time A takes to move all of B.

408. *kineisthai*, here infinitive in indirect statement after 'he says'; in the Aristotelian lemma it is governed by 'can' (*hoion te*) and is accordingly translated 'be moved'.

409. Here and throughout, Simplicius means that the same thing is moved the same amount.

410. The difference between these two consequences is slight: it does not cause motion (*kinei*) and it is not motion-causing (*kinêtikê*).

411. Simplicius quotes Aristotle's word *kinêsei*, but makes it transitive.

412. *touto paskhein*, where *touto* is object of *paskhein*.

413. Simplicius means 'from any time no matter how small'.

414. The final two sentences of this paragraph emphasize that *no* part of a whole that possesses infinite power can possess infinite power. In fact, this point is made well enough at 1326,29-31.

415. i.e. John Philoponus, Simplicius' Christian rival whom Simplicius never names. Simplicius here criticizes at length some of the arguments in a work by Philoponus which is different from *Against Aristotle on the Eternity of the World*. See P. Hoffman, 'Simplicius' Polemics', pp. 57-83, and C. Wildberg, 'Prolegomena to the Study of Philoponus' *Contra Aristotelem*', p. 198 and n. 5, both in *Philoponus and the Rejection of Aristotelian Science*, ed. R. Sorabji, London, 1987.

416. Simplicius *in Phys.* 1117,15-1118,11.

417. viz. the lay people.

418. At *Cael.* 1.12, Aristotle argues that the heaven is ungenerated and indestructible.

419. Simplicius *in Cael.* 25,23-38,5; 42,17-49,25; 56,26-59,23; 66,8-91,20; 119,7-144,4; 156,25-201,10.

420. Simplicius *in Phys.* 1129,29-1152,19; 1156,30-1169,8; 1171,30-1182,39.

421. Literally, 'the power being taken' (*hê … lambanomenê dunamis*). This notion is in accord with Aristotle's definition of the infinite, for which see below, 1328,12-30.

422. This odd expression reflects the definition of the infinite (see just below).

423. i.e. the capacity to be moved ad infinitum.

424. Above, 1180,20-6.

425. *Phys.* 3.6, 206a27-9 (approximate quotation) (translation taken from the Revised Oxford Translation, ed. J. Barnes).

426. *Phys.* 3.6, 206a26-7.

427. *Phys.* 3.3, 202a13-b22.

428. *khorêgos*, 'chorus leader', 'benefactor', one who provides the financial backing of a play, or contributes money for civic purposes.

429. In the sense of having power that is infinite all at once together.

430. *Phys.* 8.1.

431. *Phys.* 8.8.

432. *Cael.* 1.12.

433. Simplicius *in Cael.* 59,15-18 together with context 56,26-59,23.

434. Placing the comma after *ekeinôn*.

435. *atopôteron*, lit. 'more out of place'; note the play on words.

436. Just as the components are joined together in a particular way, and the creator knows how they are joined together, so he knows how the compound can be taken apart. What he knows is the account (*logos*) of their union and also of their undoing.

437. Closing the parenthesis after *esti* (1331,13) (with the authority of MS A), instead of *apeiron* (1331,15).

438. Philoponus.

439. Plato, *Tim.* 41b4-6, discussed below at 1338,16-30.

440. *Cael.* 1.4.

441. Removing the parentheses in 1332,9-10.

442. i.e. the time it takes for the ladlefuls of water to perish.

443. Simplicius seems to misunderstand Philoponus' argument, or at least to interpret it uncharitably; Philoponus does not make the mistake Simplicius charges him with, of failing to apply the principle (1332,16-18) that larger amounts of water last longer than smaller ones.

444. This principle was stated at 1332,7-8, but as Simplicius reads the argument, it is not used until now. Alternatively, *hôi*, translated as 'a <principle> that', can be translated 'which', referring to 'the entire world', and the point could be that until now Philoponus has been discussing the celestial and the sublunary bodies, but not the world as a whole.

445. *ta ourania*, normally translated 'the heavenly bodies'; that translation is not possible here.

446. *logos phusikos*. This expression is apparently synonymous with 'definition of their nature' (*logos tês phuseôs*, above 1333,16).

447. Punctuating differently from Diels.

448. viz., that bodies are divisible ad infinitum.

449. i.e. parts that have extension and are divisible into parts.

450. As Aristotle explains below, he uses this expression to indicate that the different parts of a body have different spatial locations, even if the endpoints of adjacent parts coincide.

451. Plato, *Tim.* 41b1-6, discussed below.

452. See n. 415 above.

453. In this, Philoponus is following a chronographic tradition traceable to Sextus Julius Africanus (third century AD). Sextus places the creation of Adam 5500 years before the birth of Christ (which Sextus places in 2/1 BC by our reckoning). Further, according to the *Epistle of Barnabas* 15, God will bring the world to an end in its six-thousandth year (J. Finegan, *Handbook of Biblical Chronology*, Princeton, 1964, pp. 139-47).

454. i.e. in the sublunary realm.

455. Omitting the full stop after *holên* (1336,9) and adding parentheses.

456. *Cat.* 7, 7b15-22. The qualifications introduced at 7b22-8a12 do not apply to the case of part and whole.

457. The following attempt to reconcile Aristotle's doctrine of the prime mover with Plato's account of the fashioning of the world in the *Timaeus* is heavily neoplatonic in thought and expression. Many of the ideas and much of the terminology can be found in Proclus' *Elements of Theology*. The edition, translation and commentary of that work by Dodds (Oxford, 1963) is a valuable tool for understanding the present discussion.

458. Heaven is corporeal: Plato, *Tim.* 36e5. At 29e1 it is not the cause (i.e. the demiurge) but the model after which the demiurge fashions the heaven or world (*ouranos ê kosmos*: 28b2-3) that Plato describes as being always in the same condition and unvarying. For the distinction between cause and model, see 28a4-8.

459. For 'procession', see Proclus, op. cit., 25-30. That procession is continuous is claimed in Proclus, op. cit., 112.

460. Reading *diestos* in 1337,11, with MS A.

461. For causation by the self-constituted, see Proclus, op. cit., 40. For the thesis that what is caused by the self-constituted is always in a process of coming to be, see Proclus, op. cit., 55.

462. Simplicius (along with some Plato MSS and some other ancient authors who quote this passage; see Burnet's (OCT) app. crit. ad loc.) has *emou ge thelontos* where Burnet reads *emou ge mê thelontos* ('as long as I am unwilling [sc. to undo them]'). A.E. Taylor argues for keeping *mê* (*A Commentary on Plato's Timaeus*, Oxford, 1928, pp. 250-1).

463. Plato, *Tim.* 41a7-b7 (approximate quotation).

464. *tôn holôn*. See Proclus, op. cit., 112, 120 for this use of the word. Simplicius is describing the model after which the demiurge fashions the world (*Tim.* 30c3-31a1).

465. i.e. next in order after the creator.

466. i.e. by that stage of creation.

467. i.e. the heavenly bodies.

468. Evident *to us*, that is. See *An. Post.* 1.2, 71b33-72a5 for the distinction between what is better known to us and what is better known in nature or per se.

469. Simplicius offers the same incorrect etymology of *theos* (god), from *thein* (run) that Plato gives at *Crat.* 397d.

470. i.e. the heavenly gods'.

471. Plato, *Tim.* 41a7.

472. Plato, *Tim.* 31c2-4.

473. embracing, and so holding together.

474. i.e. in the mind of.

475. i.e. next in order to the creator.

476. i.e. the intellective goodness of the creator.

477. *epeisakton*, i.e. furnished by a separate cause. The word occurs in Proclus, op. cit., 201 and is explained in Proclus, op. cit., 188.

478. In Simplicius' terms, Plato's account of motion is based on theology, whereas Aristotle's begins with considerations of motion, which is the subject matter of physics. See *Phys.* 8.1, 251a8-9.

479. *Phys.* 8.1.

480. The reasoning given in this sentence is not found in Aristotle. However, see *Phys.* 8.6 259b32-260a1 for the claim that what is moved by the primary mover must be eternal, and *Phys.* 3.3, 202a13-b22 and 5.1, 224b25-6 for the claim that motion is in what is moved.

481. *Phys.* 8.4.

482. *Phys.* 8.10.

483. *Phys.* 8.10, 267a21-5, b5-7, with 8.8.

484. *Phys.* 8.8.

485. *Phys.* 8.9, 265a29-b1.

486. Simplicius bases this interpretation on *Tim.* 39e3-40b6, 27d6-28a2, 28b2-c3.

487. Plato, *Tim.* 41a7-b6.

488. *tou paragontos*. In the previous discussion of Plato *paragein* means 'create' (1337,13, 16, 17, 21), but Aristotle's prime mover is not a creator. In the next paragraph (1339,40) it occurs in the expression 'creates motion' and perhaps this is what Simplicius means here. On this interpretation *kai* (1339,22) is epexegetic.

489. Simplicius has *eneinai* where the Aristotle MSS (followed by Ross) have *einai* (b8).

490. Ross's app. crit. reports that Simplicius has *pasês de* where the Aristotle MSS which Ross follows have *ei ge pasês* (b15-16), but I find no evidence for this claim in the text of Simplicius

491. Simplicius has *tônde* where the Aristotle MSS (which Ross follows) have *toutôn* (b26).

492. According to Ross's app. crit., Simplicius omits *dunamin* (b24) (which all the Aristotle MSS have), but I find no basis in the text that Simplicius did omit the word.

493. 'Kindred' translates *homogenes*, 'of the same kind' translates *homoeides*; Aristotle has *tên autên tôi genei*, which I render 'the same in kind'.

494. Ross follows Simplicius as in printing *kai* (which all the Aristotle MSS omit) before *kineisthai* (b33).

495. Simplicius has *hoion kinein* where Ross, following some Aristotle MSS has *hoion te kinein* (a2).

496. Simplicius along with all the Aristotle MSS has *toiouton*, which Ross omits (a4).

497. Simplicius and some Aristotle MSS omit *toiouton*, which Ross prints (a4).

498. Simplicius omits *holên* (a12).

499. At *in Phys.* 75,12-14 Simplicius associates the expression 'having no place' (*khôran mê ekhon*) with 'impossible'. In the present passage, then, the claim is that unless the axiom is accepted as necessary, the argument based on it must fail.

500. The reference is presumably to Plato (*Tim.* 80a1).

501. Simplicius quotes Aristotle's word *kinei*, but makes it transitive.

502. Either *alla hama pan kekinêsthai kai pepausthai* or *alla hama panta kai kineisthai kai pepausthai*; Ross agrees with Simplicius that the latter reading is preferable. See Ross's app. crit. ad loc.

503. i.e. the first ring, the ring next to the primary mover.

504. viz., the second ring.

505. viz., able to cause motion.

506. Omitting *apodidôsi pros* with MS a. Diels prints these words, which he labels a 'corrupt reading of manuscript A'.

507. *Cat.* 7, 7b15-22.

508. I translate following Diels' tentative suggestion to emend the MS reading.

509. The reference is presumably to *Cael.* 3.2, 301b22-30, cf. *Cael.* 4.5, 312a25.

510. *Phys.* 8.5, 257b28-258a27.

511. *Phys.* 8.4, 255b29-31 (approximate quotation).

512. *Cael.* 4.5, 312a23-7, cf. 4.4, 311b8-12.

513. *Phys.* 5.3, 226b34-227a1.

514. *Phys.* 5.3, 227a7, 6.1, 230a22-3; but whereas Simplicius says that things in contact are next to (*ekhomena*) one another, Aristotle says that their endpoints are together (*hama*).

515. *kineisthai*, *kinein*, and *pauesthai* are quoted from Aristotle but occur in a different syntactical construction. Also Simplicius uses *pauesthai* transitively, where it is in intransitive in Aristotle.

516. i.e. the air.

517. For references, see following notes.

518. *Phys.* 8.4.

519. Plato, *Tim.* 57e2-6. Simplicius agrees with some Plato MSS in reading *einai*, where Burnet in the OCT follows other MS authority in reading *eneinai*.

520. Plato, *Tim.* 57e6-58a1 (approximate quotation).

521. Plato, *Tim.* 58b6-c4 (approximate quotation).

522. Plato, *Tim.* 59a1-4 (approximate quotation).
523. The reference may be to *in Phys.* 668,24-669,15, above.
524. Simplicius has *de* instead of the Aristotle MSS' *gar* (b2).
525. *Phys.* 8.1.
526. *Phys.* 8.7, 260a23-6.
527. *Phys.* 5.4, 228a20-2.
528. *Phys.* 8.6, 259a18-19.
529. *Phys.* 8.5, 256a4-b3.
530. *Phys.* 8.5, 256b4-257a27.
531. *kinein* is quoted from Aristotle but is used in a different syntactical construction.
532. Placing a comma instead of a full stop after *kinein* (1353,13) and a semicolon instead of a comma after *diamenein* (1353,14).
533. Placing a comma instead of a full stop after *einai* (1353,16) and a full stop after *estai* (1353,17).
534. *Cael.* 1.12.
535. Simplicius along with several Aristotle MSS has *de*, instead of *dê* (which Ross prints, b6).
536. Simplicius omits *en* (b6).
537. Simplicius has *to* (b8) where the Aristotle MSS have *ta* (which Ross adopts).
538. In the system of Philolaus the celestial bodies move around a central fire. The present text, whose origin Burkert thinks is 'ultimately Aristotle' (W. Burkert, *Lore and Science in Early Pythagoreanism*, Cambridge, Mass., 1962, p. 340) is the only source for the view that the central fire provided the celestial bodies with their 'motive power'.
539. Lit., 'the great circle that goes through the poles'. Eudemus, fr. 122a, p. 50,21-9 (Wehrli[2]).
540. Reading *toutôn* (1354,13) (with MS A) for Diels' *toutôi*.
541. The great circle has two poles.
542. Simplicius refers to the present passage.
543. For discussions of Pythagorean cosmology, see W.K.C. Guthrie, *History of Greek Philosophy*, vol. 1, Cambridge, 1962, pp. 282-301 and W. Burkert, op. cit., pp. 299-368.
544. *Phys.* 8.9, 265b10-11; but there Aristotle says only that *circular* motion is the measure of other motions, and says nothing about being included or containing an image.
545. i.e. Alexander's suggestion.
546. Eudemus, fr. 123a, p. 50,35-7 (Wehrli).
547. Eudemus, fr. 122b, p. 50,30-4 (Wehrli).
548. Simplicius adds the definite article.
549. Simplicius seems to have read *to auto* (cf. 1356,7) where the Aristotle MSS (followed by Ross) have *hauto* (b11).
550. Simplicius agrees with several Aristotle MSS in writing *dei*, where Ross, following other Aristotle MSS, has *dei aei* (b12).
551. Simplicius has *kai to hudôr*, which Ross omits (b14).
552. i.e. this latter way of causing motion
553. Simplicius has *allos aei kai allos* where Aristotle has *allos aei*.
554. *Phys.* 8.10, 267b8.
555. Eudemus, fr. 123b, p. 51,1-12 (Wehrli).
556. *Phys.* 8.10, 267b6-9.
557. *Phys.* 8.10, 267b9-15.

558. *Phys.* 3.5, 204b4-206a8.

559. Simplicius makes this claim elsewhere too. For references and discussion, see W.D. Ross, *Aristotle's Physics*, Oxford, 1936, pp. 1-11. Note however the different division, in *in Cael.* 226,19-20: four books *On Natural Principles* and four *On Motion* (cf. *in Phys.* 802,8-11, which attributes this latter division to Porphyry). I am indebted to Han Baltussen for these references.

560. The word 'theology' does not occur in *Phys.*, and 'god' (*theos*) occurs only once, as a throwaway example (*Phys.* 8.8, 262a3). This claim is founded on Aristotle's identification in *Metaph.* 12.7 of the prime unmoved mover with god and Aristotle's characterization of theology as the science whose object is 'the eternal, unmoved, and separate', as opposed to physics, whose objects are subject to movement (*Metaph.* 6.1, 1026a10-22).

561. Plato, *Tim.* 27d6-28a4.

562. Plato, *Tim.* 28a5-6 (approximate quotation).

563. The mention of births of the gods suggests that Simplicius is thinking of Hesiod, whom Aristotle calls a theologian (*Metaph.* 3.4, 1000a9). But the reference could be more general.

564. *Phys.* 3.6, 206a21-3 (approximate quotation).

565. *Metaph.* 12.9.

566. *Metaph.* 12.7, 1072b29.

567. *Metaph.* 12.7, 12.9

568. *Metaph.* 12.7.

569. Plato, *Tim.* 29d7-e1 (approximate quotation).

570. Plato, *Tim.* 30b4-6 (approximate quotation).

571. Plato, *Tim.* 41a7.

572. Plato, *Tim.* 41b7-8 (approximate quotation).

573. Plato, *Tim.* 47c1-4 (approximate quotation).

574. *Phys.* 2.3, 194b29-31.

575. *Cael.* 1.4, 271a33.

576. *Cael.* 1.9, 279a27-30 (approximate quotation).

577. *GC* 1.3, 318a1-5 (approximate quotation). 'The account *On Motion*' refers to *Phys.* 8.6.

578. Simplicius refers to this book as 'big alpha'.

579. *Metaph.* 1.3, 984b15-18 (approximate quotation).

580. *Metaph.* 1.3, 984b18-20.

581. *Metaph.* 1.3, 984b20-2.

582. Simplicius incorrectly says 'shortly above'; the reference is *Metaph.* 1.8, 989b15.

583. *Phys.* 2.2, 198a2-3 (approximate quotation).

584. The manuscripts of Simplicius here omit some words from the text of Aristotle's *Physics*.

585. *Phys.* 2.2, 198a5-13 (approximate quotation).

586. This work of Ammonius has not survived.

587. Above, 1359,30.

588. *Phys.* 8.1, 250b11-15.

589. *Phys.* 8.1, 250b15-251a5.

590. The reference is to *Phys.* 8.1, 251a5-b10, b28-252a4, but Aristotle does not commit this *petitio principii*.

591. *Phys.* 8.1, 251b10-13.

592. *Phys.* 8.1, 251b19-26.

593. At *Phys.* 8.1, 252a5, Aristotle says that it seems contrived to say that sometimes there is motion and sometimes there is not. At 252a5-6, he says that it is likewise contrived to say 'that is how it is by nature'. Only later, at 252a14-16, does he object to the proposal that there is an infinite period of rest prior to motion, and the objection is based on the grounds that nature does not work like that.

594. *Phys.* 8.2, 252b9-12.

595. *Phys.* 8.2, 252b12-16.

596. *Phys.* 8.2, 252b17-28.

597. *Phys.* 8.2, 252b28-253a2.

598. *Phys.* 8.3, 253a22-30.

599. *Phys.* 8.3, 253a32-b6, 254a23-33.

600. *Phys.* 8.3, 253b6-254a1, 254a33-b4.

601. *Phys.* 8.3, 254a3-15.

602. The subject is broached at *Phys.* 8.3, 254a15-b6. The reference is 254a15-16, but Aristotle does not prove this point directly.

603. *Phys.* 8.3, 254a16-b4.

604. This possibility is stated at *Phys.* 8.3, 254b5-6 as 'what we must prove'.

605. *Phys.* 8.4, 254b7 is the first time this distinction occurs.

606. *Phys.* 8.4.

607. *Phys.* 8.5, 256b14-27, 257b26-258b9 proves that the first mover is immovable; that some things are sometimes in motion and sometimes at rest is obvious (*Phys.* 8.3, 254a5-8, a35-b4); from this it follows that there is something that is always in motion (*Phys.* 8.6, 259b32-260a5).

608. *Phys.* 8.5, 256a4-b3.

609. This seems to refer to *Phys.* 8.5, 258b4-5 with the supporting argument at 256b26-258a4.

610. Simplicius abruptly changes from the plural to the singular.

611. *Phys.* 8.4, 255a30-b12.

612. *Phys.* 8.4, 256b24-8.

613. *Phys.* 8.4, 255b29-256a3.

614. *Phys.* 8.5, 256a4-21.

615. *Phys.* 8.5, 256a21-b3.

616. *Phys.* 8.5, 256b4-13.

617. *Phys.* 8.5, 256b13-24.

618. *Phys.* 8.5, 256b27-257a26.

619. *Phys.* 8.5, 257a26-7.

620. *Phys.* 8.5, 257a27-31.

621. *Phys.* 8.5, 257a31-3 raises this question.

622. *Phys.* 8.5, 257a33-258a20; but Aristotle does not there speak of soul and body.

623. *Phys.* 8.5, 258a22-7.

624. *Phys.* 8.5, 258a23-4, cf. 257b26-258a1.

625. *Phys.* 8.5, 258b4-9.

626. *Phys.* 8.6, 258b11-259a6.

627. *Phys.* 8.6, 259a6-20.

628. *Phys.* 8.6, 258b10.

629. In *Phys.* 8.8-9, Aristotle proves that circular motion is the only kind of motion that can be continuous and eternal, without reference to the heavenly bodies. Alternatively, the reference may be to *Cael.* 1.3, 270b1-16.

630. *Phys.* 8.6, 259a20-b28.

631. *Phys.* 8.6, 259b32-260a10.

632. *Phys.* 8.8-9.

633. *Phys.* 8.7, 260a20-6.

634. *Phys.* 8.7, 260a26-261a25.

635. *Phys.* 8.7, 261a27-b26 proves that only locomotion can be continuous. *Phys.* 8.8 proves that among locomotions only circular motion can be continuous. In particular, *Phys.* 8.8, 262a12-b8 proves that rectilinear motion cannot be continuous for the reason given here.

636. *Phys.* 8.8, 263b9-26.

637. *Phys.* 8.8, 263b26-264a6.

638. *Phys.* 6.1, 231a24-b18.

639. *Phys.* 8.8, 264a7-21.

640. *Phys.* 8.8, 264a21-33.

641. *Phys.* 8.8, 264b1-6.

642. *Phys.* 8.8, 264b9-265a2.

643. *Phys.* 8.8, 265a7-8.9, 265a27.

644. *Phys.* 8.9, 265b10-11.

645. *Phys.* 8.9, 265b8-10; but Aristotle says nothing there about circular motion being simpler.

646. *Phys.* 8.9, 265b11-16: circular motion alone is uniform.

647. *Phys.* 8.9, 265b17-266a1.

648. *Phys.* 8.9, 266a6-9.

649. *Phys.* 8.10, 266a10-11.

650. *Phys.* 8.10, 266a24-b6 proves that a finite magnitude cannot contain infinite power; 267b20-1 states that there cannot be an infinite magnitude.

651. *Phys.* 8.10, 266b6-24.

652. *Phys.* 8.10, 266b25-267a20.

653. *Phys.* 8.10, 267b17-26.

Additional note

At 1359,1-3, Simplicius claims that Alexander knows that the sphere of fixed stars gets its eternal being (*einai, ousia*) from God in finite dollops at a time. His actual quotations from Alexander discuss its power of undergoing and causing motion, not its being. But the view about being was that of the late Neoplatonists, starting from Proclus, and something similar had even been attributed to Aristotle by Simplicius' teacher Ammonius. So Simplicius may be inferring the ascription to the Aristotelian Alexander. Philoponus, as a Christian, had not endorsed the view, but had used it *ad hominem* in a dialectical argument against Proclus. Here Philoponus is accused of not recognising Alexander's endorsement.

Taieb Farhat has kindly drawn my attention to the relevance of his finding of July 1993 at the Centre d'Etudes médiévales of the University of Paris–I. Averroës is another person who cites Alexander as making the eternity of the heavens due to the external influence of the divine Unmoved Mover. But Farhat showed that Averroës was misled by a passage in an Arabic text wrongly identified as a translation of Alexander, rather than of Philoponus (*Contra Proclum*, ed. Rabe pp. 238-42). The passage is pp. 89-91 of *De regiminibus orbium*, edited as D15 by H.J. Ruland, *Die arabischen Fassungen* … (1976). As regards Farhat's new suggestion that our Simplicius passage may be relevant, my inclination is to think that the tradition of Ammonius, reflected here in Simplicius, may have led to Averroës' attribution to Alexander. [Ed.]

Bibliography

Blumenthal, H.J., 'Dunamis in Simplicius', in F. Romano and R. Loredana Cardullo (eds) *Dunamis nel Neoplatonismo*, Florence, 1996.

Burkert, W., *Lore and Science in Early Pythagoreanism*, Cambridge, Mass., 1962.

Cameron, A., 'The last days of the Academy at Athens', *PCPS* 15, 1969, 7-29.

Cornford, F.M., 'Aristotle, *Physics* 250a9-19 and 266a12-24', *CQ* 26, 1932, 52-4.

Diels, H., 'Zur Textgeschichte der Aristotelischen Physik', in *Abhandlung der Königlichen Preussischen Akademie der Wissenschaften zu Berlin, Phil.-hist. Kl. I*, 1882, 1-42; also in H. Diels, *Kleine Schriften zur Geschichte der antiken Philosophie*, ed. W. Burkert, Hildesheim, 1969, 199-238.

Dodds, E.R., *Proclus, The Elements of Theology*, Oxford, 1970.

Dodds, E.R., 'Simplicius', *OCD*, Oxford, 1970.

Finegan, J., *Handbook of Biblical Chronology*, Princeton, 1964.

Guthrie, W.K.C., *A History of Greek Philosophy*, vol. 1, Cambridge, 1962.

Hadot, I. (ed.), *Simplicius, sa vie, son oeuvre, sa survie* (= Peripatoi 15), Berlin-New York, 1987.

Hadot, I., 'La vie et l'oeuvre de Simplicius d'après des sources grecques et arabes', in id. (ed.), *Simplicius, sa vie, son oeuvre, sa survie*, Berlin-New York, 1987, 3-39; also translated in a slightly revised version as 'The life and work of Simplicius in Greek and Arabic sources' in R. Sorabji (ed.), *Aristotle Transformed*, London and Ithaca NY, 1990, 275-303.

Heath, T.L., *The Thirteen Books of Euclid's Elements*, Cambridge, 1925.

Hoffman, P., 'Simplicius' polemics', in R. Sorabji (ed.), *Philoponus and the Rejection of Aristotelian Science*, London and Ithaca NY, 1987, 57-83.

Hülzer, K., *Die Fragmente zur Dialektik der Stoiker*, Stuttgart-Bad Cannstadt, 1987.

Hussey, E., *Aristotle's Physics, Books III and IV*, Oxford, 1983.

Lloyd, A.C., 'Simplicius', in P. Edwards (ed.), *Encyclopedia of Philosophy*, vol. 7, New York and London, 448-9.

Long, A.A., *Hellenistic Philosophy*, London, 1974.

McDiarmid, J.B., 'Phantoms in Democritean terminology: *peripalaxis* and *peripalassesthai*', *Hermes* 86, 1958, 291-8.

McKirahan, R.D., *Philosophy before Socrates*, Indianapolis, 1994.

Morrison, D., 'Philoponus and Simplicius on tekmeriodic proof', in E. Kessler (ed.), *Method and Order in Renaissance Philosophy of Literature: The Aristotle Commentary Tradition*, Aldershot, 1997, 1-22.

Mueller, I., 'Aristotle and Simplicius on mathematical infinity', *Proceedings of the World Congress on Aristotle*, vol. 1, Athens 1981, 179-82.

Praechter, K., 'Simplicios', *RE* 3A, 1: (zweite Reihe), 1927, cols 204-13.

Ross, W.D., *Aristotle's Physics: A Revised Text with Introduction and Commentary*, Oxford, 1936.

Sambursky, S., *The Physical World of Late Antiquity*, London, 1962, ch. 1.
Solmsen, F., *Aristotle's System of the Physical World*, Ithaca NY, 1960.
Sorabji, R., *Necessity, Cause and Blame*, London and Ithaca NY, 1980.
Sorabji, R., *Time, Creation and the Continuum*, London and Ithaca NY, 1983.
Sorabji, R. (ed.), *Philoponus and the Rejection of Aristotelian Science*, London and Ithaca NY, 1987.
Sorabji, R., *Matter, Space and Motion*, London and Ithaca NY, 1988.
Sorabji, R. (ed.), *Aristotle Transformed: The Ancient Commentators and their Influence*, London and Ithaca NY, 1990.
Sorabji, R., 'Simplicius', *OCD*, 3rd edn, Oxford, 1996.
Tarán, L., 'The text of Simplicius' commentary on Aristotle's *Physics*', in I. Hadot (ed.), *Simplicius, sa vie, son oeuvre, sa survie*, Berlin-New York, 1987, 246-66.
Taylor, A.E., *A Commentary on Plato's Timaeus*, Oxford, 1928.
Vasoli, E., 'Simplicius', *Enciclopedia di Filosofia*, vol. 5, col. 1386.
Verbeke, G., 'La Physique d'Aristote et les anciens commentaires grecs', *Proceedings of the World Congress on Aristotle*, vol. 1, Athens, 1981, 187-92.
Verbeke, G., 'Some later Neoplatonic views on divine creation and the eternity of the world', in D.J. O'Meara (ed.), *Neoplatonism and Christian Thought*, Norfolk, 1982, 45-53
Verbeke, G., 'Simplicius', *Dictionary of Scientific Biography* 12, 1975, 440-3.
Wehrli, F., *Die Schule des Aristoteles*, vol. 7: *Herakleides Pontikos*, Basel and Stuttgart, 2nd edition, 1969.
West, M.L., *Ancient Greek Music*, Oxford, 1992.
Wicksteed, P.H. and Cornford, F.M. (tr.), *The Physics with an English Translation* (= Loeb edition), vol. 2, London-Cambridge, Mass., 1934.
Wieland, W., 'Die Ewigheit der Welt (der Streit zwischen Joannes Philoponus und Simplicius)', in D. Heinrich, S. Schulz and K.H. Volkmann-Schluck (eds), *Die Gegenwart der Griechen im neueren Denken: Festschrift für H.-G. Gadamer zu 60. Geburtstag*, Tübingen, 1960, 290-316.
Wildberg, C., 'Prolegomena to the study of Philoponus' *Contra Aristotelem*' in R. Sorabji (ed.), *Philoponus and the Rejection of Aristotelian Science*, London and Ithaca NY, 1987, 197-209.
Wildberg, C., *John Philoponus' Criticism of Aristotle's Theory of Aether*, Berlin, 1988.
Wildberg, C., *Philoponus: On Aristotle on the Eternity of the World*, London and Ithaca NY, 1987.
Wildberg, C., *Simplicius: On Aristotle on the Eternity of the World*, London and Ithaca NY, 1991.

Appendix
Notes on the Text of Aristotle's *Physics*

Discrepancies between Simplicius' text of the *Physics* and the text as given in Ross's edition:

258b15 Simplicius has *kath' hauto* (1251,17) for *haplôs*, which Ross prints, following the Aristotle MSS
258b16 Simplicius (agreeing with the Aristotle MSS) has *de* (1251,26), which Ross emends to *dê*
258b16 Simplicius has *bouloito* (1251,26) for *bouletai*, which Ross prints, following the Aristotle MSS
258b28 Simplicius has *tôn men* (1252,26) where the Aristotle MSS have *tôn aei*, which Ross emends to *tôndi*
259b29 Simplicius (agreeing with the Aristotle MSS) has *kinountôn* (1252,26), which Ross omits, following Philoponus
258b29 Simplicius has *sunekhôs* (1252,27, 31) for *sunekhous*, which Ross prints, following the Aristotle MSS
258b31 Simplicius has *aïdion ex anankês* (1253,6) for *aïdion kai ex anankês*, which Ross prints, following the Aristotle MSS
258b32 Simplicius (agreeing with the Aristotle MSS) has *arkhai* (1253,19), which Ross omits
259a1 Simplicius (agreeing with many Aristotle MSS) has *kinousôn* (1253,19) for *kinountôn*, which Ross prints, following one Aristotle MS
259a16 Simplicius has *to gar* (1255,35) for *kai gar to*, which Ross prints, following the Aristotle MSS
259a22 Simplicius (agreeing with the Aristotle MSS) has *tôn kinountôn* (1256,34), which Ross omits
259b4 Ross reports that Simplicius has *ginesthai*, but I find no reason to suppose that Simplicius does not agree with the majority of the Aristotle MSS (which Ross follows) in having *engignesthai*
259b18 Simplicius has *heautou* (1259,10) for *hautou*, which Ross prints, following the Aristotle MSS
259b28 Simplicius' lemma has *kineisthai* (1261,12), but Simplicius' commentary has *to kineisthai* (1261,20, 1262,11), in agreement with most of the Aristotle MSS, which Ross follows
259b31 Simplicius omits *de* (1261,13), which Ross prints, following the Aristotle MSS
259b32 Simplicius has *esti ti aei toiouto to kinoun* (1262,14) for *estin ti aei toiouton, kinoun*, which Ross prints, following the Aristotle MSS
260a4 Simplicius, followed by Ross, omits *tên autên* (1262,32), which some Aristotle MSS have

260a4 Simplicius omits *aei* (1262,32), which Ross prints, following the Aristotle MSS
260a6 Simplicius (agreeing with some Aristotle MSS) omits *hupo kinoumenôn men* (1262,33, 1263,16), which Ross prints, following other Aristotle MSS
260a6 Simplicius (agreeing with one Aristotle MS) has *ê* (1263,16) for *de*, which Ross prints, following the other Aristotle MSS
260a26 Ross reports that Simplicius has *gar*; however, Simplicius's lemma (1265,8) has *de*, while Ross (following most of the Aristotelian MSS), prints *d'*
260b19 Simplicius has *kai tôi khronôi* (1267,37) for *kai to tôi khronôi*, which Ross prints, following the Aristotle MSS
261a13 Simplicius has *ei phainetai* (1271,19), where Ross (following the Aristotle MSS) omits *ei* (which I do not translate)
261a34 Before *mê on horoi* Simplicius (agreeing with some Aristotle MSS) omits *to* (1274,9), which Ross prints, following other Aristotle MSS
262a4 Simplicius takes *touto esti* with 'that in which' (1278,37), whereas Ross takes it with the following clause and punctuates accordingly
262a4-5 Simplicius has *megethos ê eidos* (1278,39-1279,1) for *eidos ê megethos*, which Ross prints, following some Aristotle MSS
262a12 Simplicius has *eis ta plagia* (1279,18) for *eis to plagion*, which Ross prints, following the Aristotle MSS
262b2 Simplicius omits *kai allôi* (1283,19), which Ross prints, following the Aristotle MSS
262b5 Simplicius (agreeing with the Aristotle MSS) has *de* (1283,24), which Ross emends to *dê*
262b15 Simplicius (agreeing with some Aristotle MSS) has *oukh hama ara* (1285,31), where Ross prints the conjecture *ou gar hama* (which for reasons obscure to me he reports to be supported by Simplicius)
263a5 As Ross observes, there is no trace in Simplicius of *kai axiountas* (1288,37), which the Aristotle MSS contain, but which Ross omits. But given the nature of the commentary, this is inadequate reason to suppose that Simplicius' text of Aristotle did not contain these words
263a16 Ross reports that Simplicius has *peperasmenôi khronôi* where the Aristotle MSS (which Ross follows) have *peperasmenôi*, but I see no clear evidence (e.g. 1289,28) that this is the case
263b21 Ross reports that Simplicius apparently read *tou husterou* (the reading which Ross prints), which is absent from the Aristotle MSS
263b21 Simplicius (agreeing with some Aristotle MSS) omits *ou* (1295,26), which Ross prints
263b21 Ross reports that Simplicius has *ei*, which Ross (following some Aristotle MSS) omits, but the text of Simplicius (1295,26) does not warrant this interpretation
263b22 Before *leukon* Ross adds *to*, which is not found in Simplicius (1295,27) or the Aristotle MSS
264a26 Simplicius has *estai* (1304,35) for *estin*, which Ross prints, following the Aristotle MSS
264b14 Simplicius' *ep' eutheias* (1307,32, in a paraphrase, not a quotation) agrees with some Aristotle MSS, where Ross prints *kat' eutheian*, following other Aristotle MSS
264b18 Simplicius' lemma has *heautou* (1308,33), which accords with the *hautou* of most Aristotle MSS (which Ross prints). In the commentary, Simplicius quotes the text as *apo tou autou* ('from the same <point>', 1309,13-14)
264b19 Simplicius' lemma gives *heauto* (1308,33) (a conjecture of Diels), which

accords with *hauto*, a conjecture which Ross prints. In the commentary, Simplicius quotes the text as *eis auto* ('to it', 1309,14-15)

264b19 Simplicius omits *aph' autou* (1308,34), which Ross prints, following the Aristotle MSS

265b14 Simplicius (followed by Ross) omits the redundant *pleion* (or *pleiô*) of the Aristotle MSS (1317,23).

265b26 Simplicius omits *kai hôs* (1319,7), which Ross prints, following the Aristotle MSS

266a2 Simplicius (agreeing with one Aristotle MS) has *kata topon* (1320,12), which Ross prints, where one other Aristotle MS has *tên kata topon* and the rest have *tên kata topon kinêsin*

266a19 Simplicius (agreeing with several Aristotle MSS) has *tôi D* (1323,6) for *têi D*, which Ross prints, following other Aristotle MSS

266b8 Simplicius has *eneinai* (1340,25) for *einai*, which Ross prints, following the Aristotle MSS

266b26 Simplicius has *tônde* (1343,13) for *toutôn*, which Ross prints, following the Aristotle MSS

266b24 Ross reports that Simplicius omits *dunamin* which Ross prints, following the Aristotle MSS. I do not think that this claim, which will be based on 1343,24, is certain

267a2 Simplicius has *hoion kinein* (1345,14) for *hoion te kinein*, which Ross prints following some Aristotle MSS

267a4 Simplicius (agreeing with the Aristotle MSS) has *toiouton* (1345,28), which Ross omits

267a4 Simplicius (agreeing with some Aristotle MSS) omits *toiouton*[2] (1345,30), which Ross prints, following other Aristotle MSS

267a12 Simplicius omits *holên* (1344,5), which Ross prints, following the Aristotle MSS

267b2 Simplicius has *de* (1353,10) for *gar*, which Ross prints, following the Aristotle MSS

267b6 Simplicius (agreeing with several Aristotle MSS) has *de* (1353,35) for *dê*, which Ross prints, following other Aristotle MSS

267b6 Simplicius (agreeing with one Aristotle MS) omits *en* (1353,35), which Ross prints, following the other Aristotle MSS

267b8 Simplicius has *to* (1354,7) for *ta*, which Ross prints, following the Aristotle MSS

267b11 Simplicius seems to have read *to auto* (1356,7) for *auto*, which Ross prints, following the Aristotle MSS

267b12 Simplicius (agreeing with several Aristotle MSS) has *dei* (1356,7) for *dei aei*, which Ross prints, following other Aristotle MSS

267b14 Simplicius (agreeing with one Aristotle MS) has *kai to hudôr* (1356,12), which Ross omits. The other Aristotle MSS have *ê to hudôr*

267b14 Simplicius has *allos aei kai allos* (1356,13) for *allos aei*, which Ross prints, following the Aristotle MSS

English-Greek Glossary

able to be got completely through: *diexitêtos*
able to be undone: *lutos*
absurd: *atopos*
absurdity: *apemphainon, to atopon*
accident: *sumbebêkos*
accomplished by the intellect: *noêtos*
account: *logos*
accurate: *akribês*
acquire an additional attribute: *prosgignesthai* (+ dat.)
active: *energêtikos, energos*
activity: *energeia*
actual: *to energeiâi* (dat. s.), *kat' energeian*
actuality: *energeia, to energeâi* (dat. s.), *entelecheia*
actually, in actuality: *energeiâi* (dat. s.), *kat' energeian*
add: *epagein, epipherein, proskeisthai, prostithenai*
additional increase: *prosauxêsis*
additional premise: *proslêpsis*
add together: *suntithenai*
ad infinitum: *ep' apeiron*
adjacent: *prosekhês*
adjacent parts: *ta ephexês*
admit as causes: *aitiasthai*
advance: *epagein*
adventitious: *epeisaktos*
affection: *pathêma, pathos*
affective: *pathêtikos*
affirmation: *kataphasis*
affirmative: *kataphatikos*
agree: *homologein, sumphônein*
agreement: *sumphônia*
alter: *alloioun*
alteration: *alloiôsis*
amount: *megethos, plêthos*
analogous: *ana logon*
analysis: *analusis*
answer (v.): *apokrinein*
antecedent: *to hêgoumenon*
application: *epibolê*
apply: *harmottein*
apply to: *epharmottein, epharmozein, harmozein*
approach (n.): *ephodos*
approach (v.): *plêsiazein, prosagein, proskhôrein*
appropriate: *epitêdeios, oikeios, prosêkon*
appropriately: *oikeiôs*
arc of a circle: *periphereia*
argue: *epikheirein*
argument: *epikheirêma, epikheirêsis, logos*
argumentative: *antilogikos*
assumption: *lêpsis, to prolêphthen*
author: *hupostatês*

be consistent: *prosâidein*
be hurled about: *peripalassesthai*
be interrupted: *dialeipein*
be joined together with: *suneinai*
be joined with: *proskrinesthai*
be left over: *hupoleipesthai, peritteuein*
be less: *elleipein*
belief: *doxa*
be located in: *eneinai*
belong to: *huparkhein, proseinai*
be moved: *kineisthai, pheresthai*
be mutually implied: *antakolouthein*
be pedantic: *akribologeisthai*
bring into creation: *dêmiourgein*
by-product: *parupostasis*

calculate: *apologizesthai*
case: *tmêma*
categorically: *katêgorikôs*
causative: *aitiôdês*
cause (n.): *aitia, aition*
cause (v.): *parekhein, poiein*
cause change: *metaballein*
cause existence: *huphistanai*
cause motion: *kinein*

cause motion primarily: *prôtôs kinein*
cause to combine: *sunkrinein*
cause to stop: *histanein*
cease: *epileipein, pauesthai*
cease to be: *phtheiresthai*
ceasing to be (n.): *phthora*
centre: *kentron*
chance: *automaton*
change (n.): *metabolê*
change (v. trans.): *ameibein, metaballein*
change (v. intrans.): *metaballesthai*
change in truth value: *metaptôsis*
change (place): *allassein*
change position: *metapheresthai*
change together with: *summetaballein*
change truth value: *metapiptein*
change truth value before: *prometapiptein*
circle: *kuklos*
circular: *kuklikos, kuklôi, peripherês*
circularity: *kuklismos*
circular motion: *kuklophoria*
circumference: *periphereia, perix (to)*
cite: *paratithesthai*
comment: *hupomnêma*
comment on: *paragraphein*
complete (adj.): *entelês, holikos, teleios, teleos*
completeness: *teleiotês*
components: *ex hôn sunkeitai*
composite: *sunthetos*
composite whole: *sustêma*
composition: *sunthesis*
compound: *sunkeimenon, sunkrima, suntheton*
comprehend: *katalambanein, periekhein, sumparateinesthai*
comprehension: *perilêpsis*
comprehensive: *periektikos*
conceive: *ennoein*
conceive as well: *sunepinoein*
conclude: *lêgein, sumperainesthai, sunagein*
conclusion: *to deikhthen, to sumbainon, sumperasma, to sunagomenon*
condition: *katastasis*
conditional, conditional premise: *sunêmmenon*
confirm: *bebaioun, pistousthai*
confirmation: *pistis*
consecutive, consecutively: *ephexês*
consecutivity: *to ephexês*
consequence: *akolouthoun, to hepomenon, sumperasma*
consequent: *to epagomenon, to hepomenon, to lêgon*
consistently: *akolouthôs, sumphônôs*
constitute: *sumplêroun*
constituted by nature: *phusikos*
constitutive of: *hupostatikos*
construct: *poieisthai*
continue: *diarkein, epagein*
continuity: *sunekheia, to sunekhes, to sunekhôs*
continuous: *sunekhês*
continuum: *to sunekhes*
contradict: *antilegein, ta enantia legein*
contradiction, contradictory state: *antiphasis*
contradictory: *antiphatikôs, kata antiphasin*
contrariety: *enantiôsis*
contrary: *enantios*
contrary to nature: *para phusin*
contrast: *antidiairein*
contribute: *suntelein*
converse: *anapalin, antistrophon*
corporeal: *sômatikos*
correspondingly: *sustoichôs*
create: *dêmiourgein, paragein, poiein*
creation: *dêmiourgêma, dêmourgia*
creative, creator's: *dêmiourgikos*
creator: *dêmiourgos, ho sunistas*
credible: *pistos*
criticize: *aitiasthai*
customary use: *sunêtheia*
cut (n.): *tomê*
cut (v.): *temnein*
cut off: *apotemnein*
cuttable: *tmêtos*

decay (v.): *phthinein*
decrease (n.): *meiôsis, phthisis*
decrease (v.): *phthinein*
deduce: *sullogizesthai*
deduction: *sullogismos*
deficiency: *elleipsis*
define: *horizesthai*
definite: *horistheis*
definition: *horismos, logos*
demand (v.): *apaitein*
demonstrate: *apodeiknunai*

demonstration: *apodeixis*
demonstrative: *apodeiktikos*
deny: *aperkhesthai, apophanai, apophaskein*
depart: *apogignesthai, existasthai*
departure, *ekstasis*
descend: *hupobibazesthai*
descent of level: *huphesis*
destroy: *apollunai*
destructive: *phthartikos*
determinate: *hôrismenos*
determinately: *diôrismenôs*
determine: *diorizein*
determined: *hôrismenos*
dialectical: *dialektikos, logikos*
dialogue: *dialogos*
diameter: *diametron*
dichotomy: *dikhotomia*
differentia: *diaphora*
difficult to divide: *dusdiairetos*
difficult to perceive all at once: *dussunaisthêtos*
digress: *ektrepein ton logon*
diminution: *elattôsis*
direct (v.): *eisagein*
directed: *pheromenos*
discover: *aneuriskein, heuriskein*
discovery: *heuresis*
discussion: *logos, skholê*
disposition: *euphuia*
disproportional: *asummetros*
dissimilar: *anomoios*
dissolution: *dialusis*
distance: *apostasis, diastêma*
distant: *diestêkôs, diestôs*
distinct: *aphôrismenos, diôrismenos*
distinction: *diorismos*
distinguish: *diastellein, diistanai, diorizein*
divide: *diairein*
dividing point: *diairesis*
divine: *theios*
divinity: *theotês*
divisibility: *to diaireton*
divisible: *diairetos*
divisible into parts: *meristos*
division: *diairesis*
draw a conclusion: *sumperainesthai, sunagein*
draw a consequence: *epagein*
draw distinctions: *diorizein*
earth: *gê*
ecliptic: *ho loxos kuklos, zôidiakos*
effect: *aitiaton*
efficient: *poiêtikos*
element: *stoikheion*
eliminate: *anairein, anairetikos einai*
eliminate along with: *sunanairein*
eliminating x eliminates y: '*x sunanairei y*'
elimination: *to anairein*
embracing (n.): *periokhê*
end: *peras, phthora, teleutê, telos*
endpoint: *to teleutaion*
engage in activity: *energein*
engage in cosmogony: *kosmopoiein*
ensouled: *empsukhos*
entailment: *akolouthia*
entity: *(to) on*
equal, equivalent: *ison*
equally: *homoiôs, ison*
essence: *to einai*
establish: *deiknunai, kataskeuastikos einai, kataskeuazein, sunistanai*
eternal: *aïdios*
eternally: *aei*
eternity: *to aïdion, aïdiotês, aiôn*
evidence: *marturion*
evident: *enargês, phaneros, prophanês*
evidently: *enargôs, ex tês enargeias*
exact: *akribês*
examination: *exetasis*
examine: *exetazein*
example: *paradeigma*
exceed: *huperballein, huperekhein*
excess: *huperbolê*
exchange (n.): *antallagê*
exchange (v.): *antimetalambanein*
exchange of locations: *metakhôrêsis*
exchange of place: *antimetastasis*
exchange of places: *metastasis*
exchange places: *antimethistasthai*
exhaust: *analiskein, katanaliskein*
exhaust completely: *apartizein*
exist: *einai, huphistasthai*
exist afterwards: *methuparkhein*
existence: *to einai, huparxis, ousia*
existing: *kathestêkos*
existing thing: (*to*) *on*
explain: *apodidonai, exêgeisthai*
explanation: *paramuthia*
expound: *exêgeisthai, saphênizein*
extend: *ekteinein, parateinesthai*

extended: *diastan, diestôs*
extension: *diastasis, to diastaton, paratasis*
external: *exôthen*
extreme: *akron*
extremity: *akrê, akron, eskhaton*

fifth body (*aither*): *to pempton sôma*
figure: *skhêma*
final: *telikos*
finite: *peperasmenos*
fire: *pur*
first: *proteron, proteros, prôtos*
first mover: *to kinoun prôton, to prôton kinoun*
fixed: *aplanês*
fixed stars: *ta aplanê*
flux: *rhoê*
follow: *akolouthein, ekdekhesthai, epakolouthein, hepesthai, katakolouthein, sumbainein, sunakolouthein, sunepesthai*
force (n.): *bia, tasis*
force (v.): *anankazein, biazesthai*
forcible: *biaios*
form: *eidos*
formless: *aneideos*

general: *katholikos, koinos*
generalize: *koinopoiein*
generalized term: *anabebêkos*
generally: *holôs, katholou, koinôs*
generate: *gennan*
the generated: *to gennêthen, to gennômenon*
generation: *genesis, to gignesthai*
generative: *gennêtikos*
generator: *to gennêsan*
genus: *genos*
geometrize: *katageometrein*
get entirely through, get through all of: *diexerkhesthai*
give an indication: *endeiknunai*
give in: *endidonai*
give in addition: *prostithenai*
go away: *aperkhesthai*
go completely through: *diexienai*
god: *theos*
go through: *diienai*
go to infinity: *ienai ep' apeiron*
grammarian: *grammatikos*
great circle (on a sphere): *ho megistos kuklos*
growth: *auxêsis*

halt: *stasis*
happen: *sumbainein*
harmony: *harmonia*
have affective qualities: *paskhein*
have a natural tendency to: *pephukenai pros*
have both attributes: *epamphoterizein*
having equal speed: *isotakhês*
having extension: *diastatos*
having or possessing infinite power: *apeirodunamos*
having to do with nature: *phusikos*
heaven: *ouranos*
heavenly: *ouranios*
heavenly beings, heavenly bodies: *ta ourania*
hold fixed: *ekhein*
hold of: *huparkhein, sumbainein*
hold up as support: *proïskhesthai*
homoeomery: *homoiomereia*
homonymously: *homônumôs*
hypothesis: *to hupothemenon, hupothesis*
hypothesize: *hupotithenai*
hypothetical: *hupothetikos*

idea: *ennoia*
immobility: *to akinêton*
immovably: *akinêtôs*
impart: *endidonai*
impassive: *apathês*
impede: *empodizein*
imperceptible: *anepaisthêtos*
imperishable: *aphthartos*
impinge: *prospiptein*
impossibility: *to adunaton, to mê endekhesthai*
impossible: *adunatos, ou* (*mê*) *dunatos*
impulse: *hormê*
in a circle: *kuklôi*
incidental, incidentally: *kata sumbebêkos*
inclination: *enklisis*
include: *periekhein, perilambanein, sunairein, sunarithmein*
incomplete: *atelês*
incompleteness: *ateleia*
incorporeal: *asômatos*

increase (v.): *auxanesthai, auxesthai, prosauxanesthai*
increase (n.): *auxêsis*
increase in magnitude: *auxanesthai*
indefinite: *aoristos*
indeterminate: *aperigraphos*
indeterminately: *adioristôs*
indicate: *dêloun, endeiknunai*
indivisible: *adiairetos, ameristos, atomos*
inequality: *anisotês*
inexhaustibility: *to anekleipton*
inexhaustible: *anekleiptos*
infer: *epagein, sunagein*
inference: *sunagôgê*
inferring from effects (adj.): *tekmêriôdês*
infinite: *apeiros*
infinite power: *apeirodunamia, to apeirodunamon*
infinitude: *to apeiron*
infinity: *apeiria, to apeiron*
instant: *nun*
instruction: *didaskalia*
integral state: *holotês*
intellective: *noeros*
intellectually: *noerôs*
intelligence: *nous*
intelligible: *gnôrimos*
intermediary: *meson*
intermediate: *mesos*
interpreter: *exêgêtês*
interrupt: *diakoptein, dialambanein*
interruption: *diakopê*
introduce: *eisagein, paragein*
introductory: *protetagmenon*
inversely proportional: *kata tên antistrophên tês analogias*
investigate: *zêtein*
investigation: *zêtêsis*
irrational: *alogos*

join: *epizeugnunai, sunaptein*
join together: *sunaptein*

kind: *eidos, genos*
kindred: *homogenês*
knowledge: *gnôsis*

lateral: *eis ta plagia, epi ta plagia, plagios*
lead (v.): *anagein*
learn: *katamanthanein, manthanein*
leave: *apoleipein, kataleipein*
leave aside: *paraleipein*
lemma: *lêmma*
length: *mêkos*
letter: *stoikheion*
limit (n.): *horos, peras*
limit (v.): *perainein*
line: *grammê*
line of proof: *agôgê tou logou*
living: *empsukhos*
living thing: *zôion*
location: *stasis*
locomotion: *phora*
luck: *tukhê*

magnet: *lithos*
magnitude: *megethos*
main point: *kephalaion*
major (premise): *meizôn*
make continuous: *sunekhizein*
make determinate: *diorizein*
maker: *poiêtês, to poioun*
manuscript: *antigraphon, biblion*
mark out: *horizein*
material (adj.): *hulikos*
matter: *hulê, pragma*
mean (v.): *sêmainein*
measure (n.): *metron*
measure (v.): *metrein*
measure out: *katametrein*
meet with: *prostunkhanein*
mention: *mimnêskein, mnêmên poieisthai, mnêmoneuein*
method: *methodos*
middle: *meson*
mind: *nous*
minor premise: *elattôn, proslêpsis*
mixture: *migma*
model: *paradeigma*
mortal: *thnêtos*
motion: *kinêsis*
motion-causing element: *to kinoun*
motion-imparting: *kinêtikos*
motionlessness: *akinêsia*
motive: *kinêtikos*
move (trans.): *kinein*
move (intrans.): *kineisthai, metabainein, pheresthai*
move along with: *sunkineisthai*
move away from: *aphistasthai*

moved by something else: *heterokinêtos*
movement: *kinêsis*
move oppositely: *antikineisthai*
mover: *to kinêtikon, ho kinôn, to kinoun*
moving cause: *to kinoun aition*
moving thing: *to kinoumenon*
mutual elimination: *to anairein allêla*
mutually entail: *antistrephein kata tên akolouthian*
mutual replacement: *antiperistasis*

natural, in accordance with nature: *phusikos, kata phusin*
natural capacity: *phusis*
naturally: *phuseôs, phusikôs, kata phusin*
natural philosopher: *phusikos, phusiologos*
nature: *phusis*
near the earth: *perigeios*
necessarily: *ex anankês*
necessary: *anankaios, anankê*
necessitate: *anankazein*
necessity: *anankê*
negation: *apophasis*
negative: *apophatikos*
non-existence: *to mê einai*
non-uniform: *anômalês, anômalos*
non-uniformity: *anômalia, anômalos phusis, anomalotês*
not being (n.): *to mê on*
notionally: *epinoiâi* (dat.)
number: *arithmos, plêthos*
numerically one: *hen* (*tôi*) *arithmôi*

object (v.): *enistasthai*
objection: *enstasis*
observable: *theôrêtos*
observation: *theôria*
obvious: *prodêlos, prokheiros, saphês*
occupy: *katalambanein, katekhein*
occur: *einai, gignesthai*
occur in: *engignesthai*
of different kinds: *anomoeidês*
offspring: *gennêma*
off the point: *parergôs*
omit: *parienai*
one's own proper: *oikeios*
opinion: *doxa*
oppose: *anthistasthai*
opposite: *antikeimenos*
opposite motion: *antikinêsis*
opposition: *antithesis*
order: *taxis*
origin: *arkhê*
original: *ex arhkês, tên arkhên*
originally: *en arkhêi, ex arhkês*
originate: *gignesthai*
originative: *arkhikos*
overcome: *katalambanein*

paradox: *paradoxologia*
part: *meros, morion*
participate in: *metekhein*
participation: *methexis*
particular: *kathekasta, merikos*
partition: *merizein*
partlessness: *to ameres*
part-like: *merikos*
passage: *lexis, rhêsis*
pass into: *parienai*
passive: *pathêtikos*
past the prime: *parakmastikos*
pay attention to: *parakolouthein*
penultimate: *pro tou eskhatou, pro tou teleuatiou*
perceive: *aisthanesthai*
perceptible: *aisthêtos*
perception: *aisthêsis*
period of rest: *êremia, êremisma*
period (of time): *aiôn*
perish: *apollusthai, huphistasthai phthoran, phtheiresthai*
perishability: *to phtharton*
perishable: *phthartos*
perishing: *phthora*
permanently: *epimonôs*
perpetual: *endelekhês*
perpetual generation: *aeigenesia*
perpetually causing motion: *aeikinêtos*
per se: *kath' hauto*
persist: *diamenein, hupomenein, menein*
persistence: *to monimon*
persisting: *epimonos, monimos*
philosopher: *philosophos*
philosophy: *philosophia*
physical: *sômatikos*
Physics: *Phusika*
place (n.): *hedra, khôra, topos*
place (v.): *tithenai*
planet: *to planômenon*

planetary: *planômenos*
plausible: *pithanos*
point (n.): *sêmeion*
point out: *paradeiknunai*
point out as well: *prosupomimnêiskein*
points on a diameter: *ta kata diametron*
pole: *polos*
pose: *tithenai*
pose additional puzzles: *epaporein*
pose a (or the) puzzle: *aporein*
posit: *hupotithenai, tithenai*
position: *thesis*
possess: *ekhein*
possessed of real being: *ontôs on*
possible: *dunaton*
posterior: *husteros*
posteriority: *to husteron*
postulate: *axioun*
potentiality: *to* (or *hê*) *dunamei, dunamis*
potentially: *dunamei*
power: *dunamis*
powerful: *dunamikos, dunatos*
preceding: *prokeimenos*
predecessors: *hoi proteroi*
predicate (v.): *katêgorein*
pre-exist: *proüparkhein*
premise: *protasis*
present (adj.): *enestôs, prokeimenos*
present (n.): *to nun*
presuppose: *proüpotithesthai*
previous: *proteros*
previously: *prosthe, proteron*
primacy: *to prôtên einai*
primary: *prôtos*
primary mover: *to kinoun prôtôs, to prôtôs kinoun*
primary thing that is moved: *to prôtôs kinoumenon*
prime: *akmê*
principal kind of thing: *proêgoumenon*
principle: *arkhê*
prior: *proêgoumenos, progenomenos, proteron, proteros, prôtos*
priority: *to proteron*
privation: *sterêsis*
problem: *problêma*
problem under investigation: *zêtêsis*
proceed: *anerkhesthai, hodeuein, proerkhesthai*
procession: *proödos*
process of work: *ergasia*
proclaim: *proeipein*
produce: *apodidonai, poiein*
productive: *poiêtikos*
progress, progression: *proödos*
project (v.): *proballesthai*
proof: *to deiknun, deixis, pistis*
proper: *oikeios*
property: *idion*
proportion: *analogia*
proportionally: *kata ton logon*
proposal: *problêma*
propose: *proballesthai, protithesthai*
proposition: *axiôma*
prove: *anankazein, deiknunai, proskataskeuazein*
prove in advance: *prokataskeuazein*
prove to be: *gignesthai*
proximate: *prosekhês*
pull (v.): *helkein*
pull apart: *apospan*
pure: *eilikrinês*
purification: *katharotês*
purpose: *prothesis*
put (an argument): *erôtan*
put before: *protithesthai*
put down (in writing): *anagraphein*
put forward: *proballesthai, proïstanai*
put in the middle: *mesolabein*
put to use: *gumnazein*
puzzle (n.): *to aporêthen, aporia, to aporoumenon*
puzzle under consideration: *ho nun aporêtheis logos*
puzzling: *aporos*

quality: *poion, poiotês*
quantity: *poson* (*to*), *posotês*
question (n.): *to erôtan, erôtêsis*

radius of the circle: *hê apo tou kuklou grammê*
radius: *hê ek tou kentrou*
raise (one or more) objections: *enistasthai*
rarefaction: *araiôsis, manôsis*
rarity: *araiotês*
ratio: *logos*
reach completion: *teleiousthai*
reading: *graphê*
receive: *dekhesthai, hupodekhesthai, katadekhesthai, paradekhesthai*

reception: *metalêpsis*
receptive: *dektikos*
reciprocally: *antistrophôs*
reciprocity: *antistrophê*
rectilinear: *ep' eutheias, kat' eutheian*
refer: *anagein*
refutation: *lusis*
refute: *elenkhein, luein*
relation: *to pros ti, skhesis*
relational entity: *to pros ti*
relationship: *skhesis*
remove: *aphairein*
replace with: *anteisagein*
reputable: *endoxos*
responsible: *aitios*
rest (n.): *anapausis, êremêsis, êremia, paula*
rest (the): *loipa (ta)*
rest (v.): *anapauein*
result: *to dedeigmenon*
reverse (adj., adv.): *anapalin*
revolve: *kuklophoreisthai, peripolein*
rotate: *peripheresthai*
rotation: *periphora*
ruler: (*to*) *arkhon*

satisfy: *arkein, autarkês einai*
science: *epistêmê*
seed: *sperma*
segment: *tmêma*
self-constituted: *authupostatos*
self-moved, self-moving: *autokinêtos*
self-mover: *autokinêton*
self-sufficient: *autarkês*
semicircle: *hêmikuklios*
sense organ: *aisthêterion*
separate (adj.): *kekhôrismenos, khôristos*
separate (v.): *diakrinein, khôrizein*
separate off: *apokrinein*
separately: *idiâi* (dat. s.)
separation: *diakrisis*
sequel: *to hexês*
set out: *apokhôrein, ektithesthai, horman*
settling: *hidrusis*
show: *dêloun, deiknunai, endeiknunai, epideiknunai*
sign: *sêmeion, tekmêrion*
significant: *axiologos*
signification: *sêmainomenon*
signify: *dêloun, sêmainein*
signifying an objection: *enstatikos*
similarity: *homoiotês*
similarly: *homoiôs*
simple: *haplous*
simultaneously: *hama*
sketch (a proof): *endeiknunai*
skip over: *paraleipein*
sleep (v.): *katheudein*
slow down: *hêsukhazein*
solution: *lusis*
solve: *luein*
soul: *psukhê*
sound (adj.): *hugiês*
soundly: *ararotôs*
special nature: *idiôtês*
species: *eidos*
specifically: *tôi eidei, idiôs*
specify: *horizein*
specify in advance: *prodiorizesthai*
sphere: *sphaira*
sphere of the fixed stars: *hê aplanês, to aplanes*
spin: *peristrophê*
square (n.): *tetragônon*
stability: *stasis*
stand: *histasthai*
stand in the same relation: *homoiôs ekhein, hôsautôs ekhein*
star: *astêr, astron*
start (n.): *endosimon*
start (v.): *arkhesthai*
start again: *dianistasthai*
starting point: *arkhê*
state: *hexis*
statement: *rhêton*
state of completion, state of perfection: *teleiotês*
state of rest: *êremia*
stationary: *hestôs*
stationary state: *stasis*
stop (v.): *ephistanein, epileipein, histanai, histasthai, pauein, pauesthai, teleutan*
stop when something else does: *sumpauesthai*
stopping: *stasis*
straight line: *hê eutheia*
structure: *sustasis*
study: *theoria*
subject (n.): *to hupokeimenon*
subject to change: *metabolikos*

subject to generation: *en genesei, genêtos*
subject to motion: *kinêtos*
subject to perishing: *phthartos*
sublunary: (*ta*) *hupo selênên*
subsequent: *to hepomenon*
subsistence: *hupostasis*
substance: *ousia*
substitute: *metalambanein*
substrate: *to hupokeimenon*
subtract: *aphairein*
subtraction: *aphairesis*
subtract previously: *proapheirein*
succession: *diadokhê*
successive: *diadekhomenos*
suffer: *paskhein*
sufficient: *autarkês*
suggest: *hupoballein*
suitability: *epitêdiotês*
suitable: *epitêdeios*
superfluous: *perittos*
supervene upon: *episumbainein*
supply (in thought): *prosupakouein*
supporting argument: *kataskeuê*
suppose: *hupolambanein, huponoein*
surface: *epiphaneia*
surround: *periekhein*
susceptibility: *eupatheia*
susceptible: *eupathês*
syllogistic mode: *tropos*

take (as): *lambanein*
take away: *aphairein, exairein*
take in: *eisagein*
take into account: *logizesthai*
take place: *epiteleisthai, gignesthai*
takes for granted: *hôs hômologoumenôs khrômenos*
take [something's] place: *antiperiistasthai*
take up: *prokheirizesthai*
tangible: *haptos*
teacher: *kathêgemon*
temporal: *khronikos*
temporally: *en khronôi, enkhronôs*
tendency: *rhopê*
term: *onoma*
terminate: *peratoun*
testimony: *marturia*
text: *graphê*
theologian: *theologos*
theology: *theologia*
theorem: *theôrêma*
theoretical interest: *theôria axiologos*
think: *hupolambanein*
think of: *epinoein*
thought: *ennoia, epinoia, noêsis*
throw (v.): *rhiptein*
thrower: *ho rhiptôn, to rhipton*
throwing (n.): *rhipsis*
thrown object: *to rhiptoumenon*
time, time interval: *khronos*
topic: *problêma*
total amount: *holotês*
train (v.): *gumnazein*
train of thought: *heirmos*
transcendent: *exêirêmenos*
transfer (n.): *metadosis*
transfer (v.): *metalambanein*
transformation: *tropê*
transmission: *diadosis*
transmit: *pherein*
traverse: *dierkhesthai, peripolein*
treatise: *pragmateia*
treatment: *diaitê*
true: *alêthês*
truly: *ontôs*
truth: *alêtheia, to alêthes*
turn back: *anakamptein*
turning back (n.): *anakampsis*
turning point: *kamptêr*
turn (v. trans.): *peristrephein*
type: *tropos*

unbounded: *aperatôtos*
unceasing: *apaustos*
unchangeable: *ametablêtos*
unclarity: *asapheia*
unclear: *asaphês*
uncontaminated: *amigês*
under construction: *episkeuastos*
undergo: *huphistasthai, hupomenein, paskhein*
undergo alteration: *alloiousthai*
undergo change: *metaballein, metaballesthai*
undergo locomotion: *pheresthai*
undergo locomotion over a circle: *kuklon pheresthai*
undergo motion: *kineisthai*
undergo mutual replacement: *antiperiistasthai*
undergoing circular motion: *kuklophorêtikos*

underlie: *hupeinai, hupokeisthai*
understand: *akouein, ekdekhesthai, noein, sunienai*
undifferentiated: *adiaphoros*
undivided: *adiairetos*
undo: *apoluein, luein*
undoing: *lusis*
unequal: *anisos*
unevident: *aphanês*
ungenerated: *agenêtos*
unification: *henôsis*
uniform: *homalês, homalos*
uniformity: *to homales, homalotês*
unintermittent: *adiastatos*
unit: *monas*
unitary: *heniaios*
universals: *ta katholou*
universe: *ta hola, to pan*
unmixed: *akraiphnês*
unmovable, unmoved: *akinêtos*
unnatural, unnaturally: *para phusin*
unqualifiedly: *haplôs*
unrelated: *askhetos*
unsatisfactorily: *adokimôs*
unsound: *sathros*
unusually: *kainôs*
unvaryingly: *hôsautôs*
use: *khrêsthai, lambanein, poieisthai*
used to prove: *deiktikos*

variety: *diaphora*
various, varying: *diaphoros*
via: *dia mesês*
visible: *horatos*
void: *kenon*

wander: *planasthai*
water: *hudôr*
withdraw: *aphistasthai*
with equal speed: *isotakhôs*
word: *logos, rhêma*
work: *ergon*
world: *kosmos*
world-maker: *kosmopoios*
wrestle around: *peripalaiesthai*
write in opposition: *antigraphein*

zoophyte: *zôiphuton*

Greek-English Index

References are to the page and line numbers of the Greek text which appear in the margins of the translation.

* An asterisk after a citation indicates that the word occurs in a quotation of Aristotle.

† A dagger after a citation indicates that the word occurs in a quotation of someone other than Aristotle.

adiairetos, undivided, 1282,15; indivisible, 1283,30; 1287,13; 1358,16*; 1366,19; not divisible, 1333,10†

adiaphoros, undifferentiated, 1279,9

adiastatos, without spatial extension, 1334,7; without intermission, 1359,12; unintermittent, 1359,25

adiexitêtos, unable to be got completely through, 1314,4.8

adioristos, not distinct, 1316,19

adioristôs, indeterminately, 1284,22

adokimôs, unsatisfactorily, 1344,20

adunatos, impossible, 1255,15; 1261,5.15; 1278,7; 1283,10; 1286,18; 1289,18.19; 1291,33; 1293,12; 1297,20; 1302,23*.23; 1303,24.27.29†; 1307,18.20; 1310,25; 1314,3.6*.8; 1322,37; 1324,19; 1325,3.4.36; 1330,5; 1335,33†.36†; 1344,32*; 1351,10*; 1352,2†.4†; 1356,21; 1359,18†; cannot, 1260,9; 1274,5.22; 1275,5.12; 1280,20*; 1303,7; 1310,37; 1321,17; 1324,3; 1326,28†; 1327,39; 1328,31; 1336,3†; 1344,10.12; 1358,4*; could not, 1307,25; ***adunaton (to)***, impossibility, 1296.12; 1307,16.19; ***eis adunaton apagôn***, by reductio ad impossibile, 1324,17

aei, always (sometimes translated 'eternally'), 1251,16 etc.; at each stage, 1257,26

aeigenesia, perpetual generation, 1361,26

aeikinêtos, perpetually causing motion, 1259,3; 1339,24

aêr, air, 1258,19; 1344,31*.33(bis).36; 1345,2.5.7.24.28*.34.35; 1346,10.30.37†; 1347,27†.33†(bis); 1348,12†.17.20.22.28.30.33.37(bis). 38; 1349,6.27.28.30; 1350,4.19*.28*; 1351,24.26; 1352,15†; 1356,12*.13

agenêtos, ungenerated, 1258,4; 1269,18; 1270,22; 1273,25; 1320,26; 1360,27; 1361,33†; 1362,15†

agnoein, be ignorant, 1334,30; 1328,9; 1329,14

agnoêsas, in his ignorance, 1328,4.33

agôgê tou logou, line of proof, 1325,39

aïdios, eternal, 1250,35 etc. [219 occurrences]; ***to aïdion***, eternity, 1273,32; 1335,22; 1339,10

aïdiotês, eternity, 1260,35; 1327,7; 1329,8; 1353,31.32.33

aiôn, period (of time), 1327,4; 1364,1; eternity, 1360,25; 1361,20; ***en aiôni***, for eternity, 1359,12.22

aisthanesthai, perceive, 1360,9

aisthêsis, perception, 1257,32; 1266,27; 1268,29*.30*; 1280,35*.35

aisthêterion, sense organ, 1258,21

aisthêtos, perceptible, 1312,32*.38; 1313,1.7; 1338,2

aitia, reason, 1254,15; 1285,31; 1298,12; 1310,37; 1359,20; cause, 1256,12; 1257,1; 1258,20.25; 1272,34(bis); 1273,3; 1310,29; 1314,23; 1315,9; 1319,28*.32; 1320,28; 1330,14†.22; 1332,38;

1333,25†; 1338,1; 1347,16†.20†; 1350,37; 1351,15.17.32; 1352,6†; 1354,38; 1356,24; 1359,17.23.26; 1360,2.32†; 1361,26*.28*.35; 1362,6*.23*; 1363,7; 1364,9; 1366,20; ***dia tina aitian***, why, 1263,37

aitiasthai, criticize, 1312,32; admit as causes, 1363,2

aitiaton, effect, 1268,12; 1279,26

aitiôdês, causative, 1338,10

aition, cause, 1251,31; 1252,11.14.21(bis).25*.27.30.32*.32. 34.35.38.40; 1253,22.23.28(bis).29. 30.31.32.33*.39; 1254,4.12.17.26; 1255,12.28.30; 1257,10; 1258,17; 1259,2.9*; 1260,7.12.35; 1261,34†; 1262,20; 1263,1*.6.7.10.32; 1264,3*; 1268,11.12; 1270,22*.25.31.32.35; 1279,26.31; 1309,4; 1316,5; 1318,20.23.30.31; 1319,18; 1333,11†; 1336,37.38; 1337,14; 1338,29; 1339,24; 1347,19†.25†; 1351,16.25.39; 1354,32†.35; 1356,11; 1359,18†.18.29; 1360,19.21.26.28.30.33; 1361,10.13.14.18.23.25(bis).33†; 1362,1.3*.8.12.16.17.18.21.25*.26*. 29*(bis).32; 1363,1.7.9.13.17; 1365,20; reason, 1258,29; 1275,31; 1317,10.12; ***to kinoun aition***, the moving cause, 1255,32; 1353,30; 1354,13

aitios, responsible, 1345,24

akatalêptos, cannot be apprehended, 1314,8

akhôristos, not separate, 1262,4

akhronos, not in a time interval, 1284,2; atemporal, 1339,26.39; 1340,3

akinêsia, motionlessness, 1339,22

akinêtos, unmoved, 1250,34; 1251,4*. 5.7.9.10*.12.14.17*.20.21.22.28.33; 1252,23.25*; 1253,19*.20; 1254,27*; 1255,26.32; 1256,33.35; 1257,3.8*.23*.25*.29*.30; 1258,3*.5.40; 1259,7.31.34*; 1260,5.9.11*; 1261,14.16.33; 1262,19.21.30.31*.33*; 1263,8.12.16*.17; 1264,4*.6*.8.9; 1269,29; 1320,34*; 1321,3; 1323,37; 1328,19.21; 1339,11.20.29.31.38; 1340,2; 1344,7; 1348,32.35; 1349,17; 1350,7; 1352,33*; 1353,4.5*.7.17.30; 1354,9.15.38; 1356,3.23*.24.25.30.32; 1357,19.33; 1358,2.4*.36; 1359,25.29; 1360,20; 1361,10.29*.31; 1362,18.20; 1363,7; 1364,26.38; 1365,4.7.11.12.13.19.20; without motion, 1272,9*; unmovable, 1359,36; ***akinêtôs***, immovably, 1355,27†; ***to akinêton***, immobility, 1357,34

akmê, prime, 1335,4.5

akolouthein, follow, 1268,24*; 1284,35; 1296,12; 1297,3; 1299,14; 1300,23.1307,19; 1323,27.29; 1325,4; 1326,11.15; 1346,35; be a consequence of, 1302,24; 1325,2; be subject to, 1258,39

akolouthoun, consequence, 1326,24

akolouthia, entailment, see *antistrephein*

akolouthôs, consistently, 1322,12; 1365,19

akouein, understand, 1254,1; 1281,31; 1293,15; 1307,7†; 1354,27; hear, 1318,13; 1334,17; 1360,27; listen, 1339,26; ***hoi akouontes***, audience, 1360,5.8

akraiphnês, unmixed, 1274,23.26

akrê, extremity, 1285,37*.38; 1286,2

akribês, exact, 1273,27; 1361,22*; accurate, 1352,18

akribologeisthai, be pedantic, 1276,26

akron, extreme, 1257,22; extremity, 1280,38; 1281,3.21.26.33.37; 1282,3.6; 1283,39; 1284,6; 1286,9*; 1287,2.3.4(bis); 1288,33

akron daktulon, fingertip, 1332,13

alêtheia, truth, 1289,31*; 1290,19

alêthês, true, 1255,5; 1263,2; 1270,16; 1285,12; 1289,32.33; 1296,22†; 1297,18; 1299,10†; 1300,1†(bis).2†.3†.4†.5†(bis).7†.8†. 17.20(bis).23.24.26.27.29(bis); 1302,10.11.14; 1314,37; 1325,12†; 1326,26; 1327,26; 1335,19; 1336,17; 1340,29(bis); ***to alêthes***, the truth, 1293,19

allassein, change (place), 1260,29†; 1354,21
alloiôsis, alteration, 1258,36; 1265,12.15*.16.20.25.26*; 1266,2.13.15.16.19.32.38; 1267,2.7; 1270,15.29; 1271,11.12.13; 1272,23; 1273,4; 1274,10*; 1278,28.39; 1302,11; 1304,13; 1305,35; 1311,27.30.36; 1313,2.3*.6
alloios, different, 1290,8; 1319,12.15
alloioun, alter (tr.), 1265,36; 1266,2
alloiousthai, be altered, 1265,29.30.31.35.36.37; 1266,1; 1319,11†.15†; alter (intr.); 1269,25; 1312,35; 1319,13; 1320,13.14; undergo alteration, 1270,20; 1313,3; 1319,4*
to alloioun, something that alters 1265,30; what alters, 1265,31; what causes alteration, 1265,35
to alloioumenon, what is altered, 1265,30.35
ta alloioumena, things that are altered, 1320,17
allotrios, alien, 1317,28
alogos, irrational, 1283,19
alutos, unable to be undone, 1331,29; 1337,23†; 1338,26; 1339,16
amegethês, without magnitude, 1321,12.14.23; 1340,15; 1344,8; 1352,27*; 1358,1; 1366,11
ameibein, change, 1320,19; 1355,20†
amerês, without parts, 1252,2.6*.15.17*.18; 1283,6; 1294,6; 1295,17; 1297,3.8.9.11.29; 1298,9.18.19.21.23.24.31; 1299,30; 1321,4*.12; 1323,37*; 1338,35; 1340,15; 1344,7; 1354,27; 1355,1.29†.30; 1357,21†.23†; 1358,1.2.16*; 1360,21.22; 1365,31(bis); 1366,11.18; ***to ameres***, partlessness, 1355,11
ameristos, indivisible, 1355,26†; 1359,12; (in) no (way) divisible, 1321,23
amesôs, immediately, 1261,3
ametablêtos, unchangeable, 1251,30; 1262,32; 1337,6.7; 1353,28.33; 1356,25; 1359,20.23.25.29.36; 1360,18.20.23
amigês, uncontaminated, 1362,10
amphisbêtein, dispute, 1361,12
anabebêkos, generalized term, 1320,4
anagein, lead, 1360,18; refer, 1364,14
anagraphein, put down (in writing), 1288,3
anairein, eliminate, 1252,32; 1259,3; 1276,18; 1279,28.29(bis).30; 1280,15; 1289,13; 1290,37.40; 1314,20.21; 1321,13; 1322,4; 1335,22; 1344,26; 1356,4; ***to anairein***, elimination, 1279,31; ***to anairein allêla***, mutual elimination, 1279,27
anairetikos einai, eliminate, 1280,9.10; 1308,27†
anakampsis, turning back, 1278,32; 1280,6*.7; 1287,6; 1306,1; 1312,16.17
anakamptein, turn back, 1278,4.18.18*.21.23*; 1279,5.10.19; 1280,21.22*.23.25.26.28.32.34; 1281,6.7.8.12.15.16; 1283,27.28; 1284,8.11.12.15; 1286,24.32.35; 1287,26; 1288,10.17.18.21(bis).24.25.26*; 1290,26; 1296,38; 1297,1; 1301,9(bis).11.14; 1302,25.35; 1303,1.16.31†; 1308,9.14.28.30; 1309,12.31.34(bis).37; 1310,4.5.6.13.24(bis).28.31.35; 1311,5.19.22.24; 1312,4.19.23.25; 1314,11.12(bis).14*.30.36; 1365,25.26; ***anakamptôn***, backward-turning, 1304,23; 1309,4
analambanein, review (v.), 1352,21. 31; 1357,31; 1364,19; 1366,10
analêpsis, review (n.), 1353,24; 1357,36
analiskein, exhaust, 1323,16.19.26.27
analogia, proportion, 1322,9.16; 1338,6; 1341,15; 1342,35*.39; see s.v. ***antistrophê***, 1342,14*.17*; 1343,11*
analusis, analysis, 1272,14
anankaios, necessary, 1252,5*.34; 1255,12; 1277,25; 1280,24; 1325,25; 1340,18; 1365,16; ***anankaiôs***, by necessity, 1340,11.15
anankazein, compel, 1305,8; 1315,20; 1316,22; 1318,12; 1349,29; necessitate, 1334,22; prove, 1341,9; force, 1363,2
anankê, necessity, 1286,25; 1329,31;

1334,23; must, 1251,3* etc. [137 occurrences]; necessary, 1255,2(bis).3.4.10; 1285,10; 1306,30; 1344,14; 1363,6; 1365,2; need, 1255,21; 1269,26; 1275,36*; 1276,6*; 1285,27†; 1304,40; 1306,28; 1309,27; 1314,30; 1328,2; 1330,23.25.35; 1344,17; 1353,9*.10*; ***ex anankês***, necessarily, 1256,14; 1257,31; 1277,30; 1281,11; 1286,34; of necessity, 1252,33*; 1253,6*.7; 1306,10; 1329,18

anapalin, reverse, 1280,9.15; 1308,3; in reverse directions, 1308,24†; in reverse, 1310,13; the converse, 1325,14†.26; 1326,27

anapauein, rest, 1356,19

anapausis, rest, 1255,23; 1258,40

anaplêroun, supply, 1267,33; fill in, 1343,16

anapnoê, breathing, 1256,37; 1258,17.18

anarkhos, without beginning, 1352,24

anatithenai, ascribe, 1319,6.32

anathumiasis, exhalation, 1258,21.26

aneideos, formless, 1333,7†

anekleiptos, inexhaustible, 1256,15.18.19.24; 1313,18; 1337,18; 1339,1.13; ***to anekleipton***, inexhaustibility, 1253,24; 1256,11; the property of being inexhaustible, 1256,16

anempodistôs, without being impeded, 1302,3

anepaisthêtos, imperceptible, 1313,13

anepidektos, incapable of admitting, 1250,36

anerkhesthai, proceed, 1359,26

aneuriskein, discover, 1359,10

anisos, unequal, 1276,33

anisotês, inequality, 1351,33; 1352,6†

anô, upwards, 1259,22; 1276,10; 1279,6; 1302,16; 1303,17; upward, 1274,22.25; 1349,31.33; 1357,27†; up, 1278,25; 1279,36; 1280,13; 1288,15; 1302,16.19; 1303,13; 1308,5.11.19; 1330,29; 1352,10†; at the top, 1317,26; ***to anô***, the top, 1281,34; 1288,15; 1302,21(bis).31; 1303,4.5.6.9.10(bis).14.15.18.21.22; 1304,25.26.36.38.39.40; 1317,27; ***ta anô***, things that are up, 1330,28

anômalês, non-uniform, 1317,14.17*

anômalia, non-uniformity, 1317,15; 1351,33

anômalos, non-uniform, 1353,16; ***anômalos physis***, non-uniformity, 1352,6†

anomalotês, (state of) non-uniformity, 1352,5†.12†

anomoeidês, of different kinds, 1343,22

anôthen, from above, 1279,6; 1288,15; up, 1281,34

antakoloutheîn, be mutually implied, 1317,1

antallagê, exchange, 1350,32

anteisagein, replace with, 1335,22

anthistasthai, oppose, 1308,27†

antidiairein, contrast, 1359,33

antigraphein, write in opposition, 1327,8

antigraphon, manuscript, 1317,6†

antikeisthai, be opposed, 1276,20*.22(bis).25(bis).28*.32.33; 1277,5†.9†.13†; be opposite, 1308,13; 1326,1; 1341,4

antikeimenos, opposite, 1266,27; 1274,1*(bis).2.3.4.7(bis).8.10.14.15. 16; 1275,12.16.20.25.28.31.34.36; 1276,2.5.12.14.28; 1277,17(bis).21(bis).25.27.28; 1278,29; 1279,30; 1280,11(bis).14; 1289,11; 1301,27 (ter); 1304,11(bis).14.18*.20.29*.37; 1305,5.9.16.32; 1306,34; 1307,26*.29*.30.31*; 1308,3*.9.15.20†.26†.28.29; 1309,1.11.38*; 1310,1.27.36; 1311,21; 1312,5.7.24; 1313,12; 1321,13; 1324,28; 1325,35; 1331,36; 1341,3(bis); 1356,4; 1365,33.38

antikineisthai, move oppositely, 1279,21.33.34.35; 1280,6.7; 1308,24†

antikinêsis, opposite motion, 1280,8; 1308,25†

antilegein, contradict, 1329,5; 1331,39

antimetalambanein, exchange, 1351,31
antimetastasis, exchange of place, 1351,39; 1352,14
antimethistasthai, exchange places, 1351,22
antiperiistasthai, undergo mutual replacement, 1351,21; take (something's) place, 1351,24.26
antiperistasis, mutual replacement, 1350,31*.31.36.37*; 1351,6.12.13.14(bis).16.17.18.29.30.32
antiphasis, contradictory state, 1275,37*; 1276,6*; 1312,22; contradiction, 1296,17; ***kata antiphasin***, contradictory, 1275,25.31.33; 1276,2; ***antiphatikôs***, tr. 'contradictory', 1275,36; 1276,5
antistrephein kata tên akolouthian, mutually entail, 1268,10; 1269,24
antistrophê, reciprocity; ***kata tên sun antithesei antistrophên***, by contraposition, 1334,29; ***kata tên antistrophên tês analogias***, inversely proportional, 1342,14*.17*; 1343,11*
antistrophon, converse, 1340,19.29; 1366,13
antistrophôs, reciprocally, 1317,2*
antithesis, opposition, 1276,21; 1277,11†.14†.25; see s.v. *antistrophê*, 1334,29
antitithesthai, be opposed to, 1281,32
aoristos, indefinite, 1300,9†; 1342,5.7
apagein: ***eis adunaton apagôn***, by *reductio ad impossibile*, 1324,17
apaitein, demand, 1260,31†; 1299,7†; 1360,12
apallassesthai, get rid of, 1318,3
apantan, meet, 1267,25†; 1273,7†; 1280,3
apantêsis, reply, 1289,35
apartizein, exhaust completely, 1343,24(bis)
apathês, impassive, 1266,35; 1362,10; that cannot be affected, 1334,14
apaustos, unceasing, 1260,13*; 1263,15
apeikazein, compare, 1259,15
apeiria, infinity, 1290,14; 1328,20; 1336,11
apeirodunamia, infinite power, 1328,23; 1329,10
apeirodunamos, having infinite power, possessing infinite power, (*estin apeirodunamos*, tr. 'has or possesses infinite power'), 1327,32; 1329,22.23.29; 1331,9.14†.23†; 1333,18†.19†.20†.28†; 1335,18.20.23.25†.29†.31; 1336,2†; 1338,35; 1339,28; 1340,12; 1358,28; ***to apeirodunamon***, infinite power, 1335,22
apeiros, infinite, 1252,36* etc. [229 occurrences]; ***apeira***, an infinite number of, 1289,7.8.9.12.13.15.18.19.21.22.24.25.26.28.29(bis).38*.40; 1290,3(bis).15.18.34.35.37; 1291,2.5.9.11.28*.29.31.37†; 1292,1†.2†.3†.4†.6†.10†.13*.17†.18†.24*. 26.31.33(bis).36; 1293,1.7*.9.10.13.15.16.17.21; 1341,33; infinite in number, 1312,18; ***to apeiron***, infinitude, 1339,10.20; 1340,5; infinity, 1341,16; ***ep' apeiron***, *ad infinitum*, 1252,37; 1253,15.18.35; 1263,11; 1289,7.25; 1290,1.39; 1291,2.11(bis).16.37†; 1292,19.21.23.37; 1293,3.15.17.19.28.31; 1313,17; 1327,33.37.39.40; 1328,5.9.10.22.32.37; 1329,33; 1331,33; 1333,5†.6†.12†.36.37.40; 1334,2.9.22; 1335,20; 1336,10.26; 1337,20; 1338,27; 1339,12.15.17; 1340,6.7; 1341,13(bis).15; 1342,1.4.7.27; 1353,4; 1358,19.29(bis); 1359,2.3.19; 1360,13.15; 1364,34; ***ienai ep' apeiron***, go to infinity, 1257,28.32; 1320,29
apekhein, be a distance away, 1317,33
apemphainon, absurdity, 1330,6
aperatôtos, unbounded, 1278,3
apergazesthai, accomplish, 1360,36†

aperigraphos, indeterminate, 1299,38†; 1300,9†.10
aperkhesthai, go away, 1282,18; 1287,12; depart, 1285,15*.30†
aphairein, take away, 1294,24; subtract, 1323,9.13.18.20; 1324,35; 1326,17.29†.31†.32†; 1331,20†; 1341,22.27.30; 1342,1.35; 1343,1.3.4.5(bis).7.9.10.33†.37†; remove, 1364,31
aphairesis, subtraction, 1323,3.23; 1341,34; 1343,8.38
aphanês, unevident, 1257,11
aphistasthai, withdraw, 1262,34; 1263,20; 1349,8.14; move away from, 1317,19.23*; (perf.) be far from, 1306,29; 1307,34; 1316,20; ***ison aphestanai***, be equidistant from, 1317,29
aphoran, keep in view, 1276,21; 1313,9; look to, 1343,3
aphôrismenos, distinct, 1260,34†
aphthartos, imperishable, 1273,25; 1314,17; 1320,27; 1329,26.27.38†; 1330,17†; 1331,30; 1333,31†; 1334,28.31
aphupnizesthai, wake up, 1258,24
aplanês, fixed, 1254,25; 1261,18; 1263,18; 1355,4.13; ***ta aplanê***, the fixed stars, 1358,33; ***hê aplanês*** (sc. ***sphaira***), ***to aplanes***, sphere of the fixed stars, 1261,23.25.34†; 1262,1†; 1263,18; 1355,7.10; 1356,34; 1357,2.6†.14; 1358,37(bis).40
aplatês, breadthless, 1292,7†.8†
apoballein, lose, 1306,5; 1334,23
apodeiknunai, demonstrate, 1251,2; 1253,27; 1269,18; 1271,25; 1273,1.27.34; 1278,15; 1285,12; 1294,33; 1304,22; 1316,38; 1320,25; 1327,3.9.16; 1328,34; 1329,4; 1330,34; 1331,8.27; 1332,1; 1340,14.20; 1341,18.19; 1344,2.15; 1353,34; 1358,18; 1364,5; 1365,23; prove, 1329,4; 1330,34
apodeiktikos, demonstrative, 1279,23.25.27; 1301,18
apodeixis, demonstration, 1262,23; 1272,19; 1278,33; 1279,12.19; 1280,19; 1295,5; 1304,2.31; 1318,10.11.12.14; 1327,12; 1329,5.13; 1331,39; 1333,31†; 1340,22; 1342,28; 1343,14.38; 1351,35; 1352,22; 1358,3; 1360,17; 1364,22; 1365,35
apodidonai, produce, 1254,22.23; 1266,37; give, 1272,29; assign, 1318,20; 1319,3; 1334,31; 1362,1; explain, 1319,23
apogignesthai, depart, 1275,3; 1281,2.5; 1282,13.17.20.27*; 1283,8.9.11.12.16.23; 1284,7.8.9.11.13.20.24.27.29.33.34. 35; 1285,2.5.9.10.13.24†.32*.33*; 1286,10.17.24.25.27.33.35; 1287,9.11.12.13.15.20(bis).24.28*; 1288,35; 1290,28; 1294,7; 1297,2.6; 1303,10.19; 1309,3.25.26.30(bis).37; 1313,9
apographein, copy, 1347,38
apoios, without qualities, 1333,8†
apokhôrein, set out, 1285,23†.28†
apokrinein, answer, 1291,27.32; separate off, 1319,5
apolambanein, attain, 1268,3; 1272,16*
apoleipein, leave, 1305,28; 1311,10; 1331,3; 1351,32
apollunai, destroy, 1273,11†; 1332,12; (pass.) perish, 1261,1†
apoluein, undo, 1334,1
apologizesthai, calculate, 1312,7; reckon, 1319,19
apophainesthai, declare one's opinion, 1318,17; 1339,18
apophanai, deny, 1329,11
apophasis, negation, 1274,10; 1279,29
apophaskein, deny, 1335,20
apophatikos, negative, 1253,9
apophthengesthai, pronounce, 1313,15
aporein, be puzzled, 1255,34; 1348,7.16; pose a (or the) puzzle, 1258,1; 1261,15; 1285,11; 1286,8.11; 1290,12.14; 1292,24; 1293,18.20; 1354,12.25; 1356,33; 1357,5.17.18; ***to aporoumenon***, ***to aporêthen***, puzzle, 1263,35; 1289,35; 1296,18†; 1346,3; 1351,10*; 1357,6
aporia, puzzle, 1256,6; 1258,4; 1261,20; 1284,16.21.22(bis);

1287,19.23; 1290,5.21; 1296,19†; 1298,4.11; 1299,4.9†; 1344,30; 1345,23; 1346,30; 1347,38; 1348,18; 1354,17; 1357,3.4.23
aporos, puzzling, 1263,36; 1298,10
apospan, pull apart, 1334,3.4.5
apostasis, distance, 1316,14
apotelein, bring to completion, 1338,3
apotemnein, cut off, 1334,11
apotithesthai, confer, 1336,20
apsukhos, that lacks a soul, 1263,26†; that has no soul, 1364,6.9
araiôsis, rarefaction, 1266,18
araiotês, rarity, 1266,24*.25.26
ararotôs, soundly, 1332,40
aretê, virtue, 1276,24.33
aristeros, left, 1278,26; 1280,12; 1302,31; 1308,6.12; 1334,11
arithmein, count, 1289,16.18 (ter).38; 1290,2.4.5.8.15.23.29.36; 1291,6.15.18*.22
arithmos, number, 1255,32; 1316,35; 1332,22†; 1336,2†; 1366,6; ***hen*** (***tôi***) ***arithmôi***, numerically one, 1255,16(bis).17(bis); 1281,37*; 1282,11*; 1283,29.34*; 1294,25
arkein, satisfy, 1251,33; 1252,22; be sufficient, 1275,39; 1328,6.30; 1361,13; 1362,16; last, 1332,19†; 1336,13.20
arkountôs, well enough, 1363,11
arkhê, beginning, 1250,38; 1252,3; 1256,7.27; 1258,5; 1267,33; 1268,20*; 1280,37*.38; 1281,1.4.6.11.23.24.25(bis).27.28.29.34.35; 1282,4.5.8.14; 1283,25.28.31.36.38; 1284,1.3.7.14; 1286,6.33*; 1287,5.6.7.8.29; 1288,12(bis).14.16; 1290,25*.28; 1291,18; 1294,5.11.14.18.21.24.26; 1295,12.13.16(bis).19.23.32; 1296,1.2.8.10.25†; 1299,18; 1300,34.36; 1302,25; 1308,10(bis); 1309,7.12.13.14.16; 1310,11.31.32; 1311,1*.2.3.5(bis).6.8.19.33.34; 1312,26; 1314,31; 1315,10.12.14.16.17.19.21.22.24.26.28.29*.30.35*; 1316,5.6.7.8.9.12*.13.17.18; 1317,11.17*.19.20.28.29.31*; 1318,29; 1320,26; 1323,20; 1326,38; 1327,9; 1328,11.39; 1335,4; 1338,38.40; 1339,1; 1356,18; 1357,22†; 1363,29; 1365,30; principle, 1251,32.33; 1252,2.3.8*.9.11.23; 1253,19*.34; 1254,28*; 1257,10; 1266,19.21; 1269,28.29; 1271,35*.35; 1272,1.7*; 1319,17.20; 1348,25; 1354,4.5; 1359,5; 1362,6.9; 1364,14; 1365,18; 1366,21; outset, 1252,3; origin, 1256,34*.35(bis).37; 1257,2.3.5.9*.28.36.37.38*; 1258,15.25*.30*.40; 1259,5.6; 1260,36*.37*.38(bis); 1261,1†.3*.3.4.6(bis).7.11; 1262,2†; 1266,15*; 1268,14*.22*.25*; 1272,31.33.34; 1318,20*; 1319,31*.33; 1320,28*; 1338,1; 1347,3†.14†.34†; 1348,2†; 1349,6; 1353,5; 1354,31†.32†; 1359,40; 1360,3.4.10.12; 1361,14.15*.27*; 1362,17.19.20.23*; 1363,15; 1365,5; starting point, 1264,26.28; 1268,27; 1271,21; 1275,16; 1345,23; 1360,17; ***ex arkhês***, original, 1323,18; 1331,26; 1341,14; ***ex arhkês***, ***en arkhêi***, originally, 1265,21; 1273,25; ***tên arkhên***, in the first place, 1291,33; 1340,21; 1347,11†; original, 1347,1†
arkhesthai, start (intrans.) 1255,20; 1300,1†; 1302,1.6.7.18; 1303,2.13; 1307,22; 1309,32.34; 1310,25; 1311,8; 1315,13; 1318,1(bis); 1320,27; 1325,9†.26; begin (intr.), 1258,10; 1279,35.37; 1280,1; 1284,26; 1285,1.16†.17†.18†.19†.38; 1338,31; 1352,23; 1354,11
arkhikos, originative, 1337,36
arkhon (***to***), the ruler, 1268,23
asapheia, unclarity, 1285,35
asaphês, unclear, 1284,23; 1285,14
askhetos, unrelated, 1334,27; 1355,25†.26†.31; 1359,8
asômatos, incorporeal, 1252,15; 1260,34†; 1321,12; 1328,24; 1334,6; 1337,32.34; 1338,1; 1354,28; 1359,7; 1363,7
astêr, star, 1330,11†
astron, star, 1263,20
asummetros, disproportional, 1276,34
ateleia, incompleteness, 1274,13*
atelês, incomplete, 1268,2;

1271,34.37; 1272,7; 1274,14; 1312,14; 1313,32; 1314,14*.15(bis).16(bis).21.22.25.27; 1361,4†
atelesteros, less perfect, 1271,30 [see also n. 173]
ateleutêtos, without end, 1352,24
athanasia, immortality, 1338,20.27
athanatos, immortal 1251,12; 1260,13*; 1278,35; 1300,7†.30; 1337,25†; 1338,19†.26; 1339,15; 1361,21*
atomos, indivisible, 1297,21*.22.23.27*.30.32.35.39*; 1298,2.5.11.12.15*.15.17; 1299,1; 1318,34; 1319,10*.16; 1333,38.40; ***hê atomos***, atom, 1319,12.13
atopos, absurd, 1257,27; 1277,30*; 1284,26.27.28; 1285,18†; 1287,22; 1289,20.27.30*; 1294,27.32; 1295,18.37; 1302,38.39; 1323,15; 1324,26; 1325,1.3.4.39; 1326,3.11(bis); 1330,31; 1332,4.5; (***to***) ***atopon***, absurdity, 1298,23; 1302,34; 1306,7.19; 1323,26; 1325,2.30
autarkês, sufficient, 1289,28.32; 1336,33; 1355,17; self-sufficient, 1329,22(bis).28; 1335,25†.29†; ***autarkês einai***, satisfy, 1252,2
authupostatos, self-constituted, 1328,24; 1329,30; 1337,3.5.7.11.12(bis); 1338,18; 1359,17
autokinêtos, self-moving, 1261,16; 1263,10; self-moved, 1339,4; 1349,18; 1365,10; ***autokinêton***, self-mover, 1251,8.12.21.32; 1252,15.16.18.23; 1253,19.20.32.33; 1254,1; 1257,29.31.39; 1258,35.38; 1259,6.7.9.31; 1260,5.6.8; 1263,12; 1269,29; 1272,33; 1319,29; 1320,2.35.36; 1347,7†; 1348,5†.16.18.20.27.27†.31; 1349,9†; 1350,3†; 1353,11; 1364,26; 1365,6; what is self-moved, something that is self-moved, 1253,35; 1353,7; 1364,36; 1365,4.5.10.11; what moves itself, 1272,30
automaton, chance, 1362,21.24.25*.28*.29*; 1363,1
auxanesthai, increase in magnitude, 1267,27†; increase, 1269,25; 1273,9†; 1274,13; 1319,4*; 1320,13.14; 1326,9
auxein, make something grow, 1265,24; make something increase, 1324,33
auxesthai, grow, 1257,35; 1265,21(bis).27(bis).28; 1267,9.10.11(bis).12.16.18; 1270,15.20; increase, 1267,8; 1271,12; 1274,19; 1312,36; 1324,33.38; 1326,18.19; 1341,16
auxêsis, growth, 1256,37; 1258,16.19; 1265,27; 1266,32; 1270,29; 1278,28; increase (n.), 1265,11.15*.20; 1266,13; 1267,23.27; 1271,11; 1272,23; 1273,5.11; 1274,11*.13; 1302,12; 1304,12; 1311,27; 1312,2; 1313,7; 1342,1
axiologos, significant, 1349,3; ***theôria axiologos***, theoretical interest, 1340,20
axiôma, axiom, 1274,31; 1276,18; 1297,15.20; 1301,33; 1302,10.18; 1303,29; 1324,7; 1344,12.19; proposition, 1299,37†; 1300,10†.11
axioun, think that someone should do something, 1262,13; 1281,31; 1290,5; 1327,21; 1330,3; 1350,4; postulate (v.); 1290,1; 1302,14; 1324,13; 1334,20

baros, heaviness, 1258,22
barus, heavy, 1266,22*.24
barutês, heaviness, 1318,35; 1343,18; 1347,33†
bebaios, certain, 1276,19; 1318,13
bebaioun, confirm, 1356,3; 1364,20
bia, force, 1317,18; 1347,29†; 1364,29
biaios, forcible, 1259,20; 1321,5; ***biaiôs***, by force, 1257,16
biazesthai, force, 1351,27
biblion, book, 1251,1.3; 1257,25; 1258,5; 1276,30; 1320,26; 1321,15; 1326,38; 1327,9.36; 1328,11.38; 1329,3; 1335,2; 1348,24; 1358,7; 1361,20; 1363,8.25; manuscript, 1288,4; 1292,15; 1319,9
boulê, plan, 1333,26†
boulêsis, will, 1331,25†.29.30; 1334,24.25; 1337,26†

bradus, slow, 1317,19.20; 1321,11; 1332,18†
brakhus, brief, 1284,23; small, 1332,14; 1340,27; tiny, 1340,28

daimonios, marvellous, 1359,5
deiknunai, prove, 1250,34 etc. [329 occurrences]; show, 1256,3; 1275,37; 1283,32.39; 1301,3; 1307,29; 1354,36; 1356,10; 1357,5; establish, 1289,33.36; 1290,37; ***to deikhthen***, conclusion, 1260,4; ***to dedeigmenon***, result, 1286,6; ***ta deiknunta***, the proofs, 1327,10
deiktikos, used to prove, 1272,28
deixis, proof, 1256,4; 1257,23; 1279,32; 1290,36; 1297,1; 1305,30(bis); 1306,19.23; 1313,31; 1321,37; 1343,2
dekhesthai, receive, 1329,34†; 1330,16†.22.25†; 1331,31; 1337,15.18(bis); accept, 1291,3; admit, 1311,12
dektikos, receptive, 1330,16†
dêlos, clear, 1252,18 etc. [60 occurrences]; ***dêlon hoti***, ***dêlon hôs***, clearly, 1271,10*.15; 1298,19; 1321,22; 1322,23; 1324,32; 1325,17†; 1331,20†; 1337,6; 1339,9; 1340,14; 1352,33; 1355,22†.24†; ***dêlonoti***, that is to say, 1254,4; 1317,24; 1363,5; clearly, 1298,35; 1300,22; 1323,10; 1327,2; 1332,21†; 1341,22.25; 1361,27; ***dêlos estin***, he makes it clear, 1360,14; he clearly …, 1360,33
dêloun, reveal, 1256,13.20; 1361,8; 1362,14; signify, 1260,11.36; 1278,3.35; 1325,18†; 1359,31; 1360,13.23; show, 1280,29; 1335,28†; 1340,31; 1349,3; 1351,38; 1361,17.25; indicate, 1291,35†; 1342,14; 1356,29; 1357,13
dêmiourgein, bring into creation, 1327,3; create, 1338,3
dêmiourgêma, creation, 1331,3; 1360,1
dêmourgia, creation, 1360,4; 1361,8†
dêmiourgikos, the creator's, 1337,28; creative, 1337,33; 1355,7; 1361,24
dêmiourgos, creator, 1327,2.4.21; 1330,37; 1331,5.15; 1334,34; 1337,21.23†; 1338,2.9.10.15; 1355,6; 1359,9.21; 1360,37(bis); 1361,2.2†
desmos, bond, 1331,24†; 1337,27†; 1338,6.7
dexios, right, 1278,25; 1280,12; 1302,31; 1308,6.12; 1334,11
diadekhesthai, be followed by, 1314,36; 1315,38; 1316,1; 1339,14; inherit, 1346,5
diadekhomenos, successively, 1328,2; 1356,8.11
diadosis, transmission, 1350,17
diadokhê, succession, 1255,12.20; 1356,15; ***ek diadokhês***, in succession, 1255,23.28; 1350,28; 1356,21; ***kata tên diadokhên***, as it is passed on, 1346,14
diairein, divide, 1269,12; 1282,1.7; 1283,6.30.36; 1284,5; 1287,33; 1288,36; 1290,23.24.30.31.32.34.36.38.40; 1291,7(bis).15.19.21.22.23; 1292,37; 1293,4.15.17; 1294,9.28*; 1295,9; 1297,8.20*; 1301,4; 1303,20; 1317,35; 1323,21; 1326,33†; 1332,14.21†.30.32; 1333,7†.10†.12†; 1334,1.3.4.6.7.9.11.15; 1352,29*; 1365,27.28.31; make a division of, 1264,19; 1290,33; 1364,14
diairesis, division, 1257,19; 1267,32; 1268,36*; 1269,6; 1283,2; 1293,29*; 1294,6.9; 1295,11; 1311,18*; 1314,1; 1321,14; 1328,17; 1333,15†.16†; 1364,19.21; dividing point, 1282,34*
diairetos, divisible, 1289,25; 1290,1.39; 1291,2.17.37†; 1292,20.21.23; 1333,5†.6†.12†.37; 1334,1.2.9.17; 1356,12*; ***to diaireton***, divisibility, 1334,22
diaitê, treatment, 1329,8
diakopê, interruption, 1312,18
diakoptein, interrupt, 1260,14; 1275,40; 1304,30; 1305,6; 1316,4.11; 1350,22; 1356,30
diakosmein, arrange, 1318,29; 1319,19
diakosmêsis, arrangement, 1277,33; 1337,30
diakrinein, separate, 1266,18; 1267,19.27†; 1273,9†; 1318,24.30*;

1319,10*; 1338,12.13.25; break up, 1352,9†
diakrisis, separation, 1266,17*.18.29*.30.33; 1267,1.2.5.6.16.22; 1273,3; 1277,33; 1311,22; 1318,21*; 1319,24.26; 1338,8.14
diakritikos, tending to separate, 1266,27
dialambanein, interrupt, 1274,29; 1275,34; 1276,15; 1277,18.20; 1278,27; 1310,15; 1312,12; 1365,34.39
dialeipein, be interrupted, 1255,7.10.13.14.15; 1308,31*
dialektikos, dialectical, 1301,23
dialogos, dialogue, 1360,36
dialuein, break up, 1269,4*; 1336,16
dialusis, dissolution, 1331,12†
diamenein, persist, 1273,10†; 1277,31; remain, 1331,29; 1353,27; last, 1332,19†; 1353,14
diametron, diameter, 1307,36.37*; 1308,2.3.4.19; ***ta kata diametron***, the points on a diameter, 1307,33*.34; 1308,16*
dianistasthai, start again, 1357,6
diapherein, differ, 1251,21; 1279,8.13(bis); 1280,16; 1304,36; 1305,5.10; 1331,21†; 1336,27; make a difference, 1275,35*; 1312,8*.10; be different, 1327,21; there is a difference, 1276,4; the difference is, 1280,29
diaphora, variety, 1263,29; differentia, 1279,7; 1301,21; difference, 1309,19.20; 1328,5.33.36; 1336,37; 1338,28
diaphoros, different, 1263,23.27.28; 1266,29; 1280,13; 1301,10; 1330,10†; 1336,35; varying, 1263,33; various, 1364,16
diarkein, continue, 1259,23; 1327,25.31; last, 1327,28; 1332,31(bis); 1335,40†
diarkês, long lasting, 1336,24
diarrhiptesthai, be dispersed, 1334,8
diastasis, extension, 1334,4.6.16
diastatos, having extension, 1333,6†; ***diastatos (esti)***, has extension 1333,39; 1359,24; ***to diastaton***, extension, 1331,20†; what has extension, 1333,8†
diastellein, distinguish, 1280,36
diastêma, distance, 1309,33; 1320,20; 1335,13
diatithenai, arrange, 1263,21; 1272,13
didaskalia, instruction, 1359,6; 1363,12
dierkhesthai, traverse, 1289,7.9.10.17.25; 1290,3(bis).6; 1291,29.30.31.35†; 1292,1†(bis).4†.8†(bis).17†.18†.26. 33; 1311,9
diexerkhesthai, get through all of, 1285,7; get entirely through, 1289,12.13.38*; 1291,28*.38†; 1292,17†; 1314,6
diexienai, go completely through, 1283,2; 1289,22(bis).23; 1291,30
diexitêtos, able to be got completely through, 1290,15
diienai, go through, 1284,26; 1287,22.23; 1289,28; 1290,1.2.15.18
diistanai, distinguish, 1359,10
diestêke, be distant, 1268,26*; 1285,22†; 1308,2.5; 1316,16
diestêkôs, distant, 1285,21†
diastan, extended, 1334,8
diestôs, distant, 1308,16†; extended, 1337,11; 1359,14
dikhotomia, dichotomy, 1289,7; 1293,18
diorismos, distinction, 1278,30; 1361,13
diorizein, determine, 1251,16; 1278,30; 1279,16; distinguish, 1353,10; make determinate, 1284,22; draw distinctions, 1300,12
diôristhai, be determinate, 1310,23; be distinct, 1314,32; 1315,10.11; 1334,12
diôrismenos, distinct, 1310,31.33; 1311,4; 1315,17; 1316,5.6.10.17
diôrismenôs, determinately, 1315,28
diplasiazein, double, 1322,24; 1324,32.34
diplasios, double, 1322,28; 1324,38.39; 1341,27.28.30(bis).34; 1342,35.36.38.39; 1343,1.2.4.6.10; twice, 1341,27; 1342,34*
diüpnizein, fall asleep, 1258,25

doxa, belief, 1331,26; opinion, 1351,28; 1366,8
dunamikos, powerful, 1335,37†
dunamis, power, 1268,23*(bis) etc. [239 occurrences]; potentiality, 1269,2*(bis); 1331,1; 1337,2; 1339,30; 1356,26
dunamei, potentially, 1265,1; 1270,4; 1277,19; 1280,39; 1281,7.9.38(bis); 1282,29; 1284,4; 1285,4.8.26†; 1286,15.20.26; 1287,26.29*.31.32.37; 1288,2(bis).6.7.33; 1290,21; 1291,2.4.5(bis).8.9.12.17.33.34†. 37†.38†(bis); 1292,2†.11†.14†.15†. 17†.18†.19.22.35.37; 1293,1.4.6.8†.8.9.10(bis).21(bis); 1305,1; 1310,10.22; 1315,20.22.26.35; 1327,35; 1328,3; 1333,17†; 1334,21†; ***to (***or ***hê) dunamei***, potentiality, 1265,30*; 1292,39; 1330,17†; 1364,30
dunatos, powerful, 1268,23*
dunaton, possible, 1253,3; 1259,35; 1261,24; 1269,32*.33; 1282,16; 1293,23*; 1299,36†; 1312,18.34; 1323,8; 1326,16.35†; 1342,6.8; 1348,12; tr. 'can', 1252,9*; 1262,8.10; 1264,16; 1265,29; 1273,32; 1289,16; 1293,11(bis).14; 1308,29; 1311,16.28; 1312,30; 1315,23.25; 1321,32.34; 1327,11; 1332,39; 1334,27; 1336,8; 1342,28; 1365,38; ***ou (***or ***mê) dunaton***, impossible, 1274,24; tr. 'cannot', 1275,22; 1277,7†; 1298,31; 1304,27; 1324,5; 1344,3; 1348,10.18; 1358,9; 1366,12; ***kata to dunaton***, tr. 'as best I could', 1335,2; ***hôs dunatos ên epoiêsamên***, tr. 'I have done my best', 1363,25
dusdiairetos, difficult to divide, 1333,10†
dussunaisthêtos, difficult to perceive all at once, 1272,13

egrêgorsis, waking, 1258,17.19; 1259,33
eidos, species, 1260,20; 1264,20; 1265,10; 1266,21; 1273,2; 1279,3*.9.13.14; 1280,14.16; 1301,10; 1304,36; 1305,2.3.8.10; 1311,27; 1316,33; 1320,32; form, 1262,2†.3; 1263,19*.22.23.27†.27; 1272,17; 1274,37; 1275,10; 1277,14†.31.33; 1279,1*; 1280,31; 1312,11.13.20; 1329,21.34†.35†.36†.37†; 1330,4.10†.13†.15†.16†.23.25†.32; 1331,19†.20†.36(bis).37; 1332,8.12.14.28; 1333,3.8†.10†.13†.14†.15†; 1334,23; 1354,28; kind, 1292,34; 1301,13; 1341,12; ***tôi eidei***, specifically, 1304,35*
eikêi, at random, 1362,4*
eikein, yield, 1319,1
eikos, reasonably, 1313,10
eikotôs, reasonably, 1253,29; 1255,34; 1256,27; 1305,30; 1325,39; 1338,13; it is reasonable, 1264,7; 1304,5; 1329,25; 1366,7
eilikrinês, pure, 1275,1; 1318,4.5
einai, be, exist, occur, 1250,34 etc.; ***to einai***, existence, 1253,31; 1268,10.11; 1269,24; 1271,7; 1273,24; 1289,13.20.21; 1291,1.12.14; 1293,32*; 1297,17; 1327,29; 1328,15.29; 1329,21.25.30; 1333,36; 1335,26†.38†; 1336,10.14.18.19.21.24.29.32(bis); 1337,13.18; 1338,19; 1339,5.8; 1358,32; 1359,14.15.16; 1360,14.15*; 1361,22*; 1363,22; 1364,14; essence, 1292,3†.5†.6†.7†; 1293,1; 1329,33†; being, 1252,11; 1299,10†(bis); 1337,5; 1361,20*; ***to mê einai***, non-existence, 1276,4; ***(to) on***, existing thing, 1251,29; 1257,11*.22.33; 1260,13*; 1263,32; 1273,26; 1275,21; 1318,31; 1324,2; 1327,4.5; 1340,14; 1357,32; 1364,24; thing that is, 1252,19; 1262,24; 1359,26; 1362,6*; what exists, 1252,40; 1274,9*; thing that exists, 1253,2(bis); 1257,17; that which exists, 1260,17*.18*; entity, 1340,1; ***(to) on***, being, 1275,14.20; ***to mê on***, not being, 1275,20; ***(to) mê on***, what does not exist, 1274,9*; 1275,21.37*.38.38*; what is not an existing thing, 1275,21; ***ontôs***, really, 1286,10; 1313,10; truly, 1353,27; 1359,5; 1361,7†;

ontôs on, that has real being, what has real being, 1337,1.4.6.11.19; 1338,18; 1359,10.19.21; 1360,21; possessed of real being, 1337,4
eisagein, introduce, 1255,13; 1341,3; take in, 1335,15; direct, 1353,21
ekdekhesthai, understand, 1287,2; follow, 1329,4; accept, 1350,4
ekhein, have, 1252,16 etc.; possess, 1318,7; 1339,10.39; 1340,6.12.16.17; 1341,19.21.23.28.30; 1343,6.30; 1344,11; 1346,1.6.17; 1347,2†; 1348,30; 1350,6; 1358,10.11.30.31.34; 1359,1.2.4; hold fixed, 1322,27
ekhesthai, be next to, 1283,14; 1297,4.26*.30; 1298,1*.9.14*.22.35*.37.39; 1300,25; 1350,25.30.34; 1351,3; 1352,29*
ekhthos, hatred, 1318,26†
ekluesthai, be exhausted, 1317,19; 1349,1
ekphainein, bring to light, 1313,13; 1315,9; 1338,29; reveal, 1360,8
ekstasis, departure, 1257,15
ekteinein, extend, 1341,13.14; 1365,35
ekthesis, illustration, 1286,28; 1341,19; 1342,33
ektithesthai, set out, 1264,34; 1267,35; 1268,13; 1273,2.5; 1274,8; 1288,20; 1344,20; 1355,28; 1364,3
ektrepein ton logon, digress, 1257,9
ektropê, deviation, 1262,9
elattôn, minor (premise), 1270,8; 1272,7.12; 1279,15
elattôsis, diminution, 1342,27
elenkhein, refute, 1257,19; 1329,27; 1364,19
elleipein, be less, 1342,19.21
elleipsis, deficiency, 1276,25
ellipês, short, 1326,18
empodizein, impede, 1302,3; 1317,35; 1343,2
emprosthen, above, 1255,16; ***to emprosthen***, the front, 1302,31; front, 1308,6; in front, 1308,12
empsukhos, ensouled, 1257,39; 1354,29†.32†.34†; living, 1320,4*.5.7*; with souls, 1344,34; 1364,7
enantios, contrary, 1262,33* etc. [187 occurrences]
enantiôs, in contrary ways, 1263,1
ta enantia legein, contradict, 1328,38; 1330,24
enantiôsis, contrariety, 1275,41*; 1277,11†.28; 1279,7; 1301,13
enargeia, clarity, 1299,32; ***ex tês enargeias***, evidently, 1312,34; 1316,27; because it is evident, 1364,16
enargês, evident, 1269,19; 1297,15; 1313,27; 1324,7.11; 1325,31; 1326,7.23; 1341,5.8; ***enargôs***, evidently, 1252,12; 1295,36; 1312,17; 1321,35; 1335,36†
endeiknunai, show, 1267,17; indicate, 1298,39; 1319,26; 1340,34†.37†; 1363,28; give an indication, 1335,8; sketch (a proof), 1340,22
endeixin (eis), to indicate, 1299,35
endekhesthai, can, 1252,4.7*; 1255,10; 1256,21*; 1268,35*; 1269,27*; 1270,5*.6.8.12; 1273,22*; 1276,27; 1278,2; 1288,28*; 1313,6*.22; 1314,32(bis).37.38*; 1315,2*; 1320,33; 1324,6*.21*; 1325,30*; 1326,1*.37†; 1327,28; 1339,28*; 1340,24*.33†.36*; 1341,1.2*; 1350,18*; 1356,4; 1357,4.18.19.21; 1358,6*; is possible, 1252,5; 1255,13; 1264,29*(bis).31*; 1289,38*; 1291,28*.32; 1294,6; 1357,5; be able, 1259,1; ***to mê endekhesthai***, the impossibility, 1276,1*
endelekhês, perpetual, 1313,8; 1331,7; 1332,37; 1352,13†
endidonai, concede, 1331,28; impart, 1345,19.32.37; 1346,14; 1348,28; 1349,16.19.23.25.26.27.28.35; 1350,5; 1358,15; give in, 1357,2
endosimon, start (n.), 1347,4†.14†.34†.35†; 1348,2†
endothen, from within, 1257,35
endoxos, reputable, 1301,24
eneinai, be in, 1258,10; 1290,34.35; 1340,25*.27.31.36*; 1341,5*;

1362,2*; be possible, 1326,28†; be located in, 1336,23
energeia, activity, 1255,24; 1258,28; 1335,29†; 1337,8; 1339,11; actuality, 1331,1; 1333,17†.30†; 1334,20.21†.39†; 1337,2; 1339,31.32.38.39; 1356,26; ***energeiai*** (dat. s.), in actuality, 1278,16; 1281,6.20.40; 1282,2.6.9.14; 1284,5.6; 1286,20.27; 1288,8.9; 1290,22.24.30; 1291,4.5.9.10(bis).33; 1293,4.11.21.24*; 1303,19; 1309,24; 1310,30; 1314,3; 1315,20.28.36; 1327,31; 1340,19; actually, 1287,25.28*.30.33.35.36(bis); 1288,2(bis).6.34; 1310,10; 1315,27; ***to energeiai*** (dat. s.), the actual, 1265,31; actuality, 1292,38; 1364,31; ***kat' energeian***, in actuality, 1280,39(bis); 1281,3.10.20.21.39; 1282,4.5.10; 1293,3.13; 1310,31.33; 1311,18.20; 1315,10.12.21.24.30.34; 1316,2.8; 1334,12.15; 1344,9; actually, 1281,9; 1290,34; 1291,6; 1309,29; 1328,36; 1329,32; actual, 1288,9; 1309,2.6; 1315,33
energein, be active, 1255,24; 1345,26; engage in activity, 1258,22; 1337,17
energêtikos, active, 1327,33
energos, active, 1348,29
engignesthai, occur in, 1258,11; 1332,29
enistasthai, object, 1258,4; 1276,38†; 1289,20.32; 1303,12; 1327,10; 1340,22; 1341,1; raise (one or more) objections, 1289,2; 1364,2; ***enestôs***, present, 1300,18
enkhronôs, temporally, 1339,26.27; 1340,3
enklisis, inclination, 1263,20
enkrinein, choose, 1354,22
enkuklios, that revolves in a circle, 1357,17
ennoein, realize, 1328,37; 1332,33; conceive, 1330,7
ennoia, thought, 1327,6; 1334,15; 1336,36; 1355,17; 1360,1; idea, 1333,36
enstasis, objection, 1251,31; 1252,3; 1270,17; 1271,18.21; 1275,17; 1277,3†; 1291,1; 1298,4; 1325,12†; 1340,31; 1341,10.18; 1344,18.21; 1350,12; 1351,35.36; 1364,11; 1366,17(bis)
enstatikos, signifying an objection, 1340,32
entelekheia, actuality, 1269,2*.4*(bis); 1291,2
entelês, complete, 1269,6
entimos, honored, 1268,10
entunkhanein, read, 1318,12; 1327,22; 1333,34; 1363,28
epagein, draw a consequence, 1252,35; 1302,39; 1344,18; infer, 1253,29; 1254,3.6.26; 1262,25; 1286,24; 1288,28; 1289,21; 1322,12; 1329,18; continue, 1258,31; 1273,6; 1276,6; 1323,28; 1342,35; 1345,13; 1350,28; 1362,5; add, 1272,19; 1310,38; 1314,15; 1320,4.7; 1333,1; 1350,17.19; 1353,25; 1362,24; bring up, 1284,21; advance, 1333,2; 1344,20.21; go on to say, 1357,6.33
epagomenon (***to***), the consequent, 1254,6
epakolouthein, follow, 1285,7; 1346,11
epamphoterizein, have both attributes, 1257,22; 1347,32†
epaporein, pose additional puzzles, 1354,15
epeigesthai, hasten, 1317,28
epeisaktos, adventitious, 1338,23
epharmottein, apply to, 1304,24; fit onto, 1334,7
epharmozein, apply to, 1301,23
ephesis, aim, 1354,33†
ephexês, next, 1253,22; 1259,7; 1262,17; 1264,15.35; 1270,3; 1294,8; 1299,4; 1301,18.24; 1313,24; 1319,19; 1345,18; 1346,9; 1348,34.38; 1349,10.20; 1350,4; 1351,3; 1364,37; 1365,12.29; consecutive, 1253,26; 1255,8.10.19.36*.36.39; 1256,1*.2.3.6.8.11.12.13*.14(bis).16.18.19.24.25.28; 1269,30*.31*.31; 1298,15*(bis).35*.37.39*; 1299,2; 1306,33*.33.36; 1350,23*.24.30; 1351,10; 1356,15; consecutively, 1255,9; 1309,33; 1310,24; ***to ephexês***, consecutivity, 1256,15; ***to***

(or ***ta***) ***ephexês***, what follows, 1321,21; 1322,12; 1345,21; ***ta ephexês***, the adjacent parts, 1350,7; ***kai ephexês houtôs***, and so on, 1350,16
ephistanein, remark, 1262,5; 1267,15; 1269,10; 1273,24; 1310,28; 1316,36; 1325,8; 1330,3.18; 1334,26; 1336,35; 1355,36; 1363,38; 1365,28; stop (intrans.), 1283,38; 1287,33; notice, 1292,12.21; 1327,21.29; 1328,37; 1329,7; 1332,27; 1358,27.39; make the remark, 1358,18
ephodos, approach, 1257,9; 1345,23; 1364,36; 1365,33; guide, 1363,26
epiballein, address, 1350,8
epiballesthai, set oneself (to do something), 1329,17
epiblepein, attend, 1256,34*; 1257,8*
epibolê, consideration, 1258,8; 1364,16; application, 1343,25
epideiknunai, show, 1266,22; 1350,20
epidekhesthai, also admit, 1333,16†(bis).27†; 1334,20†
epikheirein, attempt, 1255,33; 1290,35; 1330,2; 1332,3; argue, 1286,6
epikheirêma, argument, 1250,37; 1265,14.15; 1266,12; 1271,22; 1272,28; 1308,35; 1309,19; 1321,20.35; 1329,16.20; 1331,8.25; 1335,17
epikheirêsis, argument, 1254,33; 1257,6; 1260,6; 1267,31; 1277,26; 1301,24; 1303,28; 1305,23.31.33; 1333,1.2
epileipein, cease, 1341,33.35; give out, 1341,36; stop, 1356,30
epimenein, last, 1314,32; remain, 1344,28; 1349,7.9.16
epimonos, persisting, 1349,24; ***epimonôs***, permanently, 1288,21; 1310,32
epinoein, think of, 1333,23†; 1355,16
epinoia, thought, 1333,17†.21†.29†; 1334,18.19.21†.38†
epinoiai (dat. s.), notionally, 1283,35.37
epiphaneia, surface, 1268,34*
epipherein, add, 1284,2; assert, 1289,27
epirrhêma, adverb, 1340,32
episkeuastos, under construction, 1338,21
epistêmê, science, 1273,27
episumbainein, supervene upon, 1333,9†; 1362,22
epitêdeios, suitable, 1349,3; appropriate, 1355,5; ***epitêdeiôs ekhein***, be suitable, 1329,34†; 1330,13†; 1331,31; 1334,25; 1349,31
epitêdiotês, suitability, 1358,22†.29
epiteleisthai, take place, 1266,2; 1267,1; 1278,38; 1301,14; 1307,32; 1313,4; 1320,1; 1321,9; 1351,13
epitithenai, place on, 1366,21
epizeugnunai, join, 1308,22†
ergasia, process of work, 1338,3
ergon, work, 1337,23†; 1338,3; 1360,36†; 1361,3†
êremein, be at rest, 1257,12*.14.18*(bis).19.21; 1258,9; 1263,2*.33; 1264,1*(bis).2.10; 1274,33*.33.37; 1275,3.9.19.20.21(bis).24.36*.37*; 1276,6*.10; 1281,14; 1282,15.20.25; 1283,15.18; 1284,10.11.15.24.28.35; 1285,12.25†.27†.28†.29†(bis).33.35; 1286,1.3; 1287,1; 1296,38; 1297,7; 1301,11(bis); 1302,17; 1303,20; 1304,21.28*.40; 1305,4.18; 1309,3.8; 1310,11.12.28.36; 1311,20; 1312,2; 1315,15*.34.35.37; 1316,1*; 1317,23*.24; 1320,12; 1350,18*.21; 1357,24†.28†; 1364,16.18(bis).20.21.24.25; be in a state of rest, 1276,12; 1278,4
êremêsis, rest, 1361,15*
êremia, state of rest, 1257,16; 1274,29; 1275,17.27.29; 1276,7*.7.8.10.12.14.16*.16.19.20*.26.29*.31.38†; 1277,4†.5†.7†.8†; 1286,11; 1288,12; 1304,29*.38*; 1305,6.9.12.16; 1306,9.35; 1310,15; 1311,30.35; 1312,11; 1313,12; 1315,38; 1316,1; kind of rest, 1304,14.16.18*.21*; period of rest, 1317,26; 1350,22; rest, 1364,15; 1365,39
êremisma, period of rest, 1311,32
erôtan, propound (an argument),

1288,37*; put (an argument), 1289,14; 1290,8.11.17.19; ask, 1291,28*.31*; 1296,32; ***to erôtan***, the question, 1289,37*
erôtêsis, question, 1289,36; 1290,9
eskhatos, last, 1298,13*.34*; 1335,6; 1346,23.24.26.27.28; 1350,35; 1363,34; ***pro tou eskhatou***, penultimate, 1346,22
eskhaton, extremity, 1268,16*; 1300,16; 1307,2.3.4.6†; 1335,9; furthest part, 1331,5
ethos, practice, 1301,18.23; 1318,10.15; 1366,8
eudiairetos, easy to divide, 1317,35; 1356,12
eukinêtos, easily moved, 1352,16†; 1356,12
eulogos, reasonable, 1254,12.14.32; 1264,18; 1298,6; 1365,1.16
eupatheia, susceptibility, 1347,8†
eupathês, susceptible, 1347,29†
euphuia, disposition, 1349,7
euthus, straight, 1308,22†
euthus, ***euthu***, immediately, 1252,3; 1255,11.14.22; 1320,26; 1327,15; 1329,15; 1334,33; 1335,22; 1337,10; 1347,11†; 1348,14†.15†
euthus, as soon as, 1277,30*; 1344,35; 1346,35; right away, 1303,3; straightforwardly, 1343,2
euthu, right from when, 1303,2
eutheôs, immediately, 1310,8
eutheia, ***(hê)***, straight line, 1278,14 etc. [107 occurrences]; ***ep' eutheias***, in (on) a straight line, 1266,7; 1278,6.8; 1279,35; 1288,24; 1302,34; 1307,32; 1308,21†.24†.27; 1309,25; 1310,13; 1312,36; 1315,36; 1365,24.35; rectilinear, 1273,19; 1278,8.13.26; 1281,14; 1304,13; 1305,23; 1308,30; 1309,2; 1310,34; 1311,15.16; 1313,26*.28.33.34; 1314,1.10.15.22.23.39; 1315,1.8.9*; 1317,15.16*; 1365,39; ***kat' eutheian***, rectilinear, 1273,31.33; 1278,14.20; 1279,5.12; 1301,5.7. 12.21.28; 1302,29*; 1303,36; 1304,23.25*.31; 1307,16.24; 1309,4.20.21.25*; 1316,3; 1320,32
exairein, take away, 1295,22
exairetôs, by the special term, 1358,7
exêgeisthai, expound, 1288,4; 1292,11.16; 1319,11; 1335,2; 1361,32; explain, 1304,34
exêgêtês, interpreter, 1329,7
exêirêmenos, transcendent 1320,35; 1330,14†.22; 1333,25†; 1355,2.25†.26†.31
exetasis, examination, 1330,2; 1336,34
exetazein, examine, 1262,4; 1352,17
existasthai, depart, 1316,22
exôthein, expel, 1350,32.33(bis).34(bis).36.38; 1351,1.2(bis)
exôthen, external, 1257,2.27.33; 1258,11.25*.31*.33.36.39; 1259,2.6.16; 1354,32†; from without, 1333,9†; 1344,23

gê, earth, 1259,22; 1338,5; the earth, 1357,26†.28†
genesis, generation, 1250,36; 1251,5.19.34; 1252,4*.15.19.20.21; 1253,22.39; 1259,4; 1262,24.26.28.29(bis); 1263,3.7.14.37; 1266,14.31*.32.34.38; 1267,1.3*.4.7.27; 1270,22*.25.27.27*.32.34(bis); 1271,2(bis).4.6.8.9.10.11.15*(bis).17. 21.27*.32*; 1272,2.5*.11*.22; 1273,5.11.35; 1274,3.9*; 1275,6.26.38; 1276,11; 1277,18*.19.27.29; 1278,28; 1279,2; 1297,17.19; 1302,13; 1304,3.12; 1306,4.10.13; 1311,28; 1312,3*.22; 1313,1.2*.9; 1314,22; 1318,31; 1319,18.23; 1328,16; 1330,36; 1331,5.7(bis); 1332,38(bis); 1339,15; 1352,12†; 1359,38.39; 1360,20; 1363,16.19; 1365,13; coming to be, 1295,25; 1297,31; 1328,15; 1336,9; 1359,18†.31; 1360,32†; 1363,22; birth, 1360,8; ***en genesei***, subject to generation, 1263,10; 1270,13.18; 1271,14.26*.29; 1272,8*.13*; 1294,1*
genêtos, subject to generation, 1253,9.11.28; 1258,1; 1263,9; 1270,9.11; 1271,7.12; 1327,2;

1331,27; 1339,6.15; 1359,16; 1360,11; 1363,14
gennan, generate, 1266,36; 1270,25.37*; 1272,2; ***to gennêsan***, the generator, 1270,24*; ***to gennômenon***, ***to gennêtheis***, the generated, 1270,24*.25
gennêma, offspring, 1340,28
gennêtikos, generative, 1262,34
genos, kind, 1268,14; 1361,3†
gignesthai, come to be, 1252,37 etc. [296 occurrences]; be generated, 1252,24*.24; 1253,15.16.21.25.38; 1255,3.9.11; 1261,2†; 1262,16.30; 1263,6.32; 1264,9; 1270,15.20.22.23.25.28*.31.31*.32. 35.36; 1271,5.6.34(bis); 1272,4; 1274,20; 1275,13(bis).14; 1277,30*.31; 1312,37; 1319,4.17; 1321,6; 1337,11.19; 1340,27; 1359,35.40; 1360,2; 1363,15(bis).18.23(bis).29; 1364,6; take place, 1252,30; 1259,4; 1260,28†; 1280,12.21; 1286,18; 1297,14; 1302,15; 1304,1; 1306,15.36; 1307,33; 1308,3.8.21†.31; 1309,6; 1312,8; 1314,11; 1316,17.23; 1323,30; 1335,12; 1343,8; 1345,36; 1350,36; 1351,5; 1357,11; 1364,4; become, 1253,22; 1257,16*; 1264,3.27; 1267,25†; 1273,8†; 1274,35.36.37; 1280,36; 1281,24.25.26.28.32.33; 1283,33(bis); 1293,5; 1295,25.26*.28; 1300,6†; 1304,2; 1305,38*; 1311,5; 1318,4(bis); 1332,9; 1347,6†.17†.22†; 1348,5†; 1349,9†; 1350,3†; occur, 1255,20; 1257,16; 1265,12.16.35; 1267,13; 1270,19; 1271,13; 1273,12; 1274,39; 1275,12.29.30.40; 1289,16; 1293,27*; 1294,9; 1295,3; 1297,6; 1300,9†; 1304,15; 1312,16; 1316,20; 1321,6; 1323,20; 1346,37; 1350,19*.28; 1351,12.14.29.39; 1352,14; arise, 1258,7; 1265,7; 1274,17.22.23; 1290,4; 1301,24; 1361,11; turn out, 1279,9; 1306,14; 1310,6; turn out to be, 1274,24; 1307,9†; 1312,27; 1349,19; 1354,32†; 1356,11; originate, 1303,17; 1304,23; prove to be, 1348,20; involve, 1260,27†; 1311,24; proceed, 1273,29; start, 1303,5; is based, 1305,32; undergo, 1311,27; go, 1312,13; gets, 1317,34; generates, 1321,5; made, 1335,14; is, 1309,10; 1356,19; have, 1362,25*; ***to gignesthai***, generation, 1262,32; 1269,26; 1313,3; ***dêlon gegonen***, make it clear, 1327,26; ***gignomenon***, generation, 1331,4; ***to genomenon megethos***, the magnitude that results, 1332,11; ***kinêtikon gignesthai***, to cause motion, 1350,30
glukus, sweet, 1266,26
gnômon, gnomon, 1265,28
gnôrimos, intelligible, 1343,32†
gnôsis, knowledge, 1268,28*
gnôstikos, cognitive, 1277,33
Grammatikos, grammarian (referring to Philoponus), 1326,38; 1358,26.39
grammê, line, 1268,34*; 1273,20; 1283,6.8.21; 1284,4; 1285,13; 1290,22; 1291,36†.38†; 1292,2†.4†.5†.6†(bis).7†.8†.9†.13*. 16†.19.20.22.23*.33.36.39; 1293,2.7*.9.10.16; 1308,22†.23†.25†; 1311,9.24; 1314,31.34; ***hê apo tou kuklou grammê***, radius of the circle, 1316,14
graphê, reading, 1288,3; text, 1317,3; 1319,9; 1345,9; 1355,34
gumnazein, train, 1288,38; put to use, 1302,10

hama, simultaneously, at the same time, 1252,36* etc.; see also ***holos hama***
haplous, simple, 1264,7*; 1301,6; 1303,23; 1313,28.32(bis).33; 1316,32.35.38; 1317,1; 1332,8; 1333,21†; 1338,8.9.24.25; 1364,28; 1366,4; ***haplôs***, without qualification, 1250,37; 1251,14; 1258,27; 1268,22*; 1307,30; 1320,11.14*.15.16; 1326,4; 1348,27; 1364,26; simply; 1339,36; 1358,23†; unqualifiedly, 1349,19
haptesthai, touch, 1268,4†; 1345,15;

1355,29†.30; 1357,20†.22†(bis).25†; 1362,5; be in contact, 1344,27*.28(bis); 1345,6*.16; 1346,8; 1350,23.24
haptos, tangible, 1338,5
harmonia, harmony, 1338,16
harmottein, apply, 1290,12; 1363,19
harmozein, apply to, 1305,34; fit together, 1331,16; 1337,24†; 1338,14.15
hedra, place, 1352,16†
hedrastikos, that establishes, 1355,6
hêgeisthai, govern, 1267,2; consider, 1290,17; 1319,7; 1330,9†; 1362,16; 1363,9; ***to hêgoumenon***, the antecedent, 1254,5; 1255,5; 1277,21; 1279,11; 1284,27
heirmos, train of thought, 1257,6
hêliakos, solar, 1331,19†
hêlios, sun, 1263,19; 1270,37*; 1330,11†.29
helkein, pull, 1317,20.21; 1339,35; 1356,6.8.19.20; 1357,20†; 1364,29; attract, 1345,32(bis)
helkuein, attract, 1345,16.17; 1346,7
helxis, pull, 1356,18
hêmikuklios, semicircle, 1310,30; 1311,23
hêmisu (***to***), the halfway point, 1303,17.18.20(bis)
heniaios, unitary, 1337,30
henôsis, unification, 1338,12.22.23.25
henousthai, be unified, 1338,11
hepesthai, follow, 1255,35; 1289,31; 1297,4; 1298,23; 1302,29; 1303,26; 1307,16.18.20; 1309,26; 1329,18; 1357,9†; be a consequence of, 1279,24.28; 1282,9; 1289,19(bis); 1302,34; 1325,30; come after, 1296,25†; ***hepomenos***, subsequent, 1356,21; ***to hepomenon***, the consequent, 1255,5; 1284,14; 1289,11; consequence, 1279,23.32; 1294,27; 1295,18; 1324,26; 1325,37
hêsukhazein, slow down, 1302,17
heterokinêtos, moved by something else, 1253,36(bis); 1339,6.12.19; 1363,3
heuresis, discovery, 1273,28
heuriskein, discover, 1255,33; 1257,10; 1273,20.28; 1359,20; 1364,37.38; find 1257,31; 1273,23; 1317,3.7†; 1355,34; 1362,31
hexês, in sequence, 1299,1; following, 1321,35; ***to hexês***, the latter, 1295,25; the sequel, 1346,2; ***en tois hexês***, in what follows, 1261,9; ***kai tôn hexês***, 'and following', 1272,10; 1291,36†
hexis, state, 1273,10†; 1277,6†.10†; 1333,33
hidrusis, settling, 1351,5
hidrusthai, be established, 1339,39; 1353,15; 1355,4; 1359,22
histanai, stop (trans.), 1280,3.7; 1291,7
histasthai, come to a stop (pres.), stop (fut., aor. and plupf.) (intrans.), 1257,33; 1280,5.23.25.26.34*; 1281,1.5.7.9.11.13.15.16.19; 1282,2.12; 1283,9.10.11.26*.27*.31; 1284,21; 1287,20; 1288,18.21.22.25.25*.27; 1309,25.27.29.36.38; 1310,2.4(bis); 1328,29; 1365,26; stand, 1350,34
hestôs, stationary, 1267,24†; 1273,7†
histanein, cause to stop, 1257,34
hodêgein, guide, 1318,17
hodeuein, proceed, 1271,34; head, 1335,8
holikos, complete, 1336,29.30(bis)
holos, whole (adj.), 1269,3*; 1288,20; 1291,19; 1295,15; 1296,6; 1299,24*.29; 1300,5†; 1304,35*; 1307,28(bis); 1315,25; 1321,10; 1322,10(bis).21.26.37; 1327,17†.20.23; 1328,23; 1329,13; 1336,3†; 1343,29.37†; 1355,15.38; entire, 1279,11; 1304,31; 1310,16; 1313,14; 1329,4; 1330,26; 1333,4; 1344,28; 1357,10.12.14; 1359,6; 1360,36; 1363,8; in their entirety, 1260,29†; as a whole, 1268,32*; 1348,33; 1360,22; whole amount of, 1332,22†; entire amount of, 1332,33
kath' hola, as a whole, 1261,28; ***holos hama*** (also ***hama holos***), all at once together, 1327,30.32.34.38; 1328,4.8.20.23.27.28.29.31; 1329,11.32; 1331,32; 1332,5.7; 1335,19; 1336,6.9; 1337,13;

1338,20; 1339,6.8.9; 1340,6; 1358,32; 1359,4
holon, whole (as a noun), 1260,24†.27†.32†.35†; 1261,27(bis); 1268,31*; 1269,3*; 1290,2.3.4; 1307,7†; 1320,18; 1326,34†(bis); 1332,31; 1334,7(bis); 1335,26†.27†.31†.35†.36†.38†; 1336,1†.7.11.12(bis).14.16.18.19.20. 21.26.28.32; 1348,18(bis).19(bis); 1355,25†; 1365,8(bis).9; ***ta hola***, the universe, 1337,30
holôs, in general, 1251,34; 1252,1; 1257,34.39; 1270,23; 1280,12; 1324,6*; 1328,12*; 1359,35; 1361,16*; at all, 1252,16; 1259,4; 1269,8; 1270,36; 1271,31; 1286,8.13; 1291,28; 1296,31†.35; 1307,26.29; 1309,25; 1328,28; 1323,33; 1324,2.27; 1325,34; 1326,13; 1340,14.19; 1344,9; 1350,25; generally, 1268,12; 1270,23; 1278,39; 1302,12; 1311,24; entirely, 1272,9*; 1289,13; as a whole, 1332,20†
holotês, total amount, 1332,3.15†.26†.40; 1336,22.24; integral state, 1336,16
homalês, uniform, 1317,12*.14.30; 1353,16.17.22.23.26.27; 1356,22.28; 1357,1; 1366,7; ***to homales***, uniformity, 1317,16
homalos, uniform, 1352,4†
homalotês, (state of) uniformity, 1351,40†; 1352,5†; 1353,21
homoeidês, of the same kind, 1306,35; 1343,17.20.22.37†
homogenês, kindred, 1343,16.36†
homoiomereia, homoeomery, 1318,29
homoios, similar, 1277,23; 1317,36; 1352,7; 1353,26*; 1354,25†; 1356,29; like, 1268,6†; 1334,40; ***homoiôs***, in a similar way, 1260,31†; similarly, 1261,19; 1263,21; 1264,9; 1265,33; 1285,8; 1289,24; 1298,7; 1324,33; 1330,4; 1344,32*; 1356,28; like, 1266,26; likewise, 1275,34; 1281,29; 1292,10†; 1302,8*.11.27; 1308,5; 1312,16; 1343,20; 1350,30; 1366,4; equally, 1294,10; 1336,11.26; 1352,24; in the same way, 1284,2; 1289,29*; 1290,10.13; 1326,35†; 1346,13; 1357,26†; evenly, 1317,33; as, 1317,36; just as, 1327,1; just like, 1341,36; ***homoiôs ekhein***, be in a similar state, 1264,6; be similarly related, 1347,26†; 1356,27*.29*; stands in the same relation, 1353,18; be similar, 1277,23*; 1356,25*; be related in the same way, 1265,32*
homoiotês, similarity, 1277,24
homologein, agree, 1262,24.28; 1264,14; 1303,9; 1313,31; 1324,19; 1336,28; 1343,12
homologoumenon, for granted, 1321,34
homônumôs, homonymously, 1327,38; 1358,20
horatos, visible, 1338,4
horismos, definition, 1274,15; 1306,38; 1363,33
horizein, mark out, 1253,24; 1308,6.18; 1311,30.32; 1312,10.22; specify, 1281,23; 1300,8†
horizesthai, define, 1293,3.4.5; 1359,11; 1362,17
hôrismenos, determinate, 1268,14.26*; 1302,2.3; 1312,21; 1341,25; 1342,10.11.12.13.21.22. 23.24.32*; determined, 1268,16*
horistheis, definite, 1342,6
horman, set out, 1267,31; 1285,14*; 1299,36†; 1303,29; 1309,7
hormê, impulse, 1257,1; 1258,6.27; 1260,1; 1302,1.4.17.18.20.36; 1307,23; 1310,9; 1312,15
horos, limit, 1274,8.9*; 1277,33; boundary, 1294,11; 1295,11; 1307,1.2
hôsautôs, in the same way, 1261,4; likewise, 1312,4*; unvaryingly, 1337,17; ***hôsautôs ekhein***, be unvarying, 1330,39; 1337,1; 1339,2; stand in the same relation, 1353,28; be in the same condition, 1359,11.22.32; 1360,23; 1361,10
hou heneka (***to***), (that) for the sake of which, 1271,35.37; 1277,31
hudôr, water, 1332,10.12(bis).13.14.18†.19†.20†.22 †.30(bis).33.35(bis).37; 1335,15.38†; 1336,22; 1345,28*.34.36;

1347,17†.18†.21†(bis).32†; 1348,31; 1349,30; 1350,19*.28*; 1356,12*
hugiês, sound, 1254,3; 1267,1; 1300,11
hugros, moist, 1352,16†
hugrotês, moisture, 1258,22
hulê, matter, 1269,3; 1329,20.21(bis).24.26.33†.35†.36†.37†(bis).39; 1330,4.7†.9†.10†.13†.15†.17†.19.20.21.23.24.27.31.35; 1333,7†; 13†; 1361,28*
hulikos, material, 1361,35
hupagein, get, 1326,39
hupantan, confront, 1258,5; 1289,32; 1327,10
huparkhein, be, 1251,24; 1252,26; 1294,25; 1338,5; be present, 1254,18*; hold (abs.), 1256,9; 1277,24.25.27; belong to (w. dat.).1256,16; 1261,22*.29; 1270,5.6.29; 1271,13.26*.29.32; 1272,8*.12*.15*.17; 1273,32; 1276,2; 1277,12†; 1292,35; 1293,2.9; 1301,15.26.29; 1312,31; 1327,39; 1328,14; 1339,19; 1350,21; hold of (w. dat.), 1277,29; 1320,5; 1336,26; be an attribute of, 1262,7; 1279,22; 1327,35; 1334,10; ***x huparkhei tôi y***, y has the attribute of x, 1319,14; 1327,32.40; ***hama huparkhein***, coexistence, 1276,1*; ***to huparkhon***, attribute, 1268,5†; 1305,25.26(bis).34
huparxis, existence, 1292,27
hupeinai, underlie, 1320,20
huperanekhein, be beyond, 1251,24; 1254,26; 1359,37; 1365,14
huperballein, exceed, 1326,17; 1335,35†; 1341,38; 1342,3.22(bis).24.31*.32; be greater, 1342,20
huperbolê, excess, 1276,24
huperekhein, exceed, 1276,23*(bis); 1322,11; 1341,39
huperphuês, above nature, 1359,6; 1366,19
hupertithenai, defer, 1258,7; 1264,31
huphesis, descent of level, 1337,10
huphistanai, cause existence, 1337,7; 1361,1; ***huphistasthai***, exist, 1252,22; 1263,29; 1267,22.28; 1270,28; 1271,11; 1316,14; 1320,21; 1332,8.10; 1333,14†; 1338,26; 1339,21; 1340,2; 1360,2.22; undergo, 1337,2.8; be constituted, 1338,23; ***huphistasthai phthoran***, perish, 1332,17†
hupnos, sleep, 1258,17.19.28.40; 1259,33.37
hupoballein, suggest, 1359,40
hupobibazesthai, descend, 1337,11
hupodekhesthai, receive, 1338,23; 1349,32
hupokeisthai, be posited, 1252,26; 1253,31; 1269,34; 1332,18†; underlie, 1301,22; 1329,39; 1330,7†.35; ***to hupokeimenon***, subject, 1268,36*; 1269,1; substrate, 1287,37; 1310,18; 1329,26.27; 1330,9†; 1331,19†.35
hupolambanein, suppose, 1269,32*; 1285,16†; think, 1362,6*
hupoleipesthai, be left over, 1323,3.22
hupomenein, persist, 1294,1; 1336,17; undergo, 1329,6; 1333,30†; 1337,10; keep, 1349,22
hupomimnêiskein, remind, 1263,34; 1289,3; 1320,24; 1348,23; recall, 1277,4†; 1278,30; 1302,29; 1313,16; 1352,22; 1357,32; 1358,3(bis)
hupomnêma, comment, 1327,36
huponoein, suppose, 1326,3
hupostasis, subsistence, 1337,16; 1359,12.25; 1363,3
hupostatês, author, 1327,5
hupostatikos, constitutive of, 1363,6
hupothesis, hypothesis, 1256,11.19; 1273,24; 1286,3; 1306,11; 1332,30.33; 1335,34†; 1341,22
hupothetikos, hypothetical, 1254,33; 1272,15
hupotithenai, hypothesize, 1251,29; 1252,13; 1255,38; 1256,21*; 1260,33†; 1263,37; 1273,21*; 1285,24†.25†; 1293,11.12; 1295,9.13; 1307,17; 1322,4.8; 1324,28; 1325,8†.16†.36; 1331,34; 1332,34; 1342,37; 1351,7.12; 1360,4; 1365,15.21; 1366,14; posit, 1254,21; 1266,35.36; ***to hupothemenon***, the hypothesis, 1326,11
hupsêlos, high in the sky, 1263,21
husteros, posterior, 1255,39; 1266,6*;

1268,15*.25*; 1269,4; 1271,16.27*.33*; 1272,5*.11*; 1279,25; 1291,18; 1313,28; 1362,27*; ***husteros*, *husteron***, later, 1258,8; 1265,23; 1270,13; 1273,35; 1294,13(ter).17.20. 23(bis).28*.29.31*.34*.35; 1295,1*.2(bis).9.10.18; 1297,16; 1363,36; 1364,1.7.8; 1365,17.30; ***husteros***, after, 1272,9; 1363,35; last, 1363,35; ***husteron***, below, 1269,33; next, 1343,9; afterwards, 1363,30; ***to husteron***, posteriority, 1268,13; 1269,6.7

idiai (dat. s.), separately, 1310,15; 1342,29
idion, property, 1281,14; 1344,35
idiôs, specifically, 1261,7; 1268,3†; 1272,32; 1297,11
idiôtês, layperson, 1326,39; special nature, 1338,17.30
iskhuros, strong, 1286,11
iskhus, strength, 1358,21†.25†
ison, equal, 1276,22*.32; 1317,33; 1322,25; 1323,19.22; 1325,12†; 1326,32†; 1335,4; same, 1283,19; equivalent, 1282,34; 1283,29; 1294,34; equally, 1285,21†; 1316,15.20; ***ison aphestanai***, be equidistant, 1317,29
isotakhês, having equal speed, 1284,36; 1285,6.29†; 1287,21.22; ***isotakhôs***, with equal speed, 1284,25.28.30; 1285,19†

kainôs, unusually, 1295,3
kamnein, grow weary, 1255,24(bis).25; 1259,22.25; 1358,21†.26†
kamptêr, turning point, 1278,26
katadekhesthai, receive, 1328,27.29; 1329,35†; 1330,13†.14†.32; 1332,12; 1337,14; 1338,27; 1339,7; 1346,18
katageometrein, geometrize, 1341,20
katakolouthein, follow, 1273,1
katalambanein, overcome, 1258,21; occupy, 1267,10; 1351,2.4.6.9.26; comprehend, 1333,29†; 1334,38†
katalêgein, end up, 1311,9
kataleipein, leave, 1331,21†; 1343,16.30
katamanthanein, learn, 1327,11
katametrein, measure out, 1322,39; 1323,1.18.19.24.25.27; 1332,21†; 1343,20.22.23*.26.27.28(bis).29. 32.34†
katanaliskein, exhaust, 1322,21*.38(bis); 1323,2.6*.8*.17*.22
katantan, arrive, 1312,12; 1364,35
kataphasis, affirmation, 1274,9; 1279,29
kataphatikos, affirmative, 1253,8
kataskeuastikos einai (w. gen.), establish, 1279,15
kataskeuazein, establish, 1254,5; 1263,35; 1274,31
kataskeuê, supporting argument, 1272,6.8; 1321,24
katastasis, condition, 1272,21
katatemnein, carve up, 1332,10; 1334,18.22
katêgorein, predicate, 1299,10†
katêgorikôs, categorically, 1256,2; 1271,25
katekhein, occupy, 1260,34†; 1316,3; 1354,27
kath' hauto, per se, 1251,17.22.24; 1259,31; 1260,9; 1261,17; 1262,9.19; 1282,34; 1292,5†(bis).8†.25.30.32; 1301,15.26; 1305,25.26(bis).29.33; 1310,23; 1334,6; 1348,32; 1352,27.34; 1353,7; 1362,22.27*; 1364,39; 1365,2; in its own right, 1340,20; ***auto kath' hauto***, all by itself, 1355,1
katharotês, purification, 1338,12
kathêgemon, teacher, 1267,20; 1336,36; 1360,29; 1363,9
kathekasta, particular, 1268,30*
kathestêkos, existing, 1273,10†
katheudein, sleep, 1258,20; ***katheudon***, asleep, 1258,29
katholikos, general, 1273,28; 1297,10; 1301,22.33; 1304,2.22; 1305,32; 1365,34
katholou, generally, 1281,4; 1291,27; 1304,6*; 1313,5*; 1324,6; 1365,3.36; in general, 1341,38; ***ta katholou***, universals, 1268,30*

katô, downwards, 1259,22; 1279,6; 1302,3.17.19; 1303,4.9; downward, 1274,23.26; 1349,31; beneath, 1276,10; 1277,8†.9†; down, 1278,25; 1279,36; 1280,13; 1288,15; 1308,5.11.20; 1330,28; 1352,10†; ***to katô***, the bottom, 1281,36; 1302,4.5.19.20.21.22.31.36; 1303,4.7.22.23; 1304,25.26.36.39; 1317,26.27; ***ta katô***, things that are down, 1330,29
katôthen, from below, 1279,6; 1288,15; 1302,16.18.35; 1303,3.5.13(bis).18
keisthai, be posited, 1254,19; 1283,15.33; 1285,17†; 1295,38.40; 1296,3; 1297,4.20.24; 1298,22; 1306,11; 1313,25; 1322,35; 1324,15.36; 1326,34†; ***ep' hois keimenois***, on the basis of which, 1320,25
kenon, void, 1318,32.35; 1319,6.6*; 1351,32; 1352,15†
kentron, centre, 1316,13(bis).18; 1317,30; 1355,3.5.9; ***hê ek tou kentrou***, radius, 1308,18
kephalaiôdôs, in summary fashion, 1366,18
kephalaion, main point, 1284,23; 1348,1; 1363,27; 1366,10
khôra, place, 1265,37; 1344,16; 1351,26
khorêgein, bestow, 1334,26; provide, 1341,33
khorêgos, bestower, 1328,25
khôristos, separate, 1251,23; 1336,37; 1355,1
khôrizein, separate, 1259,19; 1267,13; 1273,36; 1333,13†; ***kekhôrismenos***, separate, 1262,2†; 1333,22†; 1334,4
khreia, need, 1252,3; 1332,28
khrêsthai, treat, 1268,20*; employ, 1268,36*; 1289,27; 1317,6†; 1318,14; 1348,34.35; make use of, 1276,26.37†; 1297,12; use, 1281,5.11; 1282,14; 1283,24*.26.31.35.36.37.39(bis); 1284,3; 1286,33*; 1287,35; 1288,11.36; 1290,25*.27; 1291,21; 1299,7†; 1309,7; 1323,17; 1324,8; 1363,20; be made up of, 1336,25; ***hôs hômologoumenôs khrômenos***, takes for granted, 1321,34
khronikos, temporal, 1359,40; 1360,3.4.10.12; 1363,15
khronos, time, 1267,37* etc. [216 occurrences]; time interval, 1274,38 etc. [186 occurrences]; ***en khronôi***, temporally, 1339,32.35
kinein, move (tr.).cause motion, impart motion, produce motion, make something move, 1251,7 etc. [291 occurrences]; (***to***) ***kinoun*** (or ***kinêsan***), the mover, what causes motion, what moves, thing that moves, the thing or element that causes (imparts) motion, motion-causing element, 1251,10* etc. [178 occurrences]; ***ho kinôn***, mover, 1348,8*; 1349,12*; ***to heauto kinoun***, something that moves itself, 1257,29; (***to***) ***prôtôs kinoun*** (or ***kinêsan***), (***to***) ***kinoun prôtôs***, (the) primary mover, 1250,34.35; 1251,3.20.23.28; 1254,4.6.12.31; 1255,4; 1256,33; 1259,10.17; 1262,18; 1264,8.14.26; 1265,7; 1320,34; 1321,3.12; 1323,37; 1324,5; 1328,18.26; 1336,6; 1338,34; 1339,26.30.32.37; 1340,12.14; 1344,6.12.15; 1345,21.22*.27; 1353,32.37; 1354,35; 1355,12.18.21†.29†.33.37; 1357,14.27†.31.33.36; 1358,4*.16; 1359,29; 1360,20.24; 1361,12.18.25; 1362,14†; 1364,37; 1365,10.12.18; 1366,11; ***prôtôs kinein***, be the primary mover, 1357,28†; cause motion primarily, 1357,29†; (***to***) ***prôton kinoun*** (or ***kinêsan***), (***to***) ***kinoun prôton***, (the) first mover, 1255,14.22; 1259,8*; 1260,10.36; 1263,12; 1264,23; 1265,4*; 1268,22*; 1269,27*; 1319,34; 1345,1*.13*.34; 1346,13; 1348,28.35; 1350,15; 1358,13*.15; 1361,32†; 1365,16; 1366,18; ***to prôton***, the first mover, 1255,10
kineisthai, be moved, undergo motion, be in motion, move (intr.), have motion, 1251,2 etc. [552 occurrences]; ***to kinoumenon***, ***to***

kekinêmenon, what is (has been) moved, moving thing, what moves (has moved) (intr.).what undergoes (has undergone) motion, thing that is (has been) moved, the moved, 1251,1 etc. [338 occurrences]; moving part, 1354,20; ***to prôtôs kinoumenon***, the primary thing that is moved, 1262,21; 1264,15.25.27; 1338,35; 1339,33; 1340,5; ***to huph' heautou kinêthen***, 'the thing that it moved', 1350,15; ***epi tôn kinoumenôn***, 'when things move', 1351,30
kinêsis, motion, movement, 1250,34 etc. [616 occurrences]; kind of motion, 1264,16.18.19.20.30*.32*; 1265,14.17; 1266,4*; 1267,32.35; 1269,15.25; 1270,3.5.7.17.18.27*; 1271,1.3.5.6.10*.14.16.17.28; 1272,4.18.29.36.37; 1273,17.29.33.34; 1274,1*.6; 1275,7; 1277,36; 1295,4; 1302,9.11; 1304,6*.11*.15.17*.19*; 1305,11.14; 1307,19; 1312,30; 1313,5*.22.24.25.30; 1314,18.39; 1315,3; 1318,15.18.21*; 1319,3.8.19.22.32; 1320,10; 13.30*; 1356,9; 1365,23; 1366,1.9
kinêtikos, (something that) causes motion, 1251,18*; 1252,22; 1253,31.32.37; 1255,27; 1260,5.30†; 1327,30; 1328,8; 1346,12; 1350,30; 1358,38; 1363,5; motion-imparting, 1254,26; 1255,30; 1258,39; motive, 1325,6; 1327,25.33.38; 1328,35; 1329,10.15; 1331,32; 1345,19.25; 1351,11; 1357,7†; 1359,3; 1362,12; ***to kinêtikon***, mover, 1260,24†; 1262,21; 1344,23; 1357,8†; motion-imparting element, 1259,31; 1260,7.8; 1261,34†
kinêtos, subject to motion, 1254,1; 1305,11; 1329,11; 1331,32; 1336,7; 1347,27†; 1358,38
koinopoiein, generalize, 1275,6
koinos, common, 1250,37; 1274,13; 1294,11.22.30*; 1295,5.11.34.35; 1296,11; 1298,10; 1301,22; 1305,23.32; 1306,38; 1307,1; 1317,5†; general, 1365,33; ***koinôs***, generally, 1261,9; 1282,36; 1334,10
kosmikos, of the world, 1327,14; 1338,15; 1359,9
kosmopoiein, engage in cosmogony, 1360,6.10
kosmopoios, world-maker, 1331,2
kosmos, world, 1260,18; 1261,10; 1264,2; 1302,33; 1327,1.3.7; 1329,4; 1330,34; 1331,5; 1333,4; 1337,29; 1353,32; 1355,2.15; 1359,10.21.23; 1360,26.30.33; 1361,4; 1362,3*; 1363,10
kouphos, light, 1266,22*
kouphotês, lightness, 1343,19; 1347,33†
kreittôn, stronger, 1330,14†.22; mightier, 1331,10.24†; superior, 1363,4
kritês, judge, 1280,36
kuklikos, circular, 1280,31; 1307,31; 1310,32
kuklismos, circularity, 1280,33; 1311,3
kuklophoreisthai, revolve, 1279,4
kuklophorêtikos, that undergoes circular motion, 1253,38; 1255,31; 1258,18.33; 1264,23; 1328,40; 1354,29†; what undergoes circular motion, 1264,25; that is in circular motion, 1264,28; 1270,26; 1335,21; 1338,36; 1357,14; of circular motion, 1331,37; in circular motion, 1353,29; ***kuklophoritikê ousa***, undergoes circular motion, 1339,21
kuklophoria, circular motion, 1256,23; 1264,21; 1266,7.8; 1278,5; 1307,17.20; 1312,31; 1313,19.24.29.30; 1314,20.28.33; 1316,26.28.31.37; 1317,5†.10.16; 1320,32; 1338,39; 1365,26.38; 1366,1.3
kuklos, circle, 1279,37; 1280,2(bis).5.24.28.31; 1308,1(bis); 1309,6.22*; 1311,7.9(bis).12; 1315,18.25.26.28.30.34; 1316,11.13.14.15.16.17; 1317,31; 1354,8*; 1355,38*; 1357,16*.16
kuklôi, circular(ly), 1260,21.25†.29†; 1261,28; 1273,19.31.33; 1280,29*(bis); 1281,17; 1301,5; 1308,30.35; 1309,5.19.20.22; 1310,7; 1311,2.10.16.17; 1312,33; 1313,6*.22.26*.33.34; 1314,23.38*;

1315,3.7.17.38; 1316,9.16; 1317,11.29*; 1331,38; in a circle, 1278,6*; 1307,21.25
kuklon, over a circle, 1280,31*; ***kuklon pheresthai***, undergo locomotion over a circle, 1280,27*.29*; ***ho loxos kuklos***, the ecliptic, 1332,39; ***ho megistos kuklos***, great circle (on a sphere), 1354,10.13; 1355,33
kuriôs, in the strict sense, 1257,2.5; 1260,38; 1264,8; 1267,4.34; 1269,30; 1273,36; 1275,5; 1279,22; 1282,34; 1299,1; 1303,37; 1320,11*; 1348,31; 1353,8; 1358,28; 1359,35.37; strictly speaking, 1258,14*; 1259,5.7; 1272,32*; 1281,2.22.39(bis); 1283,4; 1301,28; 1310,21; 1353,23; 1355,13; strictly, 1277,5†; 1278,12; 1310,20; 1353,27; 1363,19; ***kuriôteros***, more basic, 1260,7; more authoritative, 1337,27†; ***kuriôtatos***, most basic, 1315,9; 1354,5

lambanein, assume, 1253,5; 1254,16*; 1256,10.19; 1262,23; 1269,35; 1273,24; 1274,26; 1284,22.29.30.31; 1285,1.6.11; 1298,2; 1300,27; 1304,10; 1322,8; 1325,15†.21†.26; 1326,2.3.4.5.21.22; 1340,37†; 1341,21; take, 1281,22.27.33.38; 1282,9; 1285,13; 1287,3.7.8; 1289,39; 1290,24.31; 1291,8.14.16.20; 1293,25*(bis).31*; 1294,19.21.22.26.31.35; 1295,6.7; 1297,35; 1298,38; 1299,35; 1302,25; 1304,34; 1311,2; 1315,29; 1319,15; 1322,7.21.33.37; 1323,1.5.11.25; 1326,17; 1328,13*(bis); 1334,3; 1336,15; 1338,40; 1339,13; 1341,31.35.38.39; 1342,7.8.12.34*; 1343,16*.20.36†; 1345,23; 1346,31; 1347,1†.2†.4†.10†.14†.34†; 1348,3†; 1349,7; 1351,11; 1361,21*; 1363,34; take (as), 1283,36; 1291,10.24; 1293,19; use, 1287,6; get, 1320,28; 1333,36; 1363,12; reach, 1342,5*
lambanesthai, be covered, 1269,9; be occupied, 1281,9; ***labôn***, on the assumption that, 1278,5.14; making (this) assumption, 1278,17; ***eilêmmenos***, based on, 1301,26.28; ***ei houtô lambanoito***, if this assumption were made, 1325,12†
lêgein, conclude, 1300,2†; ***to lêgon***, the consequent, 1277,22; 1279,11
leipesthai, remain, 1327,17†; 1336,3†
lêmma, lemma, 1321,25
lêpsis, assumption, 1326,27
lexis, passage, 1292,12.14.24; 1340,8; 1346,2; 1347,38; 1362,14
lithos, magnet, 1345,14*.15.17.20*.23.24.30; 1346,7
logikos, dialectical, 1301,19*.19; 1305,24; 1365,32
logizesthai, reckon, 1313,14; take into account, 1332,35
logos, discussion, 1251,11*; 1254,2; 1257,9; 1258,7; 1320,24; 1328,18; 1351,35; 1353,13.21; 1354,3; 1362,5; 1366,15.16; account, 1251,33; 1252,2.23; 1313,14; 1323,24.26; 1331,11†.12†(bis).13†.15.23†.35(bis); 1333,18†.20†.27†.28†(bis); 1334,35.36.38†; 1338,13; 1346,13; 1349,25; 1351,36; 1354,39; 1357,31; 1361,29*; 1363,14; 1366,18; reason, 1254,20; 1320,20; argument, 1255,13.37; 1257,21.30; 1259,36; 1260,15; 1263,36; 1266,3; 1274,3.26; 1275,6; 1276,13.35†; 1279,23; 1280,35*.36; 1281,26; 1284,13; 1285,11; 1286,11; 1288,37*.37; 1289,3.6.20.33.34; 1290,7.8.17.19; 1293,18.20; 1296,4; 1297,12.18.36; 1299,7†.36†; 1301,16*.25; 1302,10.15; 1303,12.36; 1304,1.9.24.32; 1312,24; 1321,15.25; 1325,39; 1326,27; 1327,16.22; 1329,8; 1330,28; 1339,25; 1344,12.33.34; 1349,36; 1350,12.16; 1354,36; 1364,2.14.32; definition, 1262,10; 1268,26*.29*.30*.31*.32*; 1281,36; 1282,3*.11*; 1283,33.34*; 1292,20.37; 1294,25; 1305,20; 1314,19*.24*; 1329,37†; 1333,6†.7†.9†.12†.13†.16†.23†.24†; 1334,14; ratio, 1324,22.25; 1342,14.36; 1343,7; word, 1337,33; 1358,27; ***ho nun aporêtheis***

logos, the puzzle under consideration, 1357,6; ***ana logon***, analogous, 1282,28; 1283,20; 1285,20†.23†.26†.28†; 1365,27; ***kata logon***, in proportion, 1324,34; ***kata ton logon***, proportionally, 1322,17; ***ton logon poiein***, discuss, 1316,31; ***ekhein logon***, make sense, 1330,21
loipos, left, 1324,4; 1361,3†; last, 1335,17; ***(ta) loipa***, the rest, 1320,25; ***loipon***, then, 1256,10; 1272,38; 1278,2; 1291,27; 1293,8; 1300,15; 1304,23; 1314,16; 1320,24; 1326,24; 1344,1.16; 1346,15; 1352,21; 1364,10; 1365,3.10.21.37; 1366,7; next, 1274,16; 1340,8; future, 1296,27†.30†
loxos, ho loxos kuklos, the ecliptic, 1332,39
luein, refute, 1252,3; 1263,35; 1270,17; 1271,18; 1275,18; 1289,4; 1291,1; 1298,4; 1341,10.18; 1350,12.14; 1364,10; 1366,17; solve, 1261,20; 1263,35; 1284,16.21; 1287,19.23; 1290,21; 1296,19†.28†; 1299,10; 1344,29; 1345,23; 1346,3; 1348,12.17; 1351,11*; 1354,16.26; 1357,23; 1358,35; undo, 1301,12; 1331,16; 1337,24†.26†; 1338,14.21†
lusis, refutation.1271,21; 1289,4.26; 1290,9; 12.16; 1293,21; 1344,20; 1351,35; solution, 1286,6; 1289,36; 1344,29; 1348,1; undoing, 1331,11†.12†.15.23†.34.35; 1334,34; 1338,8.13
lutos, able to be undone, 1337,24†; 1338,7.11; 1339,15

malakos, soft, 1266,22*
mallon, rather, 1254,12*.15.18*; 1259,20; 1260,30†; 1266,5; 1274,12; 1302,7*.8*; 1325,38; 1334,17; 1337,28; 1342,29; 1352,2†; more, 1272,15*.16; 1276,2; 1277,28; 1280,19; 1281,9.16; 1283,17; 1303,4; 1304,6*; 1305,30; 1315,18*; 1316,21(bis); 1330,38; 1333,34; 1338,32; 1341,10*; 1365,5; especially, 1364,9; ***oute pollôi mallon***, much less, 1314,39; ***ek tou mallon***, *a fortiori*, 1277,26
manôsis, rarefaction, 1266,16*.17*.21.28*.30; 1319,23.24.25
manousthai, be rarefied, 1266,37
manthanein, learn, 1251,1; 1302,13; 1359,32; 1360,3
marturia, testimony, 1318,11.13.14; 1320,9
marturion, evidence, 1318,18
martus, witness, 1319,28
megalomerês, having large parts, 1335,39†
megethos, magnitude, 1252,17*; 1265,10*; 1267,5.9*.18; 1274,12*.13; 1278,16.39*; 1286,18; 1289,17.23.30; 1290,5.10.11.13.32; 1293,14.29*; 1301,14.22; 1312,2*.21; 1314,3.29; 1321,13.14(bis).16.17.18.21.23.29*.30.31(bis); 1322,5; 1323,33; 1324,1.2.4.7.12.34.35; 1325,27.37.40; 1326,6.29†.30†.31†.33†; 1328,16; 1332,8.10.11.17†.29; 1333,3.11†; 13†; 1340,11.13(bis).16.17.19.23.24*.34†.35†.36*.38; 1341,5*.7.11.13.14.15.16.19.21.24.26.28.30(bis).34.37.40; 1342,1.2.6.10.20.23.25.26.29.33*.36.37(bis).38; 1343,1.3.5(bis).6(bis).7(bis).14.18*.21.23.26(bis).27.28.30.31.33†.35†.37†; 1344,3*.8.9.10.11; 1352,11†.26*; 1358,4*.5.6*.16*; 1366,6.12(bis).14.15; largeness, 1274,12*; amount, 1341,6
meiôsis, decrease, 1258,16; 1265,11; 1266,13; 1267,13.23.28; 1271,12; 1272,23; 1273,12; 1274,14; 1278,28; 1302,12; 1304,12; 1311,27; 1312,2; 1313,7; 1342,4.14
meiousthai, become smaller, 1267,9.14.16.19; 1271,12; 1274,19; 1312,36; 1324,35; 1326,9.19; 1332,16†; 1342,7; 1346,14.19
meizôn, major (premise), 1270,11; 1272,5.11; 1279,16; larger, 1267,10; 1322,17*.18; 1325,18†.23†.37; 1340,33†; 1341,1.4.7.11*.12; 1342,12.18.19; 1343,10; 1352,9†(bis); 1366,6; greater, 1276,22.33; 1321,7.36; 1324,8.16.20; 1335,32†(bis).33†;

1337,27†; 1342,15.16; longer, 1299,27.29(bis); bigger, 1322,16; more, 1342,7
mêkos, length, 1289,37*; 1291,30*.30; 1292,7†.8†; 1308,4*.21†.25†
mellon (to), the future, 1268,19; 1363,38
menein, be permanently, 1260,17.20.21.24†.25†; 1262,1†; be permanent, 1260,36*.37*.37; 1261,4; remain, 1266,1; 1274,23.26; 1275,10; 1277,12†; 1282,14; 1290,31.34; 1292,20; 1299,9†; 1316,4; 1322,32; 1324,39; 1334,16; 1338,18; 1344,23.25; 1345,5.25; 1346,30; 1347,28†; 1349,5.19; 1350,14; 1354,21; 1355,20†; persist, 1273,10†; remain fixed, 1316,22
merikos, part-like, 1336,31; particular, 1363,19
meristos, divisible into parts, 1321,23; 1333,39
merizein, partition, 1291,6; 1334,5.12.15
meros, part, 1253,1; 1260,32†; 1268,32*; 1278,35; 1279,37; 1280,2; 1281,24; 1283,4; 1290,4; 1291,18; 1310,17.22; 1316,29; 1320,19; 1322,8*.18.22; 1323,4.9.11.12. 13.26; 1327,18†; 1330,26.35; 1331,4; 1332,35.37; 1333,39; 1334,3.6.8(bis); 1335,23.25†.26†. 27†(bis).30†.31.32†.33†.35†. 37†.38.40†; 1336,2†.3†.8.11. 12(bis).14.15.18.21.22.24.25.26.27. 28.32; 1348,19(bis).20(bis); 1355,5.26†; 1363,25; 1365,7.8.9(bis)
mesolabein, put in the middle, 1272,10
meson, middle, 1268,16*; 1280,37*.38*.39; 1281,1.2.8.20.21.22.23(bis).28.30*. 32.36.39.40; 1282,2.4.5.6.9.10.14; 1283,25*; 1285,30†; 1290,31; 1303,14.15.17; 1306,9; 1315,11.13.19.20.24.27.28.29*.31.36; 1316,5.10.12*.15.17.19; 1354,2.5.9; 1360,4; intermediary, 1365,11; ***ana meson***, in between, 1312,1*; ***dia mesês***, via, 1319,33
mesos, intermediate, 1257,21; 1259,10; 1267,25†; 1273,8†; 1349,29; 1363,37; in between, 1306,34
metabainein, move, 1260,15.19; 1270,19; 1316,7.12; 1351,4.24
metaballein, (active voice), change (intr.), 1255,26; 1258,3; 1264,5*(bis); 1265,23; 1267,8*.9.12; 1274,18.19.37; 1275,11.14.16.20; 1294,15(bis).19; 1295,39; 1296,28†; 1297,14; 1299,6†.9†; 1300,13 (ter).14.15(bis).16(bis).19.21.22.26. 28.32.35; 1301,27.29; 1302,12; 1303,1; 1304,4.7.8; 1305,13.14(bis). 15.16.17.18.27(bis).28(bis); 1306,1.6.7.16.17.18(bis).20.21; 1310,3; 1311,31; 1318,2.3.5; 1330,4.27.32.36; 1331,36; 1347,17†; 1353,19; cause change, 1258,32*.38; change (tr.).1258,34; 1259,12; 1265,30.31.32.33.34; 1289,36; 1324,30.33; 1352,11†(bis); 1364,30; make a change, 1272,21*; undergo change, 1276,28; 1359,36; undergo a process of change, 1252,6*; (middle voice), undergo change, 1251,30; 1359,15.20; be changed, 1259,3; change (intr.).1265,34; 1331,1; 1337,6; 1359,13.33; 1360,18; ***metaballein eis***, make, 1347,19†
metabatikos, involving change of place, 1258,13; 1260,1; ***to metabatikon***, that which changes place, 1354,37; ***metabatikôs***, from place to place, 1258,7
metabolê, change, 1250,36; 1251,5.17*.18*.24.30.31.34.35; 1252,2.8.28.29.38; 1253,24.30; 1255,38; 1258,12.37; 1260,27†; 1263,14; 1265,25.26*; 1266,14.16.19.20(bis).33; 1267,7.17(bis); 1270,30; 1273,36; 1274,39; 1275,6.11.14.23.26.28.41*; 1276,6*.11.12.14.16*.17.27.31; 1277,2†.3†.8†.17.20.23.25.29; 1295,3; 1297,13; 1303,37; 1304,2.7.22; 1305,13.37(bis); 1306,5.12.13.14; 1311,27; 1312,8.9.22.37; 1313,17*; 1318,4; 1337,2.8; 1339,29; 1359,24.31; 1361,15*; 1364,3.12; 1365,14.34.40; kind of change, 1273,17.29;

1275,7.15.39; 1277,36; 1307,15; 1312,17.30; 1365,24
metabolikos, subject to change, 1359,22
metadosis, transfer, 1337,19
metakhôrêsis, exchange of locations, 1351,16
metalambanein, alter, 1334,36; substitute, 1344,33
metalêpsis, reception, 1337,20
metapheresthai, change position, 1352,10†
metapiptein, change truth value, 1299,37†; 1300,3†.6†.11(bis).32
metaptôsis, change in truth value, 1300,10†
metastasis, exchange of places, 1351,16
metekhein, participate in, 1251,5; 1278,8; 1332,16†; 1347,33†; 1361,6†
methexis, participation, 1355,6
methodos, method, 1256,22*; 1267,15; 1273,27; 1335,14
methuparkhein, exist afterwards, 1256,10
metienai, set out (to do something), 1278,2
metrein, measure, 1316,29.32(bis).33; 1339,32; 1340,4; 1366,5
metrêtikos, that measures, 1340,7
metrêton, thing that is measured, 1317,1
metrios, moderate, 1276,22*
metron, measure, 1316,26.27.30.31.36.38; 1317,1.2.4; 1339,31.36; 1355,9; 1366,3.4.5.7
migma, mixture, 1307,11†
mignunai, combine, 1313,28
mikrotês, smallness, 1274,12*
miktos, combined, 1273,20.31; 1278,7.9; 1301,5.6; 1365,25; with a combination, 1278,6*
mimnêskein, mention, 1318,31
mixis, combination, 1274,25
mnêmê, memory, 1363,28; ***mnêmên poieisthai***, mention, 1318,20
mnêmoneuein, recall, 1268,7; 1289,6; 1294,5; mention, 1326,39; 1344,29; 1365,26
mokhleia, leverage, 1259,12.14*.15*.26*.28; 1260,28†
mokhlos, lever, 1259,17.18
monas, unit, 1316,35; 1366,5
monimos, persisting, 1349,27.32.35; 1350,5; ***to monimon***, persistence, 1349,33
morion, part, 1260,25†.29†; 1261,27; 1269,3*; 1289,8; 1290,2.40; 1291,3; 1294,10.11; 1296,22†; 1299,2.3; 1304,35*.39; 1305,1.5.7.10; 1307,27(bis); 1322,7; 1326,29†.31†.32†.33†.35†.36†; 1332,23†; 1338,39; 1354,17.18; 1355,18†.24†(bis); 1365,9
muriakis, ten thousand times, 1253,18

Neikos, strife (in Empedocles), 1318,22*.26†
noein, understand, 1329,2; 1332,2; 1334,35; 1337,29; notice, 1335,9; do so in thought, 1283,29*.33*
noeros, intellective, 1337,31; 1338,11.12.22.25; 1359,8.10.21; ***noerôs***, intellectually, 1338,1
noêsis, thought, 1283,35; 1337,28
noêtos, accomplished by the intellect, 1338,12
nous, mind, 1318,29; 1354,38; 1360,25; 1362,1.2*.8.9.16.17.22.25*.28*.29*.32; 1363,2.3; intelligence, 1337,31; 1360,34†; ***prosekhein ton noun***, pay attention, 1363,3
nun, instant, 1282,16.19(bis).23.25.28*.29.31.33*.37; 1283,2.3.4.7.12; 13.14(bis).20.24; 1284,2.10.34; 1287,10*; 1294,5.7.8.12.15.21(bis).25.29; 1295,6.35; 1297,2.4.5.6.7.20.21; 1298,36; 1300,19(bis).20.21.23.26.27.28.31.35; 1305,29.38; 1306,2.7.10.13.20; 1363,37; 1365,28.29; ***to nun***, the present, 1268,17.18*.19*.20*

oikeios, proper, 1257,13; 1259,25; 1260,1; 1261,23; 1272,33.34; 1304,5.8; 1305,24.26.30; 1306,18.22; 1312,21; 1317,22.28.34.36; 1318,2; appropriate, 1258,37; 1262,4; 1301,14.16*.17(bis).20; one's own, 1261,24; 1280,35; 1302,37;

1329,16.36†; 1335,27†; 1347,10†.28†; one's own proper, 1268,3; of its own, 1347,2†; ***oikeiôs***, in a way appropriate, 1258,35; 1365,36; appropriately, 1319,24; 1354,9
onoma, name, 1266,29; term, 1317,5†; 1320,9; 1336,36; 1359,38; 1360,12; 1363,20; 1366,9
onomazein, name, 1271,24; identify, 1363,18
onta, ***ontôs***, see ***einai***
opisthe, back, 1278,25
opisthen, behind, 1308,13; ***(to) opisthen***, the back, 1302,31; back, 1308,6
organon, instrument, 1259,17
ouranios, of the heaven(s), 1256,17; 1263,33; 1327,14; 1335,18; 1353,29; heavenly, 1312,31; 1330,12†.13†.15†; 1333,3.19†; 1337,22.35; 1339,18; 1362,18; ***ta ourania***, the heavenly bodies, 1312,34; 1320,1.15; 1329,20.23.25.39; 1330,5.8†(bis).12†.16†.20.23.30; 1331,9.11†.14†.21†.22†.37; 1332,4; 1333,1.26†; 1334,18.23.34; 1335,23 (sing.); 1336,37.38; 1338,11.29(bis); 1361,31; 1365,17; the things in the heavens, 1333,5†.11†; 1360,38; ***hoi ouranioi***, the heavenly beings, 1360,38; 1361,1.2
ouranos, heaven, 1261,8.16; 1263,26†; 1279,4; 1327,1.7.18†.20.21; 1328,33; 1329,1.4.9.15.18; 1330,19.24.26.30.31; 1331,18†.28.32; 1334,32.36.37; 1335,5.25†; 1336,3†.24; 1337,3; 1339,23; 1353,31; 1354,2; 1355,13(bis).14.18.38; 1357,10.12; 1360,26.31; 1361,4†.4; 1362,12.15.19.29*.32; 1363,1
ousia, substance, 1260,20.34†; 1262,2†; 1266,31*; 1267,4*.19; 1269,1*.3*; 1270,28; 1279,1; 1293,32*; 1330,39; 1337,1.8.29; 1339,30.38; 1354,28; 1356,26; being, 1268,1*.3.4†(bis).35*; 1269,11; 1271,22.23.27.28.30.33.36; 1272,1.3.4.6.18*.20.22.24.25.31.37; 1292,3†; 1293,1; 1304,5; 1314,25.26; 1336,15.38; 1338,28; 1359,15; existence, 1354,29; 1359,3.9.22; 1363,4.6.7

palin, again, 1251,29 etc.; in turn, 1254,17; 1257,8*; 1263,7.8; 1267,13; 1269,19; 1270,32; 1273,18; 1290,29; 1301,11; 1317,14; 1341,29.31; 1352,14†; 1364,23
pan (to), the universe, 1260,37*; 1261,2.9; 1338,4; 1351,31; 1355,5; 1360,32†.35†; 1361,7†; 1362,1.30*
pantôs, certainly, 1262,30; 1274,38; 1335,7.8; 1338,33; 1349,25.34; 1351,37; at least, 1265,36
paraballein, compare, 1268,17; 1335,12
paradeigma, example, 1278,36; 1302,10.15; 1304,1; model, 1338,15
paradeiknunai, point out, 1308,1; 1333,33
paradekhesthai, receive, 1347,7†; 1363,8
paradidonai, present, 1267,31; 1269,6; 1273,2; 1292,34; 1293,22; 1336,38; 1343,15; 1348,2
paradoxologia, paradox, 1299,32
paragein, introduce, 1258,1; 1277,4†; 1318,11.19; 1320,10; 1337,21; 1366,17; create, 1330,39; 1337,13.16.17; 1339,22.23.40
paragignesthai, come to be present, 1360,6
paragraphein, comment on, 1288,5; offer comments, 1351,38
parakmastikos, past the prime, 1335,7
parakolouthein, pay attention to, 1327,22
paralambanein, employ, 1269,8; 1287,4; 1319,18; 1342,4
paraleipein, skip over, 1288,26; 1324,26; leave aside, 1267,33; 1357,5
paralupein, trouble, 1312,24
paramenein, keep, 1356,16
paramutheisthai, exhort, 1306,26
paramuthia, explanation, 1341,2
paratasis, extension, 1336,18.27; 1359,15
parateinesthai, extend, 1336,31

paratithesthai, cite, 1328,12(bis); 1333,33; 1362,13
parauxesthai, become large, 1332,17†
parekhein, provide, 1253,24.26; 1347,16†.36†; 1354,31†; 1363,9; cause, 1263,1*; 1352,13†; assist, 1363,28
parelêluthos (to), the past, 1268,18; 1363,37
parergôs, off the point, 1276,35†; ***mê parergôs***, attentively, 1339,25
parienai, omit, 1252,35; 1270,9; 1325,31; pass into, 1297,18
parupostasis, by-product, 1262,8
paskhein, undergo, 1257,39*; 1348,26*; be affected, 1325,20*(bis).21†.22†.23†.24†; 1347,28†; 1358,23†; suffer, 1334,34.40; have affective qualities, 1271,14; ***peponthenai***, has as an attribute, 1347,22†
patêr, father, 1268,11; 1270,23; 1327,4; 1337,23†.29; 1338,3; 1361,3†.16*
pathêma, affection, 1266,15*.21
pathêtikos, affective, 1265,12; 1271,13; passive, 1327,34
pathêtos, that can be affected, 1334,13
pathos, affection, 1265,11*; 1268,33*.34*; 1274,10*; 1278,39*.39; 1294,13.34*.36(bis); 1295,1*.2.3(bis); 1319,25; 1347,29†
pauein, stop (trans.).make cease, 1279,33.34
pauesthai, stop (intrans.).1255,20; 1288,5.19; 1309,33; 1310,25; 1314,36; 1337,15; 1344,35.36; 1345,1*.2*.3*.4.5.10*(bis).11.12*. 13*.21*.22*.24; 1346,4.6.9.19.22.23(bis).24.25(bis). 26*.35; 1348,7*.8*.10.14†(bis).15†(bis).29; 1349,2.11*.12*.15.20.22.23; 1350,1.14; 1351,1.22; 1352,24; cease, 1274,34.36; 1278,22; 1346,10.33(bis).35; 1347,9†.11†; 1351,19*
paula, rest, 1328,1
peperasmenos, finite, 1254,14* etc. [246 occurrences]
pephukenai, be of a nature, 1257,37; 1260,2.16; 1332,12; 1333,4; 1334,13; 1349,34; 1350,29; be by nature, 1364,2; ***pephukenai pros***, have a natural tendency to, 1331,33
perainein, limit, 1315,15
peras, limit, 1275,2.3; 1278,3; 1280,4.5; 1282,23; 1286,16; 1288,19; 1293,17; 1296,29†; 1297,2.5.14.28.30.31.32; 1298,17(bis).21.25.28.36.38; 1299,5†.8†.13.14(bis).17.22.23.25. 30(bis).31.35; 1300,13.14.17.22. 31.33.34; 1302,30; 1308,5; 1310,29; 1311,7; 1312,19; 1315,15*.32; 1339,18; end, 1281,3.6.10.11. 27.28.29.34; 1282,4.6.9.13; 1283,31.36.39; 1284,1.4.8.14.15; 1286,24.25.26.27.30.31.33.35; 1287,5.7(bis).29.36; 1288,12.14.15; 1290,25.28; 1294,5.10.17.25.26; 1295,12.13.14.17.19.21.22.23.27.29. 38; 1296,4.7.15.26†.31†.34; 1302,26; 1307,1.5.7†.8†.9†.10†(bis). 36; 1308,2.10(bis).17.19; 1309,7.11.12; 1310,11.33; 1311,1*.2.3.5(bis).6.8.19.34.35; 1312,25; 1314,35.36; 1315,14.19*.21.22.24.27.28.29.31. 33.35; 1316,2.6.8(bis).10; 1317,29; 1320,24.28; 1352,21; 1357,22†.23†; 1365,29
peratoun, terminate, 1314,30; 1316,9.15; 1365,27
Peri kinêseôs, *On motion* (as a title for *Physics* 6-8), 1358,8; 1361,29*
Peri phusikôn arkhôn, *On natural principles* (as a title for *Physics* 1-5), 1358,8
periekhein, comprehend, 1253,23*; 1254,25; include, 1304,3; surround, 1355,14.15; ***to periekhon***, the environment, 1258,19*
periektikos, comprehensive, 1336,29
perigeios, near the earth, 1263,22
perilambanein, include, 1251,19
perileipesthai, remain, 1364,20
perilêpsis, comprehension, 1277,33
periokhê, embracing, 1338,10
peripalaiesthai, wrestle around, 1319,1

peripalassesthai, be hurled about, see n. 379
periphereia, circumference, 1309,24; 1310,14; 1354,8.12.17.19.20.22; 1355,19†.22†.37; arc of a circle, 1310,31
peripherês, circular, 1279,34
peripheresthai, rotate, 1310,8; 1349,4
periphora, rotation, 1307,27; 1317,7†; 1338,38.40
peripiptein, fall into, 1296,17
peripolein, traverse, 1263,26†; revolve, 1337,35
peristrephein, turn (trans.), 1321,11
peristrophê, spin (n.), 1349,4
peritteuein, be left over, 1341,39; 1343,25
perittôma, excretion, 1258,24
perittos, superfluous, 1253,13
perix (to), circumference, 1354,6
phaneros, evident, 1256,22*; 1257,10; 1258,2; 1264,27; 1273,22*; 1280,20*; 1294,27; 1333,34; 1358,15*; 1363,13
phantasia, imagination, 1306,26; 1334,31; 1359,40
phantazesthai, imagine, 1333,22†; 1334,26.28†.30.32.36
pherein, bring, 1267,25†; 1279,19; 1307,22; 1318,17; 1344,19; transmit, 1345,9; carry, 1347,36†; ***pheresthai***, (middle voice) undergo locomotion, 1273,20; 1278,18; 1280,27.29.30; 1285,19†.36*.38; 1286,2; 1302,4.6.17.37; 1303,14; 1317,17*; move (intr.), 1273,8†; 1278,14; 1303,3; 1317,25*.27.32.34; 1347,16†.35†; be moved, 1278,14; ***to pheron***, what is causing locomotion, 1269,26; ***to pheromenon***, what is undergoing locomotion, that undergoes locomotion, etc., 1273,20; 1278,6*.15.17*.18.20.21; 1280,30; 1281,15.19; 1282,4.26; 1283,7.24*; 1301,8; 1302,16; 1303,2; ***pheromenos***, directed, 1284,21
Philia, Love (in Empedocles), 1318,22*
philosophia, philosophy, 1366,20
philosophos, philosopher, 1339,22; 1353,19; 1366,21
Philotês, Love (in Empedocles), 1318,25†
phora, locomotion, 1265,13*.16.17.29; 1266,2.4*.6.8.9.12; 1267,32.35; 1269,15.22.23.34*; 1270,3.6.7; 1271,15*.16.17.23.26*.27.29; 1272,8*.12*.15*.29.32; 1273,16.18.30; 1274,1.6; 1278,1; 1281,19; 1301,13; 1302,9.15; 1303,7.36; 1304,13.23; 1307,15.16.30; 1313,25.26; 1314,18.39; 1315,1; 1317,4.5†.6†; 1318,19.23; 1319,7*.19.22; 1320,10.32(bis); 1351,25; 1365,23.24; 1366,1(bis).9; kind of locomotion, 1273,19.21.23.30.32; 1278,2; 1301,3.4
phthartikos, destructive, 1279,22
phthartos, subject to perishing, 1253,10.11.13.28; 1263,9; 1271,7; 1327,3; 1331,27; perishable, 1261,29*; 1271,4*; 1314,14.15.16.17.21.25; 1327,2.16; 1328,34; 1329,2.16.18.24(bis).25.27.29.31; 1330,24.26.34; 1331,29; 1332,6; 1333,35; 1334,32.40; ***to phtharton***, perishability, 1335,22
phtheiresthai, perish, 1252,24*; 1253,21.25.38; 1255,14; 1261,6; 1262,17.30.32; 1263,7.32; 1264,10; 1269,26; 1271,6; 1274,20; 1275,13; 1276,12; 1277,30; 1295,36*; 1305,36*.38*; 1306,16(bis); 1312,37; 1319,4; 1329,31; 1330,37; 1331,4.10.24†; 1332,34.36.40; 1333,15†.21†.23†.26†; 1334,33.36; 1335,3; 1336,23; 1346,14; 1359,35; 1363,24.30; cease to be, 1295,26*.27*.29.30.31(bis); 1296,5(bis).6(bis).7.8.10 (ter).11.13(bis)
phthinein, decrease (v.), 1257,35; 1267,8.27†; 1273,10†; 1319,4*; 1320,13; be in decay, 1313,1*
phthisis, decrease (n.), 1256,37; 1273,5; 1274,11*
phthora, perishing, 1250,36; 1251,5.19.34; 1252,5*.15.20.21; 1262,24.26.28; 1263,3.8.14;

1266,14.31*.32.34.38; 1267,1.3*.4.7.28; 1272,22; 1273,5.12.35; 1274,3.9*; 1275,6.26.38*; 1276,12; 1277,18*.19.27.29; 1278,29; 1279,2; 1294,1; 1297,19; 1302,13; 1304,3.12; 1306,3.10.13; 1311,28; 1312,3*.22; 1313,1.3*; 1314,23; 1319,23; 1330,36; 1331,7.12†. 13†(bis).23†; 1332,38(bis); 1333,16†.18†.20†.27†(bis).29†.30†; 1334,20†.29.30.36.38†; 1359,38; 1363,19; 1365,13; ceasing to be, 1275,14; 1295,25; end, 1335,8; ***huphistasthai tên phthoran***, perish, 1332,17†

phulattein, preserve, 1272,24(bis); 1347,35†; 1353,26; 1362,10; keep, 1287,4; 1323,9; ***phulattesthai***, guard against, 1275,18

phunai, see s.v. ***pephukenai***

phusikos, constituted by nature, 1254,17; 1257,36.38; natural, 1257,12; 1258,27.36; 1318,33; 1333,23†; 1359,5.6; 1364,28; having to do with nature, 1269,32; in accordance with nature, 1277,22*; of nature, 1366,19.20.21; ***Phusika***, Physics (as title of *Physics* 1-5), 1358,7; ***hoi phusikoi***, natural philosophers, 1363,31; ***phusikôs***, naturally, 1257,13; 1258,15; ***phusikôteron***, more like a natural scientist, 1265,26; 1338,31

phusiologos, natural philosopher, 1312,32; 1313,8; 1318,19; 1366,8

phusis, nature, 1254,19; 1260,34†; 1262,9; 1268,14.16; 1272,16*.17.24; 1290,20; 1295,33; 1302,33; 1312,21; 1318,17.33*.34; 1329,23; 1331,9.10.14†.24†.28; 1333,16†.19†.27†; 1334,24; 1337,15; 1338,22; 1339,1.3.11.18.20.23; 1341,20; 1347,7†.28†; 1348,23; 1358,30.38; 1359,7; 1361,19; 1362,3*.22.26*.28*.30*.32; 1363,2; natural capacity, 1346,1.6

phuseôs, naturally, 1347,13†; ***kata phusin***, natural(ly), 1257,14.15; 1259,20.21.24; 1265,16; 1266,5.8; 1269,11; 1272,1; 1274,23.24; 1301,3; 1303,29†.31†.33†; 1308,7; 1317,16.21.23.25.26.33; 1333,24†; 1334,30; 1361,8†; 1364,30; in respect of nature, 1268,8; in nature, 1268,34*; 1269,16.20.22.23; 1271,24.28; 1272,18.20; in accordance with nature, 1272,17; by nature, 1333,18†; 1334,28; 1338,23; 1347,24†.26†.30†; 1360,35†; ***para phusin***, contrary to nature, 1257,14.15; unnatural, 1274,24; 1303,30†.32†.33†; 1317,17.26; unnaturally, 1317,18.24; ***(têi) phusei***, in nature, 1256,24; 1265,20; 1267,36; 1270,3; 1271,27*.33*.35.36; 1272,3.4.5*.11*.24.35.36; 1314,19*(bis); 1317,13; 1337,4; by nature, 1333,20†.31†; 1347,22†(bis); ***phusin ekhein*** (+ infin.), be by nature capable, 1334,11

phuton, plant, 1257,35; 1271,30*; 1320,5.6; 1340,26

pisteuein, be convinced, 1256,33*; 1257,7*; place confidence in, 1301,16*; ***pepisteutai***, it is a matter of conviction, 1317,22

pistis, confirmation, 1287,29; proof, 1363,9

pistos, credible, 1277,20; 1366,16

pistousthai, confirm, 1280,10; 1297,12; 1318,16; 1366,9

pithanos, plausible, 1350,8

plagios, ***eis*** (or ***epi***) ***ta plagia***, lateral, 1274,25; 1280,13; 1349,32.33

planasthai, wander, 1262,7

planômenos, planetary, 1261,18.22.30; 1263,18; ***to planômenon***, planet, 1356,34; 1358,35.36

plêsiazein, approach, 1263,19; 1317,21.22.35

plêsion, near, 1259,25; 1302,5*; 1317,33.36; nearby, 1352,15†

plêthos, number, 1293,14; 1312,7; 1341,33; 1342,33*; amount, 1342,23

poiein, make, 1254,19; 1259,13; 1267,32; 1274,27; 1282,14; 1283,28*.30.32.38; 1286,29; 1287,7.33; 1290,25*; 1294,27;

1295,24; 1298,12; 1299,29; 1304,37; 1305,5; 1306,20; 1311,33; 1312,3; 1313,14; 1316,12; 1319,29; 1326,18; 1331,7; 1333,34; 1335,10; 1343,38; 1345,13*.17.28*; 1346,12.20; 1347,20†.25†; 1350,5.15.38*; 1351,8.10; 1361,16*; 1364,22; produce, 1257,38*; 1299,32; 1348,26*; 1363,15; do, 1281,6; 1283,17; 1288,36; 1291,17.20; 1309,35; 1318,15; 1341,32; 1356,8; 1361,19; act, 1290,29; cause, 1325,29.31.34.35.36; create, 1337,34; ***poieisthai***, base, 1259,36; 1274,3; 1305,30.34; construct, 1305,35; use as, 1279,20; 1280,8; 1311,20; assign as, 1318,20; ***to poioun***, maker, 1358,21†; 1361,16*; what works as an efficient cause, 1363,15.18

poiêsis, production, 1337,33

poiêtês, maker, 1327,3

poiêtikos, productive, 1263,24; efficient, 1318,23.30.31; 1319,18; 1354,36; 1360,25.30.33; 1361,12.14.18.25.27.30.33†; 1362,1.8.13.15†.16.17.22; 1363,1.9. 13.17

poion, quality, 1272,23; 1278,39; 1312,20

poiotês, quality, 1260,21; 1265,12.25; 1266,22; 1267,5; 1271,14; 1304,4; 1318,3; 1333,9†; 1335,37†

polos, pole, 1261,23.24(bis); 1262,1†; 1354,10.15; 1355,33.35†; 1357,11

polueidôs, in many ways, 1288,17

polukoiraniê, rule of many, 1254,13

porizein, supply, 1351,35

porrôteron, farther, 1268,15*.18*

poson (***to***), quantity, 1272,23; 1279,1; 1312,1.20

posotês, quantity, 1304,5

pragma, thing, 1266,29; 1270,28*; 1294,12.13.18.20.22.24.29*.30.31. 34*.35.36.36*; 1295,1.5.8; 13.19.23; 1296,16(bis); 1300,34; 1365,30; matter, 1289,31*; 1290,20

pragmateia, treatise, 1251,13; 1254,20; 1269,5; 1273,37; 1278,16; 1289,1; 1293,14.22; 1297,10; 1304,9; 1314,4; 1316,30; 1321,17; 1328,10; 1358,7; 1360,14; 1361,34; 1366,22

pragmateiôdês, cogent, 1299,4; 1347,38

pragmatikôs, that arises from the facts, 1289,35

proagein, bring forward, 1344,1; bring, 1353,24

proairesis, choice, 1268,24*.25*

proapheirein, subtract previously, 1341,34

proapodekinunai, demonstrate previously, 1340,20; 1366,10

proballesthai, put forward, 1297,10; 1336,36; 1344,19; 1351,36; 1352,21; project, 1338,1.17; propose, 1363,29

problêma, problem, 1264,34; 1355,28; proposal, 1340,23; 1358,19; topic, 1363,27

prodêlos, obvious, 1274,15

prodiorizesthai, specify in advance, 1301,5

proêgeisthai, be prior, 1266,33; 1279,27

proêgoumenos, prior, 1321,24; ***proêgoumenon***, principal kind of thing, 1262,9.10; ***proêgoumenôs***, chiefly, 1254,19; directly, 1259,26.27; 1352,17

proeipein, proclaim, 1319,14

proêremein, be at rest first, 1305,3

proerkhesthai, proceed, 1346,18

proelthôn, further on, 1253,3; 1256,20; 1361,3

progenomenos, prior, 1338,39; 1357,27†

proïskhesthai, hold up as support, 1327,36

proïstanai, put forward, 1318,30

prokataskeuazein, prove in advance, 1321,25

prokeisthai, (his) task is, 1251,14; 1276,30; 1278,32; 1283,28; 1304,23; 1315,4; 1322,4; 1331,27

prokeimenos, present, 1257,6; 1292,12; preceding, 1340,11; ***to prokeimenon***, the present topic, 1276,36†; 1301,17; the present point, 1297,15; 1302,15; 1336,33; 1340,33†; 1365,37; (his) present purpose, 1325,27; 1340,18; 1363,11; the thing at hand, 1301,20

prokheirizesthai, take up, 1289,3
prokheiros, obvious, 1329,10
prolambanein, assume, 1269,17; 1281,4; 1297,15; 1301,33; 1313,31; 1321,36; 1324,7.11; 1344,14; 1365,22; ***to prolabon***, the immediately preceding, 1253,22; ***to prolêphthen***, the assumption, 1332,28
prometapiptein, change truth value before, 1300,24.25
proödos, progression, 1263,11; 1351,25; progress, 1289,34; procession, 1337,10
prophanês, evident, 1252,34; 1257,11.21; 1337,34; 1341,8; 1360,11; 1363,18
pros ti, relation, 1268,14; ***to pros ti***, relational entity, 1335,27†; 1336,28; 1346,34; 1349,14
prosagein, approach, 1262,34
prosâidein, be consistent, 1350,1
prosauxanesthai, increase, 1311,11
prosauxêsis, additional increase, 1311,12
prosdialegomenos, interlocutor, 1288,38
proseinai, hold, 1277,11†; belong, 1333,17†; 1334,21†; be an attribute, 1336,7
prosêkein, be an attribute of, 1301,4; 1328,6; 1336,11; 1350,18; concern, 1301,25
prosêkon, appropriate, 1330,1; 1344,18.21; befitting, 1330,38
prosêkonta (ta), the attributes, 1352,31; 1353,25
prosekhês, near, 1256,8; 1301,20; proximate, 1257,29; 1320,28; 1338,29; next to, 1345,16.18; 1350,34; next, 1345,25; adjacent, 1350,4
prosekhôs, immediately preceding, 1254,2; proximate(ly), 1261,3.5.8; 1263,17; 1264,9.24; 1265,5; 1273,3; 1319,31; 1323,36; 1328,32; 1330,38; 1337,7.12.19; 1339,2.20; 1345,27.32; 1348,33; 1353,30; 1356,31; 1359,30; 1360,38; 1362,18; 1365,11.20; next, 1328,40; 1348,6; next in order, 1337,30; 1338,22.25; immediately following, 1364,22
prosgignesthai tini, acquire an additional attribute, 1333,25†
proskataskeuazein, prove, 1297,8.11
proskeisthai, be applied to, 1259,18; add, 1292,16; 1306,26
proskhôrein, approach, 1316,21
proskhrasthai, make use of, 1274,31; 1284,17; 1289,34; 1290,6; 1297,1; 1304,3
proskrinesthai, be joined with, 1265,24; 1319,5
proslambanein, assume in addition, 1274,30; 1285,3; 1289,10; 1313,27
proslêpsis, minor premise, 1262,27; additional premise, 1300,3†
prosnemein, assign, 1294,14.23; 1295,25.32; 1296,16
prospephukos, clinging, 1271,30
prosphorôs, accordingly, 1319,12
prosphuôs, in a naturally fitting way, 1289,4
prospiptein, impinge, 1258,33.35.37.39
prosthe, front, 1278,25; previously, 1290,12
prosthen, earlier, 1258,12; 1263,35; 1280,19; ***eis to prosthen***, forward, 1351,24
prostithenai, add, 1251,4; 1253,9.30; 1262,19; 1263,16; 1266,30; 1267,3; 1282,33; 1283,32; 1290,8; 1292,14; 1293,7.30; 1298,12.34.39; 1299,27.29.34; 1300,33; 1303,28; 1322,20.23.25.27.28.32.37; 1323,6*; 1324,35; 1326,16; 1327,19; 1342,6.8; 1343,35†; 1345,30; 1352,8.32; 1353,10; 1354,3; 1356,24.29; 1357,36; 1363,27; 1365,5.9.19.30; 1366,10.13; give in addition, 1254,15; 1262,23; attribute to, 1294,18; supply, 1360,3
prostunkhanein, meet with, 1267,23†; 1273,6†
prosupakouein, supply (in thought), 1262,13; 1309,22; 1322,11
prosupomimnêiskein, point out as well, 1271,3
protasis, premise, 1253,8.9; 1270,7.11; 1272,5.10; 1274,31; 1279,16; 1288,23; 1321,26(bis)
proteros (adv. ***proteron***), earlier, 1255,28; 1270,37;

1285,14*.15*.30†(bis); 1294,17.24.28*.30*; 1295,8.10.18; 1297,18; 1300,35; 1301,21; 1302,8*; 1309,31; 1318,19; 1321,3; 1330,19; 1346,26; 1348,16; 1362,4*; 1363,14.36; 1364,7.8; previous(ly), 1256,21*; 1262,17; 1263,36; 1270,28; 1272,30; 1273,21*; 1274,32*.33*; 1278,30; 1286,30.31; 1287,2; 1290,26; 1296,28†; 1297,16*; 1304,7; 1305,13.15; 1306,19; 1308,35; 1324,8; 1333,4; 1338,25; 1348,19; 1354,26; 1361,29; 1363,23.30.38; 1364,6.10; before, 1257,37; 1258,2.9; 1265,33; 1270,23.35; 1308,18; 1328,4; 1335,16; 1352,23; 1362,5; formerly, 1287,5; prior, 1255,38; 1256,24; 1259,4; 1267,32.36*.37; 1268,2.4†.9.14.18.19(bis).21.23*.27*.30*.31*.33*; 1269,1*(bis).3.9.10; 1270,16; 1271,27*.33*; 1272,6*.11*.35; 1279,25; 1291,18; 1302,33; 1314,17.18(bis).25.27; 1315,1.1*.2.3; 1321,25; 1337,4; 1346,20*; 1362,27*.29*; first, 1270,20; 1280,26.34; 1303,22.36; 1304,28*; above, 1325,35; ***to proteron***, priority, 1267,34.35; 1268,7.8.9.13;1269,5.7; 1272,29; 1273,16; ***hoi proteroi***, predecessors, 1318,14 (conjectural reading)

protetagmenon, introductory, 1340,31

prothesis, purpose, 1331,26

protithesthai, propose, 1269,15; 1280,36; 1288,17; 1301,3; 1321,4; 1323,36; 1340,17.21; 1344,6; put before, 1272,6

prôtos, first, primary 1250,37 etc.; (w. gen.) prior, 1264,33.35; 1265,3.14*.16; 1266,7.9; 1267,34; 1268,28; 1269,19; 1270,3.9.10; 1271,1.16.23.28.36; 1272,18*.20.28.37; 1273,16.33; 1313,25.29.31.32.34; 1314,19; 1316,32.35; 1317,14; 1320,32.33; 1353,6; 1364,26(bis).36; 1365,23.40; 1366,5.9; ***to prôtên einai***, primacy, 1316,36; ***to prôtôs kinoun***, s.v. *kinein*; ***to prôton kinoun***, s.v. *kinein*

proüparkhein, pre-exist, 1251,30; 1252,1; 1254,25; 1256,10.25; 1257,31; 1265,20; 1266,12; 1267,7.16; 1270,21.32; 1271,2.9; 1272,30; 1337,30; 1338,9.14

proüpotithesthai, presuppose, 1325,40

pseudês, false, 1289,33.34; falsehood, 1289,34

pseudos, false, 1281,31; 1293,12; 1300,4†.6†.33

psukhê, soul, 1251,8.9.11.13.32; 1253,20; 1256,35; 1257,2.34; 1259,8.9.15.16.19.23; 1260,29†; 1261,31; 1262,3; 1263,25†; 1319,28*.29.32.33; 1320,3.7; 1337,31; 1348,21.32.36; 1354,30†.37; 1360,34†(bis); 1365,7

psukhein, cool, 1258,37

psukhikos, due to the soul, 1354,33†

psukhros, cold, 1263,22; 1266,23*.24; 1335,37†

psukhrotês, cold, 1274,11

puknôsis, condensation, 1266,16*(bis).17.21.28*.30; 1319,22.23.25

puknotês, density, 1266,23*.24.28

puknoun, condense, 1266,37

pur, fire, 1259,22; 1317,27; 1338,5; 1346,16.17; 1347,17†.19†.20†; 1352,15†.16†

rhein, be in flux, 1277,31.32; 1313,1*.11

rhêma, word, 1292,12; 1325,24; 1332,27; 1337,21; 1340,21; 1348,2; 1361,32

rhêsis, passage, 1279,12.15

rhêton, statement, 1285,35

rhipsis, throw, 1345,36; throwing, 1350,29

rhiptein, throw, 1317,18; 1339,35; 1347,13†; 1357,26†

rhiptôn (ho), ***rhiptoun (to)***, the thrower, what throws, thing that throws, 1317,20; 1344,24.30.31.35(bis); 1345,4.35; 1346,31.34.36; 1347,4†.8†.9†.11†.12†.15†.28†.34†; 1348,13†(bis).15†.33.36(bis);

1349,27; 1350,3.14; 1351,23; 1364,29
rhiptoumenon (to), thing that is thrown, what is thrown, thrown object, 1344,24*.30.31.33; 1345,2; 1346,11.16.36; 1347,12†. 13†.16†.24†.27†.37†; 1350,13.22.36.37; 1351,7.9(bis).11.15.19.23.29; 1356,9*.11.18; 1366,16
rhoê, flux, 1313,9.14
rhopê, tendency, 1307,23; 1310,9; 1318,6

saphênizein, clear up, 1285,15; expound, 1330,2
saphês, clear, 1250,38; 1361,17; obvious, 1270,9; 1288,26; ***saphôs***, clearly, 1258,8; 1329,4; 1340,14; 1343,15; 1347,31†; 1359,39; 1360,38; 1361,19
sathros, unsound, 1332,32
selênê, moon, 1330,11†; 1335,28†; ***(ta) hupo selênên***, sublunary, 1253,21; 1256,9.15.18.23; 1258,35; 1277,32; 1312,35.37; 1320,1; 1327,18†; 1330,1.5.8†.20.21.27.34; 1331,4; 1336,4†; 1361,1.31; 1362,19
selênêiakos, lunar, 1331,19†
sêmainein, mean, 1269,17; 1302,8; 1309,14; 1314,25; 1338,20; signify, 1337,21
sêmainomenon, signification, 1264,21; 1267,32; 1268,7.9; 1269,7.16; 1273,18; 1314,18; 1315,2; 1320,31
sêmeion, sign, 1259,25; 1279,22*.32; 1280,8; point, 1280,1.24; 1281,7.8.10.13.15.17.18.19.20.22.37; 1282,1.2(bis).21.26.28.30.33.35.37; 1283,1.6.8.9.15.17.19*.20.23.25. 27.30.39; 1284,3.4.30.32; 1285,2.4.5.8.20†.22†.26†(bis).28†. 36*; 1286,1.15.19.20.32*; 1287,26.37; 1288,9.33; 1290,23.24*.28; 1291,17*.19.21.31; 1294,28*; 1298,13*.14.34*.35; 1308,21†.23†; 1309,3.6.8.24.32.33.35; 1311,2; 1315,18.28; 1365,27
skepsis, consideration, 1251,13; 1264,28
skhêma, figure, 1256,2
skhesis, relationship, 1263,34; 1264,11; 1354,24†; relation, 1302,32; 1333,22†; 1349,15; 1353,26; 1356,27.30; 1364,15
skholê, discussion, 1326,39; 1328,11
sklêros, hard, 1266,23*.24
sôma, body, 1251,8.22.24; 1252,19; 1255,32; 1257,4.13; 1258,18.33; 1259,11.12*.13*.14.15.17.19.20.21. 27.32; 1260,2.3.4.27†.33†; 1261,32; 1262,2†.4; 1263,14; 1264,23.28; 1266,35; 1269,9; 1270,26; 1274,27; 1312,31.32; 1313,4; 1318,34; 1319,5.10*.16; 1321,6; 1327,13.14.27.29.39; 1328,8.9.22.25.31.35.40; 1329,2.9.11.14.17; 1330,12†.15†; 1331,22†; 1333,5†(bis).11†.15†.19†.22†.37.38. 39; 1334,6.8.10(ter).13.14.17.22.28; 1335,18.20.21.39†; 1338,7.36; 1339,18; 1340,27.30; 1343,18; 1347,32†; 1348,21.23.36.37; 1350,32; 1351,31; 1352,7.33.34; 1353,29; 1354,30†; 1355,1; 1357,15; 1358,9.10.11; 1359,9.25; 1360,34†; 1361,33†; 1362,14†; 1363,4.16; 1364,28; 1365,8; ***to pempton sôma***, the fifth body (*aithêr*), 1330,10†.19.21; 1331,18†.19†.34
sômatikos, corporeal, 1313,10; 1334,16; 1337,3.34; 1338,4(bis).17; 1339,19.34; 1346,32; 1348,10; 1349,38; 1355,2; 1356,5.6(bis); 1358,37; 1359,7.14; 1363,7; bodily, 1321,5.9; 1334,16
sperma, seed, 1275,10.11; 1340,25
sphaira, sphere, 1261,22.30; 1262,7; 1315,38*; 1330,12; 1354,5.19.22; 1355,4.35†.37; 1356,34; 1357,12; 1358,34; ball, 1357,26†
stasis, halt, 1255,20; stability, 1277,32; 1352,5†; stopping, 1282,5.7.8(bis).10.11; 1283,30; 1284,5; 1309,30; 1312,12.14; 1314,36; stationary state, 1310,3; location, 1352,12†; ***statikos einai*** (w. gen.), bring something to a standstill, 1279,21
stegein, harbour, 1329,36†; 1331,35
sterêsis, privation, 1276,31;

1277,5†.9†; 1304,3.5.8.22; 1305,11.12.15(bis).19; 1312,23
stoikheion, element, 1254,21; 1259,22; 1266,36; 1274,27; 1318,24; 1319,3(bis).20; 1331,12†.17†; 1332,3.15†.17†.25†.34.36(bis).40; 1336,22; 1338,6.10.11; 1348,24; 1349,30; letter, 1286,28.29; 1287,1
sullogismos, deduction, 1272,13; 1288,20; 1301,18.19; 1321,26(bis)
sullogizesthai, deduce, 1253,3; 1265,2; 1269,20; 1270,4; 1271,25; 1272,15; 1277,19; 1278,17; 1279,24.26.27; 1304,16; 1317,16
sumbainein, hold (of), 1251,35; follow, 1254,16*; 1284,25.28; 1285,1; 1287,20; 1294,32; 1295,6.35.37; 1317,2; 1330,28; 1335,33†; 1345,5; 1366,4; happen, 1275,29; 1277,13†; 1280,28; 1292,30; 1311,29*; ***sumbebêkenai***, be an accidental attribute of, 1292,7†.13*.23*.36; 1293,7*; ***to sumbainon***, the conclusion, 1277,24
sumbebêkos, accident, 1268,31*; 1270,29; 1291,37†; 1292,9†.19.22.33.34.35; 1293,6; ***kata sumbebêkos***, incidental, 1251,6.25; 1265,8; 1292,26(bis); 1362,21.27*; incidentally, 1251,18*.20*.22; 1257,4; 1259,11*.14.27.32.35*.35; 1260,3.10(bis).11*.14.23†.26†.38; 1261,15.17.18.20*.26.28.31.35†; 1262,5.12*.13.19; 1282,36; 1291,38†; 1292,1†.3†.9†.17†.18†.25.27.29(bis). 31.32; 1320,36; 1348,33; 1353,8(bis).11; 1354,14.18.20.24†; 1355,12.17.20†.25†.27†.32; 1362,26*; 1364,31.39.40; 1365,19
summetaballein, change together with, 1259,13; 1353,9*.10*.11.14
summetrios, in due proportion, 1276,24
summetros, in due proportion, 1276,34
summiktos, combined, 1335,29†
sumparateinesthai, be coextensive with, 1336,14; comprehend, 1360,2
sumpauesthai, stop when something else does, 1347,9†.12†
sumperainesthai, conclude, 1266,3; 1272,38; 1288,19; 1313,16; 1315,16; 1365,10; 1366,17; draw (one or more) conclusions, 1342,29; 1344,1
sumperasma, conclusion, 1254,2; 1272,36; 1279,14; 1288,27; 1342,28; consequence, 1344,18
sumpherein, carry along with, 1356,34
sumphônein, agree, 1267,20; 1318,11
sumphônia, agreement, 1363,31
sumphônôs, consistently, 1271,24
sumplêroun, constitute, 1278,33.36
sunagein, infer, 1253,10; 1254,34; 1256,2; 1259,34; 1274,5; 1285,11; 1302,28; 1306,19; 1313,34; 1316,38; 1321,29; 1325,10†.36; 1329,22; 1342,26; conclude, 1321,15; 1364,32; conduct, 1321,25; draw a conclusion, 1323,15
sunagomenon (to), the conclusion, 1285,18†
sunagôgê, inference, 1329,20
sunairein, include, 1288,25; 1336,31.33; 1355,10
sunakolouthein, follow, 1352,34*; 1359,8
sunanairein, eliminate along with, 1314,22; '***x sunanairei y***', eliminating x eliminates y, 1265,18; 1314,24
sunanaphainesthai, appear together with, 1264,27
sunanapherein, bring with, 1363,16.19
sunapodeiknusthai, demonstrate together, 1264,24.32; 1353,31; demonstrate as well, 1278,9
sunaptein, join, 1260,6; 1307,1; 1311,1*.3.6.7; join together, 1269,11; 1299,2; 1307,2
sunêmmenon, conditional, 1254,3; 1262,24.29; 1300,4†.6†; 1302,14; conditional premise, 1277,22; 1289,11
sunarithmein, include, 1273,35
sunartêsis, fitting together (n.), 1337,32
suneinai, be joined together with, 1259,19

sunekheia, continuity, 1255,21.24; 1287,33; 1301,12; 1304,30; 1306,27; 1312,17; 1334,1
sunekhein, hold together, 1331,10.25†; 1333,26†
sunekhês, continuous, 1252,30 etc. [299 occurrences]; ***(to) sunekhes***, continuity, 1273,31; 1290,37; 1314,29; continuum, 1290,27
sunekhôs, continuously, 1252,24* etc. [73 occurrences]; ***to sunekhôs***, continuity, 1252,27*.31*
sunekhizein, make continuous, 1306,27
sunektikos, that maintains, 1335,38†
sunepesthai, follow, 1331,8
sunepinoein, conceive as well, 1360,10
sunêtheia, customary use, 1320,9; 1366,9
sungenês, of the same kind, 1306,34; 1317,34; 1350,24; akin to, 1340,19
sunienai, understand, 1327,8
sunistanai, establish, 1333,3; 1357,35; 1360,32†.34†; ***ho sunistas***, the creator, 1360,32†
sunistasthai, be composed of, 1331,36; 1333,38
sunkeisthai, be composed of, 1255,9; 1278,10; 1279,13; 1297,3.8.11.21.23; 1298,5.11; 1303,33†; 1307,25; 1309,2; 1314,13; 1332,24†; 1333,40; 1336,2†; 1338,6; 1365,31.39
sunkeimenon, compound, 1338,8; ***ex hôn sunkeitai***, components, 1278,7.8
sunkhôrein, grant, 1252,4; 1278,11; 1293,20; 1295,20; 1314,5; 1325,13†; 1330,7†; 1333,25†; permit, 1258,22
sunkhrasthai, make use of, 1290,14; employ, 1333,4; 1360,13
sunkineisthai, move along with, 1251,9; 1257,5; 1259,32; 1346,11; 1355,28†.31; 1358,37
sunkrima, compound, 1319,13
sunkrinein, combine, 1266,18; 1267,18.26†(bis); 1273,9†(bis); 1318,23; 1319,10*.11†; cause to combine, 1352,10†
sunkrisis, combination, 1266,17*.18.28*.30.33; 1267,1.2.5.6.16.22; 1273,4; 1318,21*; 1319,23.25
sunodos, coming together, 1338,24
sunokhikos, that holds together, 1355,6
sunoran, see, 1273,1; 1300,12
suntattein, rank, 1294,29
suntelein, contribute, 1327,12; 1328,33; 1349,33; 1351,7
sunthesis, composition, 1331,22†; 1338,7.24
suntheton, compound, 1274,28
sunthetos, composite, 1278,8; 1314,13; 1316,35; 1331,10†.11†.13†.18†.34; 1332,8; 1338,17.21
suntithenai, add together, 1323,12; 1332,31; 1336,19
suntomos, brief, 1250,37; 1254,3; 1363,26; ***suntomôs***, briefly, 1252,32; 1288,20; 1310,38; 1320,25; 1336,5; 1352,22; 1358,3; 1364,19.32
sunuparkhein, coexist, 1262,22; 1275,33; 1277,10†(bis).28; 1280,14; 1336,28
sustasis, structure, 1359,7.14; 1366,19
sustêma, composite whole, 1330,37
sustoikhôs, correspondingly, 1265,5

tasis, force, 1310,9
taxis, order, 1264,34; 1268,9.25*; 1319,12.15; 1354,39; 1362,3*; 1363,27
tekmêriôdês, inferring from effects, 1279,23
tekmêrion, sign, 1258,1; 1279,20
teleios, complete, 1268,1.2; 1271,36; 1272,1.2.19.25; 1274,13; 1311,10*; 1312,11; 1313,31.33; 1314,1.17.20.22.25.27; 1317,13.14; 1318,7; 1320,35; 1355,1; 1363,11; see also n. 173.
teleiotês, state of completion, 1271,34.35.37; 1272,2; see also n. 173; completeness, 1274,12*; state of perfection, 1353,15
teleiousthai, reach completion, 1270,14.19; 1271,32; 1272,17
teleos, complete, 1361,6†; ***teleon***, wholly, 1258,9; ***teleôs***, entirely, 1251,5; 1353,27; wholly, 1258,18; 1266,35; 1314,27; 1318,3.5; 1360,23

teleutaios, last, 1255,38; 1270,18; 1271,26.29.32; 1272,2.3.8.12*; 1279,16; 1323,19.21; 1343,25; 1346,14; 1351,3; 1358,8; ***to teleutaion***, endpoint, 1284,12; ***pro tou teleuatiou***, penultimate, 1346,19
teleutan, stop, 1309,32; 1315,14
teleutê, end, 1280,37*; 1281,24.25(bis); 1283,25*; 1288,11; 1335,4
telikos, final, 1354,34; 1360,25.28.29.33; 1361,11.23; 1362,1.8.12; 1363,13
telos, end, 1271,36; 1272,14; 1274,35; 1275,16; 1280,38; 1281,1.23.35; 1285,30†; 1287,31.32; 1288,6.7(bis).8.9; 1291,18; 1296,3.9; 1300,35; 1310,31; 1314,31; 1315,10.13.18.36*; 1316,13*.14.16.19.20.22; 1317,11.17*.28.31*; 1321,15; 1327,16.36; 1338,38.39.40; 1339,13; 1360,4; 1366,21; finally, 1346,14*.19
temnein, cut, 1283,5(bis); 1290,22; 1291,11.12.13; 1334,4.13.14
tetragônon, square, 1265,28
theasthai, recognize, 1313,13; 1362,2
theios, divine, 1278,35; 1361,21*.33†; 1362,14†
theologia, theology, 1359,6
theologos, theologian, 1360,7
theôrein, consider, 1262,6
theôrêma, theorem, 1273,28
theôrêtos, observable, 1320,20
theôria, observation, 1263,27; study, 1366,20; ***theôria axiologos***, theoretical interest, 1340,20
theos, god, 1278,34*.34; 1327,4; 1330,38; 1331,25†.29.30; 1333,26†; 1334,24.25.33; 1337,22.22†(bis).32.33†(bis).33.34.35 (bis).36(ter); 1338,1.2; 1359,10.21; 1360,8.25.30.33; 1361,2†(bis).7†.12.19; 1363,10.13
theosebeia, religious zeal, 1327,20
theotês, divinity, 1338,2
thermainein, heat, 1258,37; 1292,28(bis); 1346,16.17.18; 1347,17†.18†(ter).20†.21†
thermos, hot, 1263,22; 1266,23*; 1347,19†.21†; ***to thermon***, heat, 1266,27
thermotês, heat, 1274,10
thesis, position, 1263,23; 1265,32; 1268,9; 1269,8.9.10; 1319,12.15
thnêtos, mortal, 1255,25; 1259,36; 1260,28†; 1262,6; 1361,3†.7†
tithenai, posit, 1257,24; 1262,28; 1268,8; 1270,12; 1276,9.10; 1284,15; 1286,7; 1296,7; 1307,18.20.21; 1318,19; 1319,17.22; 1348,17; 1352,6†; 1359,16; 1362,7*.8; put, 1253,36; 1272,9.10; 1284,23; 1302,26; 1321,15; 1329,20; 1352,8†; 1359,19; place, 1287,3; pose, 1299,4
tmêma, segment, 1257,20; 1290,14; case, 1364,19
tmêtos, cuttable, 1334,13
tomê, cut, 1283,4; 1287,24*; 1288,10; 1291,3.7.12; process of cutting, 1332,11; 1333,13†
topikos, in place, 1265,34; 1270,12.14.21(bis).26.34.36; 1271,2; 1320,1.3.20
topos, place, 1257,1 etc. [117 occurrences]
trephein, nourish, 1257,35; 1265,22.24(bis); 1267,12; 1270,15; raise, 1296,33
triplasiazein, triple (v.), 1322,30(bis)
trokhos, wheel, 1314,33
trophê, nutriment, 1258,20*.21.25.27.28; 1259,32.37; 1265,21(bis).23.27.37; 1267,11.13.18
tropê, transformation, 1329,6
tropos, manner, 1254,34; way, 1257,3; 1262,32*; 1267,34; 1276,20; 1283,37; 1290,35; 1301,8; 1312,26; 1344,1; 1347,6†; 1348,5†.27†; 1349,9†; 1350,3†; 1358,36; 1365,23; type, 1267,35; 1268,13; 1272,29; 1273,16; 1276,21; 1279,30; 1362,23*; syllogistic mode, 1272,14
tukhê, luck, 1362,21.24.25*.28*; 1363,1

xun- see *sun-*

zêtein, investigate, 1251,11.15; 1257,21; 1264,19.31; 1269,7; 1273,17; 1298,2; 1333,24†; 1339,25;

1346,29.37; 1354,1; 1358,33.39; 1361,26; 1363,29; 1365,6; seek, 1263,8; look for, 1298,26; inquire, 1363,12

zêtêsis, investigation, 1263,10; 1264,35; the problem under investigation, 1358,35

zôidiakos, ecliptic, 1263,20

zôion, animal, 1251,7.8.32; 1253,20; 1256,35.36; 1257,2.12.34.39; 1258,5.9.13.18.20.23.24.26.28.30.39; 1259,1.4.10.16.36; 1260,1.28†; 1262,6; 1270,14; 1271,30; 1320,2.4*.5.6*.8; 1336,13.16; 1340,26.28; 1348,21; 1350,6; 1362,2; living thing, 1261,16; 1361,8†

zôiphuton, zoophyte, 1271,31

Subject Index

References are to the page and line numbers of the Greek text which appear in the margins of the translation.

acceleration, 1317,33-1318,7
accidents, 1270,28-30
actuality, 1265,29-31; 1293,10-11; 1327,31-2; 1339,37-9; 1356,25-6
  of points, 1280,33-1281,29; 1281,40-1282,3; 1284,4-8; 1286,14-20.26-7; 1287,25-1288,9; 1288,33-4; 1290,21-1291,24; 1308,35-1309,16; 1310,9-12.29-33; 1311,17-20; 1315,9-36
  see also infinite
affirmation, 1274,9-10
air, 1345,28-9; 1348,15-1349,10; 1349,29-32; 1350,28-30; 1356,12-14
Alexander
  has different text of 258b30, 1253,6-7
  how it is possible for a mover not to be moved even incidentally, 1260,22-35 (quotation)
  the souls of the planetary spheres are moved incidentally, 1261,30-1262,2 (quotation)
  how a thing might be called prior in respect of being, 1268,3-6 (quotation)
  one thing has only one contrary but it can be opposed to many things, 1276,35-1277,14 (quotation)
  understands 262a20 differently from S., 1281,30
  interprets well 262b14-15, 1285,15-30 (quotation)
  interpretation of 262b22-8, 1286,28-1287,2
  has a different text of 262b31, 1288,3-8
  interpretation of 263b6-9, 1291,34-1293,9 (quotations)
  solution of the puzzle about when Socrates died, 1296,18-32 (quotation)
  poses and solves a puzzle, 1299,4-19 (quotation)
  on Stoic indeterminate propositions, 1299,36-1300,10 (quotation)
  argument that reversing rectilinear motion is not continuous, 1303,27-33 (quotation)
  interpretation of 264b7-8, 1307,6-12 (quotation)
  how the points on a diameter are contraries, 1308,16-17 (quotation)
  motion on non-straight lines can be opposite, 1308,20-27 (quotation)
  interpretation of 264b20, 1309,22-5
  has a different text of 265b9, 1317,3-7 (quotation)
  has a different text of 265b29, 1319,11-16 (quotation)
  criticism of Aristotle and S.'s rebuttal, 1325,8-1326,27 (quotation)
  alternative proof that a finite magnitude cannot contain infinite power, 1326,28-37 (quotation)
  the power that moves something ad infinitum is a power only homonymously, 1327,35-8; 1358,18-1359,4 (quotation)
  criticism of Aristotle and S.'s rebuttal, 1340,32-1341,9 (quotation)
  criticism of Aristotle and S.'s rebuttal, 1343,32-1344,1 (quotation)
  projectile motion, 1346,29-1348,5 (quotations), 1348,11-15 (quotation), 1348,26-1349,10 (quotations), 1350,2-7 (quotation)
  interpretation of Plato's view of

projectile motion and S.'s rebuttal, 1351,28-1352,18
location of the primary mover, 1354,12-35 (quotations), 1355,15-38 (quotation)
how the motion of the planetary spheres is continuous, 1356,33-1357,9 (quotation)
the first mover is an efficient cause of the motion of the divine body, 1361,31-3 (quotation)
the first mover is not an efficient cause of the heaven, 1362,11-15 (quotation)
alteration, 1265,11-12.15-16.20-1266,2; 1267,2; 1271,12-14; 1274,10-11; 1278,28-9; 1313,2-4
amber, 1345,31
Ammonius, 1363,8-12
Anaxagoras, 1254,20-3; 1266,33-6; 1318,29-30; 1361,33-1362,10; 1363,31-3
Anaximander, 1266,36-8; 1319,20-7
Anaximenes, 1319,20-7
animals, 1251,7-9.31-2; 1256,36-1257,3; 1257,11-12; 1258,5-12.20-3.28-9.32-40; 1259,22-3.32-3.36-1260,2; 1270,14-16.19; 1348,31-3; 1350,6-7
see also souls
antiperistasis, see mutual replacement
arguments
a fortiori, 1277,26-9
by refuting an objection, 1341,18
conditional, 1254,2-6
contrasted with perception, 1257,30-2; 1280,33-5
demonstrative, 1279,25-7
dichotomy, 1289,6-1294,2
from consequences, 1279,22-5
from exhaustion, 1322,4-1323,27
from similarity of cases, 1277,22-6
from what is better, 1269,30-2
infinite regress, 1253,33-6; 1257,23-30; 1262,29-1263,15
proper arguments based on per se attributes, 1305,24-6
proper (appropriate) arguments contrasted with dialectical (general) arguments, 1301,12-29; 1305,23-6.29-34; 1306,18-23
reductio ad absurdum, 1294,26-1296,17; 1325,1-5
reductio ad impossibile, 1324,16-1325,7
that locomotion is prior to the other kinds of motion, 1265,14-1266,9; 1266,12-1267,14; 1272,28-38
that the primary mover is eternal, 1250,34-1256,30
through an illustration, 1341,19-20
see also proofs
atomists, 1318,31-1319,8; 1320,16-19

bees, 1293,4-5
body, 1251,24; 1252,14-15; 1259,12-28; 1260,28-9; 1274,27-8; 1321,12; 1327,38-1328,3; 1328,22-30; 1329,14-16.19-1332,2; 1333,36-1334,26; 1335,19-20; 1339,18-20; 1349,38; 1352,34; 1354,28-9; 1355,1-2; 1356,5-7; 1358,9-17; 1359,6-8; 1363,4-8
natural bodies, see elements
perceptible bodies, 1312,38-1313,15
the body that undergoes circular motion, 1264,22-8; 1270,25-6.35-6; 1328,40-1329,1; 1335,20-1; 1338,35-6
see also animals, heavenly bodies
breathing, 1258,18-19
bulls, 1293,4-5

causes, 1279,25-7; 1366,19-20
of changes, 1251,30-1
of what is eternal, 1252,31-5; 1253,8; 1260,35-1261,11; 1363,7-8; 1365,20-1
of motion, 1254,25-6; 1260,6-1261,11; 1272,38-1273,12; 1318,20; 1320,28-30; 1351,28-33; 1352,4-13; 1353,30; 1354,37-9; 1356,24-32; 1362,11-12; 1363,6-7
of generation, 1270,21-6.33-6; 1332,38-9
see also efficient cause, final cause
chance, 1362,20-1363,2
change, 1251,4-6.17-19.24-5.30-1; 1273,35-7; 1274,14-16; 1275,4-23; 1303,36-1304,9; 1305,13-20; 1311,26-8; 1312,35-7
continuous, 1304,10-23; 1313,17

instant of, 1294,5-1296,35; 1297,12-14
limits of, 1274,7-15.38-1275,3; 1278,3-4; 1288,19; 1296,28-9; 1297,27-32; 1298,13-1299,19; 1299,22-35; 1300,11-36
see also motion
circle, 1280,3-5.29-32; 1307,33-1308,5; 1311,1-9; 1315,17-31; 1316,12-16.20-3; 1317,10-12.31-2
cold, 1263,22; 1266,24-5
combination, 1266,16-1267,28;
see also motion, combined
coming to be, see generation
completeness, completion, 1270,14-19; 1271,35-1272,1.19-25; 1274,14-16; 1313,31-2; 1314,1-27; 1317,14; 1336,29-33
condensation and rarefaction, 1266,15-29; 1267,2
continuous, the, 1281,19-20; 1287,25-6; 1288,33-4; 1289,7-9; 1290,21-1291,24; 1294,11-12; 1297,9-11; 1306,30.37-1307,1; 1310,22-3
contradictories, 1312,22-3
contraries, 1273,37-1274,4; 1275,24-1276,34; 1278,24-6; 1279,19-32; 1305,31-2; 1306,32-1307,12
see also opposites, motion
cosmogony, 1336,36-1338,30; 1360,9-13

decrease, 1265,10-11; 1267,8-19; 1271,12-14; 1274,11-14; 1278,28-9
definition, 1333,15-27
one in definition, 1282,10-11; 1283,28-39; 1294,25-6
prior in definition, 1314,19-27
Democritus, 1254,20-3; 1266,33-6; 1318,31-1319,8; 1320,16-19
demonstration, see arguments
density and rarity, 1266,24-6
dialectic, 1301,12-29; 1305,23-6.29-34; 1306,18-23
digestion, 1258,20-8
dimension, spatial, 1331,20-1
divisibility, 1291,1-3.36-1292,11; 1293,10-1294,2; 1328,16-17; 1333,36-1334,16; 1365,31-2
see also indivisibility

earth, 1259,22
ecliptic, 1263,20; 1332,38-9
efficient cause, 1318,22-31; 1319,17-19; 1354,34-9; 1360,24-1363,24
elements, 1254,21-2; 1257,12-14; 1259,21-2.25-6; 1274,27-8; 1317,33-1318,7; 1332,3-40; 1348,24-6; 1364,28-31
elimination, 1279,27-32; 1280,10-11
Empedocles, 1254,20-3; 1266,33-6; 1318,22-28; 1363,31-3
eternity, 1253,26; 1260,35-1261,2; 1262,7-10; 1301,3-4; 1327,5-8; 1353,31-4; 1360,24-5; 1361,20-1
Eudemus, 1262,18-19; 1354,9-12.16; 1355,28-36; 1357,17-29
existence, 1291,13-14; 1293,24-7.31-1294,2; 1310,10; 1315,22; 1328,9-17; 1336,9-29; 1339,4-8; 1358,32; 1360,13-17; 1363,4.7-8

final cause, 1271,35-6; 1354,34-9; 1360,24-1363,24
fire, 1259,22
first mover, see primary mover
flux doctrine, 1277,31-3; 1312,32-1313,15
force, 1257,15-16; 1317,17-21; 1321,4-5

generation, 1251,33-5; 1252,14-15; 1262,23.27-9; 1263,2-3; 1264,9-11; 1266,31-3; 1270,21-6.33-6; 1271,4.12-14.29-32; 1273,35-7; 1274,3-4.9; 1275,7.25-6; 1278,28-9; 1297,15-19; 1299,18-19; 1312,22-3; 1313,2-4; 1325,32; 1328,16; 1332,38-9; 1336,9-29; 1339,4-8; 1358,32; 1359,31-1360,17
geometry, 1341,20
gnomon, 1265,28
god, 1330,37-1331,7; 1331,24-5; 1334,23-6; 1360,24-5.31-1361,11; 1361,18-19; 1363,8-12
growth, 1258,19-20; 1265,27-8

heating, 1346,16-18; 1347,17-22
heaven(s), 1256,15-18; 1258,32-8; 1261,14-1262,13; 1263,32-4; 1327,1-8; 1329,3-4.14-16;

1331,31-3; 1333,1-32; 1334,21-1339,24; 1353,31-4; 1357,9-17; 1362,11-1363,8;
body of, 1327,14-1329,19; 1339,18-19; 1353,29-30
mover of, 1353,30; 1354,1-1355,38; 1362,19-20
heavenly bodies, 1312,30-2; 1331,15-16.24-5.37; 1332,3-1336,33
heaviness and lightness, 1266,24-5; 1343,18-20
Heraclitus, Heracliteans, 1257,17; 1313,8-14; 1319,20-7
Hermotimus, 1361,33-1362,7
hypothesis, 1256,10-12.18-22; 1273,24-5
see also inferences

image, 1355,11
imagination, 1306,26-7
immortality, 1251,11-13; 1255,24-5; 1260,13-14
implication, mutual, 1317,1-3
impulse, 1257,1; 1258,6-7; 1260,1-2; 1301,33-1302,21; 1307,22-5; 1310,9; 1312,15-16
increase, 1265,10-11.15-16.20-8; 1267,8-19; 1271,12-14; 1274,11-14; 1278,28-9
see also growth
indivisibility, 1283,30; 1298,17; 1366,18-19
indivisibles, 1333,39-40
inferences, 1254,32-1255,5; 1256,1-3; 1272,14-18; 1301,12-29; 1305,23-6.29-34; 1306,18-23; 1321,26
infinite, 1254,18.21; 1278,3; 1291,27-1293,9; 1293,13-14; 1313,16-17; 1314,3-4; 1321,19-21.26-32; 1322,39; 1323,16-30; 1324,4-1326,37; 1328,9-20.30-1; 1329,14-16; 1332,5; 1335,19-20; 1336,9-29; 1338,34-5; 1339,8-12.19-22; 1340,11-1344,3; 1358,9-17; 1360,15-17; 1366,12-15
ad infinitum, 1252,37; 1253,14-15; 1289,7-9; 1291,1-3.11-14.36-1292,11; 1293,10-1294,2; 1313,17; 1327,32-3.38-1328,3; 1328,4-7.22-30.32-3.35-7; 1329,32-3; 1331,31-3; 1333,36-1334,16; 1335,20-1; 1336,6-10; 1339,12-22; 1340,6-7; 1341,13-17.38-9; 1358,18-26.29-32; 1360,13-17
nothing (no magnitude) infinite in actuality, 1278,16; 1291,33-4; 1314,3-4; 1321,16-17.30-1; 1324,1-2; 1340,13-14; 1358,6; 1366,11-12
potentially infinite, 1291,13-14.36-1292,11; 1292,24-1293,5; 1293,10-1294,2
infinite regress, 1353,4-5
instants, 1282,25-31; 1283,4-7.19-21; 1286,16; 1290,20-3; 1294,5-1296,35; 1297,1-2.4.7-1299,4; 1299,36-1300,36; 1305,29; 1363,37-8; 1365,28-30

lever, leverage, 1259,12-28; 1260,28-9
lightness, see heaviness and lightness
lines, 1282,3-31; 1283,5-24; 1284,5-8; 1286,12-20.23-4.26-7; 1287,8-16.27-37; 1288,8-9.34-5; 1290,20-3; 1291,36-1292,11; 1292,24-1293,7; 1293,10-1294,2; 1309,2-5; 1310,32-3.35-6; 1311,23-5;
circular lines, 1279,33-1280,8
straight lines, 1266,6-7; 1279,19-32; 1280,4; 1281,40-1282,3; 1286,14-20; 1302,20-4; 1308,5-20; 1309,10-12.31-5; 1314,5-10; 1315,22.36-8; 1365,24-6
locomotion, 1257,1; 1258,6-7; 1260,1-2; 1264,14-1274,6; 1267,15-1279,16; 1280,29-32; 1282,3-31; 1283,7-24; 1286,12-1288,29; 1307,15-1311,28; 1313,24-5; 1315,15-16; 1318,15-1320,21; 1320,30-2; 1365,22-6
luck, 1362,20-1363,2

magnet, 1345,13-27; 1346,5-10
magnitude, 1252,16-17; 1265,10-11; 1267,5-6; 1274,11-14; 1278,16; 1293,13-14; 1312,20-1;

1314,3-4.28-9; 1321,12-24.28-33; 1323,36-1326,37; 1328,16-17; 1332,7-14; 1340,11-1344,12; 1353,25-9; 1357,36-1358,17; 1366,11-15
material monists, 1319,20-7
measure, 1316,26-35; 1317,1-2; 1322,39; 1323,17-24; 1339,36; 1340,3-7; 1343,21-5; 1355,9-11; 1366,3-7
mind, 1360,24-5; 1362,15-1363,8
motion
  animal, see animals, souls
  back and forth, see rectilinear motion
  circular, 1256,22-3; 1258,32-8; 1260,21-2.25; 1261,27-31; 1264,21-8; 1266,6-9; 1270,25-6.35-6; 1273,28-33; 1278,4-10; 1280,29-32; 1307,15-1311,12; 1311,17-18; 1312,30-1; 1313,18-19.22-4.28-1315,4; 1315,7.17-18.38-1316,3; 1316,20-3.26-31.37-8; 1317,10-1318,7; 1320,32-4; 1328,40-1329,1; 1331,33.37; 1335,20-1; 1338,35-6; 1353,29-30; 1365,24-6.37-1366,7
  combined, 1274,16-28; 1313,27-8
  components of, 1278,33-1279,4; 1321,37-1322,2; 1352,25-9
  continuous, 1253,26-7; 1254,24-5.34-1255,1; 1255,5-23; 1256,12-23; 1259,33.37-1260,2-5; 1260,13-17; 1264,6-7.14-1273,12; 1273,23.29-32; 1274,4-5.16-28; 1278,4-1279,16; 1280,19-1288,35; 1301,3-18.32-1305,20; 1306,32-1312,27; 1312,30-2; 1313,18-19.22-3; 1314,28-9; 1315,7-8; 1317,10-12; 1320,34-7; 1328,40-1329,1; 1338,35-40; 1350,20-8; 1352,24; 1353,15-20.25-6; 1356,3-1357,9; 1357,32.34-5; 1365,14-17.21-2.24-6.35-40; see also change, continuous
  contrary, 1274,4-5.14-1275,4; 1275,41-1276,1; 1276,25-34; 1278,18-29; 1279,4-8.19-32; 1301,9-10.32-1305,20; 1306,32-1307,12; 1307,24-1308,27; 1308,35-1309,2; 1314,12-14; see also contraries
  eternal, 1250,38-1251,1; 1251,29-31; 1253,26-7; 1254,7-9.37-1255,1; 1255,5-7; 1256,18-23; 1262,16-17; 1263,3-9; 1264,9-11.22-1265,7; 1269,18; 1273,21-4; 1305,6-7; 1313,18-23; 1314,19-27; 1314,35-1315,8; 1320,26-34; 1327,30-5; 1328,22-30; 1328,37-1329,1; 1332,38-9; 1338,31-8; 1352,23-4; 1357,32; 1358,13-15; 1363,29-1364,25
  forced, see natural and forced motion
  in what is moved, 1251,1; 1331,17-18; 1338,33; 1339,36-7; 1340,1-2
  incidental, 1251,21-2; 1257,3-5; 1259,5-28; 1260,8-10.14-17.22-35; 1261,28-9.31-1262,13; 1320,34-7; 1353,6-12; 1355,11-15
  infinite, 1291,13-14.36-1292,11; 1292,24-1294,2; 1314,5-10; 1323,28-33; 1327,38-1328,32; 1331,31-3; 1335,20-1; 1336,6-10; 1339,8-22; 1358,29-32
  kinds of, 1264,19-22; 1265,10-1266,9; 1271,9-14; 1272,38-1273,12; 1275,7; 1301,9-10; 1304,33-1305,10; 1311,26-9; 1312,35-7; 1313,24-7; 1359,31-8; see also alteration, decrease, generation, increase, locomotion, perishing
  lateral, 1274,25-8; 1280,12-14; 1349,29-32
  'laws of', 1321,7-1326,37; 1340,11-1344,3; 1358,13-15; 1363,4-6; 1366,12-15
  natural and forced, 1257,12-16; 1259,20-6; 1260,1-2; 1302,33-4; 1303,5-8.29-33; 1317,16-1318,7
  projectile, 1317,18-20; 1339,35; 1344,21-1351,24; 1356,9-15; 1366,16-17
  rectilinear, 1278,13-1279,16; 1279,19-32; 1280,19-1288,29; 1301,3-18.32-1304,32; 1309,2-5; 1311,2-4.16-17; 1313,28-1315,4; 1315,8-14; 1316,3-9;

1317,16-28.33-1318,7; 1320,32-3; 1365,35-7
self-motion, 1251,8-9.21-2.31-2; 1252,14-17; 1257,3; 1260,4-1261,11; 1269,28-9; 1319,28-30; 1320,2-8.28-30.34-7; 1348,15-1349,10; 1350,6-7; 1353,6-8.10-12; 1364,25-6.35-6; 1365,5-12
single, 1255,7-8.16-21; 1256,5-6.29-30; 1264,6-7; 1274,4-5.16-28; 1278,28-9; 1279,2-4; 1301,9-10.32-1305,20; 1307,24-5; 1308,35-1309,2; 1309,38-1310,2; 1310,18-19.35-7; 1312,26-7; 1315,7; 1320,34-7; 1328,40-1329,1; 1338,37-40; 1352,24-29; 1353,15-20.25-6; 1356,25-6; 1357,10-13; 1365,14-15.21-2.35-7
uniform, 1317,10-1318,7; 1353,15-29; 1356,33-1357,9; 1366,7; see also alteration, decrease, increase, locomotion
mutual replacement, 1350,31-1351,33; 1352,14-17

natural philosophers, 1265,26-8; 1312,32-1313,8; 1363,31-3; 1366,7-9
nature, 1254,19; 1257,15-16; 1271,37-1272,1; 1272,23-5; 1318,17-18.33-4; 1341,20; 1359,6-8; 1361,18-19; 1366,19-22
prior (primary) in nature, 1265,14-1266,9; 1266,12-1267,14; 1269,16-35; 1271,23-5.36-7; 1272,19-20.28-38; 1314,19-27
things constituted by nature, 1257,35-6
negation, 1274,9-10
neoplatonic metaphysics, theology, and cosmogony, 1336,36-1338,30
now, see instants
nutriment, 1258,19-28; 1259,32-7; 1265,21-8.37-1266,2; 1267,11-13.17-19

opposites, 1273,37-1276,34; 1279,30-1280,14; 1301,24-9; 1304,10-32; 1305,3-6.15-16.31-2; 1306,34-6; 1307,25-33; 1308,3-1309,2; 1309,10-12.38-1310,1; 1310,36; 1311,18-22; 1365,32-4

perception, 1257,30-4; 1280,33-5; 1312,38-1313,15
see also sensation
perfection, 1353,15; 1355,1-2
peripatetic tradition, 1362,11-13
perishing (*phthora*), 1251,19.33-5; 1252,14-15; 1262,23.27-9; 1263,2-3; 1264,9-11; 1266,31-3; 1271,4; 1273,35-7; 1274,3-4.9; 1275,7.25-6; 1278,28-9; 1312,22-3; 1313,2-4; 1330,24-7.33-7; 1332,33-40; 1359,31.33-8
Philoponus, 1326,38-1336,34; 1358,26-9.39-1359,4
arguments that the bodies of the heaven and of the world are perishable, and S.'s criticism, 1327,14-1329,19
arguments that a finite body possesses finite power, and S.'s criticism, 1329,19-1332,2
arguments that the total amounts of the elements and the heavenly bodies have finite power, and S.'s criticism, 1332,3-1336,33
criticism of Philoponus' view that what does not in the strict sense possess infinite power cannot be moved ad infinitum, 1358,26-32.38-1359,4
philosophy, first, 1366,20
place, 1260,17-22; 1278,24-6; 1279,4-7; 1317,10-12; 1354,26-7;
motion (change) in place, see locomotion
natural place, see proper place
proper place, 1257,12-14; 1259,25-6; 1302,33-4; 1308,5-8; 1317,21-8.33-1318,7
planets, see spheres
plants, 1257,35; 1271,30-1
Plato
every origin is eternal (*Phaedrus* quotation), 1261,1-2
activity of soul (*Phaedrus* quotation), 1263,24-7
agreement with Aristotle that locomotion is the primary kind of

motion (*Laws* 10 quotation), 1267,19-28
employed the concept of priority in nature and being, 1268,34-6
locomotion is the cause of the other kinds of motion (*Laws* 10 quotations), 1272,38-1273,12
the soul is the cause of motion, 1319,28-34
Timaeus 41a7-b7, interpretation of, 1327,20-1338,30
Philoponus tries unsuccessfully to follow Plato, 1331,8; 1334,34-9
indestructibility of heavenly bodies, 1331,15-16.24-5
agreement with Aristotle on the cause and nature of the heavens (*Timaeus* quotations), 1336,35-1339,24
treats the subject of celestial motion less scientifically than Aristotle, 1338,31
the cause of motion (*Timaeus* quotations), 1351,28-1352,18
agrees with Aristotle on the causes of motion, 1354,37-9
agrees with Aristotle on the nature of god (*Timaeus* quotation), 1359,8-1360,23
god is the final and efficient cause of the world (*Timaeus* quotations), 1360,28-1361,11
pluralists, 1318,21-1319,19
points, 1280,33; 1282,31; 1283,5-39; 1284,4-8; 1286,12-20.23-7.31-3; 1287,8-16.25-1288,2; 1288,8-9.14-16.33-5; 1290,20-31; 1291,5-7.14-24; 1308,35-1309,16; 1309,19; 1315,22
poison, 1340,25; 1341,6-7
poles, 1261,23-4; 1262,1; 1354,15; 1355,34-6
potentiality, 1265,29-31; 1282,29-31; 1291,13-14; 1291,36-1292,11; 1292,24-1294,2; 1305,1; 1310,22-3; 1356,25-6
of points, 1280,33-1281,29; 1284,4-8; 1286,14-20.26-7; 1287,25-1288,9; 1290,21-1291,24; 1310,9-12; 1315,9-36
power, 1321,6-11.18-1326,37; 1327,15-16.31-5; 1328,18-31.33-5; 1329,14-1336,33; 1338,34-5; 1339,6-7.16-22; 1340,11-1344,3; 1346,16-18; 1355,15; 1358,11-13.18-26.33-5; 1363,6-7; 1366,12-15
predecessors, Aristotle's use of, 1318,10-15; 1366,7-9
primacy, see priority
primary mover 1250,34-1256,30; 1262,18-20; 1263,7-8; 1264,25-6; 1265,5-6; 1320,34-7; 1321,3-5.12; 1323,36-1324,1; 1328,18-20.23-4.30-2.39-40; 1336,6; 1338,34-40; 1339,25-1340,7.12-15; 1344,6-12; 1352,33-1355,38; 1357,9-17.31-1358,17; 1360,24-31; 1361,11-1362,10; 1362,15-1363,24; 1364,37-1365,19; 1366,11-13.18-19
principles, 1251,31-1252,23; 1257,10-11; 1266,20-8; 1269,28-9; 1271,35-6; 1272,1; 1348,24-6; 1354,4-6; 1366,21-2
priority, 1264,17-1266,9; 1267,32-1273,33; 1279,25-7; 1313,24-1315,4; 1316,26-1317,16; 1320,19-21.30-3; 1321,25-6; 1352,24; 1353,6-7; 1362,26-7; 1364,25-6.35-6; 1365,22-4; 1366,3-7
privation, 1276,31-2; 1277,5-10; 1304,3-23; 1305,11-20; 1312,22-3
projectile motion, see motion, projectile
proofs
by elimination, 1321,12-14
from signs, 1279,31-2
on the basis of a division, 1321,14-23
see also arguments
propositions, 1299,35-1300,36
pushing and pulling, 1317,20-1; 1339,35; 1345,33-4; 1348,34-5; 1350,7; 1356,6-7.20-1; 1357,19-20
Pythagoreans, 1354,2-3; 1355,3-9

qualities, 1265,12; 1266,20-9; 1267,5-6; 1274,10-11; 1312,19-20
quantities, 1312,20-1

rarefaction, see condensation and rarefaction

rarity, see density and rarity
relational entities, 1349,13-14
rest, 1274,29-1277,14; 1280,22-1288,29; 1297,1-7; 1301,11-12; 1302,15-23; 1304,28-30; 1306,34-5; 1309,2-5.19-1311,12; 1311,20.29-1312,27; 1315,15-16.36-1316,3; 1350,18; 1365,32-4

sciences, 1273,27-8
seed, see sperm
sensation, 1259,20-3
separation, 1266,16-1267,28
sleep, 1258,15-30.40; 1259,32-3.36-7
souls, 1251,7-9.11-13.31-1252,23; 1256,33-1259,28; 1260,8-10.28-9; 1319,28-30; 1320,7-8; 1364,6-10; 1365,6-10
  of planetary spheres, 1261,31-1262,13
sperm, 1340,25-8
sphere, 1315,38-1316,3; 1354,5-6.15; 1355,34-6
  celestial, 1261,23-4; 1262,1
  of fixed stars, 1254,25; 1263,17-18; 1355,4.7-11.28-31; 1357,9-17; 1358,33-5;
  planetary, 1261,14-1262,13; 1263,18; 1356,33-1357,9; 1358,35-8
states, 1276,31-2; 1277,5-10; 1312,22-3
Stoics, 1299,35-1300,36; 1320,19-21
sublunary bodies, 1312,35-7
  elements, 1332,38-9
  region, 1319,34-1320,2
  things, 1256,7-10.15-18; 1263,32-4; 1362,19-20
substance, 1270,28-30; 1356,25-6
stars, see spheres
sun, 1263,19-22

Thales, 1319,20-7
Themistius, 1253,7
theology, 1336,36-1338,30; 1359,5-30; 1360,17-23
thought, 1333,15-20; 1334,18-40
time, 1282,22-3; 1283,4-5; 1286,16; 1290,20-3; 1294,11-12; 1297,1-1299,19; 1305,29; 1306,30; 1314,28-9; 1316,28; 1324,22-3; 1327,9-11; 1328,15; 1336,29-33; 1339,25-1340,7; 1342,7-19; 1365,31-2
  of motion, 1278,33-1279,4; 1321,37-1322,2; 1325,32
  prior in, 1270,3-1271,18; 1272,28-38; 1314,19-27;
  see also instants
truth, 1318,17-18

universe, 1261,9
unmoved mover(s), 1251,7-9.13-15.21-2; 1252,14-15; 1260,4-5; 1262,18-19.20-1263,18; 1269,29; 1320,34-7; 1321,3; 1340,1-2; 1344,6-7; 1353,2-8.10-12.18-19.30; 1356,3-32; 1357,17-29.33-5; 1358,1-2; 1362,19-20; 1363,6-7; 1364,26.37-1365,21

void, 1318,32-1319,8; 1351,28-32

waking, 1258,15-30; 1259,32-3.36-7
water, 1345,28-9; 1349,29-32; 1350,28-30; 1356,12-14
wheels, 1314,19-24
world, 1260,17-21; 1261,2-11; 1327,1-8.14-1329,19; 1335,3-16; 1353,31-4; 1355,1-2.15; 1360,31-1361,11; 1363,8-12

Zeno, see dichotomy argument

Index of Names

References are to the page and line numbers of the Greek text which appear in the margins of the translation.

Alexander, 1253,6; 1260,22-35 (quotation); 1261,30-1262,2 (quotation); 1262,3; 1268,3-6 (quotation); 1276,35-1277,14 (quotation); 1281,30; 1285,15-30 (quotation); 1287,1; 1288,3; 1291,34-1293,9 (includes quotations); 1296,18-32 (quotation); 1299,4-10 (includes quotation); 1299,36-1300,10 (quotation); 1303,28-33 (includes quotation); 1307,6-12 (quotation); 1308,16 (quotation); 1308,20-7 (quotation); 1309,23; 1317,3-7 (includes quotation); 1319,11 (quotation); 1325,8-24 (quotation); 1326,3.28-37 (quotation); 1327,35; 1340,32-7 (quotation); 1343,32-7 (quotation); 1346,29-1348,5 (includes quotations); 1348,11-15 (includes quotation); 1348,26-1349,10 (includes quotations); 1350,2-7 (includes quotation); 1351,28-37; 1354,12-35 (includes quotations); 1355,15-38 (includes quotation); 1356,33-1357,9 (includes quotation); 1358,18-1359,4 (includes quotation); 1361,31-3 (includes quotation); 1362,11-15 (includes quotation)

Ammonius, 1363,8-12

Anaxagoras, 1254,21; 1266,34; 1318,29; 1361,34; 1362,4.9, 1363,32

Anaximander, 1266,36; 1319,21

Anaximenes, 1319,21

Aristotle, 1262,6.11; 1263,25; 1325,8.25; 1326,21; 1327,7.13.23.24.27; 1328,10.16.34.38; 1329,6.7; 1331,39; 1332,1; 1335,19; 1336,7.35; 1338,31; 1339,17.25.27; 1340,8; 1343,37; 1348,6.16.24; 1349,11.30.38; 1351,34; 1354,15.35; 1355,4.7.9.37; 1357,3.13.17; 1358,28; 1359,5.23.31.33; 1360,9.24; 1361,5.10.11; 1362,12.32; 1363,10.13.21

Democritus, 1254,21; 1266,34; D. and his followers, 1318,32; 1320,16

Empedocles, 1254,22; 1266,34; 1318,22-8 (includes quotations); 1363,32

Eudemus, 1262,18 (quotation); 1354,10.16; 1355,28-36 (includes quotations); 1357,17-29 (includes quotations)

Heraclitus 1257,17; the followers of Heraclitus, 1313,8-12 (paraphrase); 1319,21

Hermotimus, 1361,35; 1362,5

Peripatetics, 1362,11

Philoponus, 1326,38-1336,34 (including quotations at 1327,17-19; 1329,33-8; 1330,7-17.24-5; 1331,10-15.17-25; 1332,15-26; 1333,1-2.4-32; 1334,20-1.27-9.37-9; 1335,24-30.31-1336,5); 1358,26-9.39

Plato, 1261,1-2 (*Phaedrus* quotation); 1263,24-7 (includes *Phaedrus* quotation); 1267,21-8 (includes quotations from *Laws* 10); 1268,36; 1272,38-1273,11 (includes quotations from *Laws* 10); 1319,29 (ref. to *Phaedrus*); 1331,8.15.24; 1334,35; 1336,37; 1337,20-1338,30 (includes quotation from *Timaeus*); 1339,3.5.14; 1351,28-1352,18 (includes quotations from *Timaeus*); 1354,37; 1359,8-1360,23 (includes quotation);

1360,31-1361,11 (includes quotations)
Pythagoreans, 1354,2; 1355,3-9
Stoics, 1299,37; 1320,19
Thales, 1319,21
Themistius, 1253,7
Timaeus (the character in Plato's dialogue), 1313,10
Zeno, 1288,36-1291,24 passim

Index of Passages Cited

References are to the notes to the translation. As a result of constraints of space, references in the notes to Simplicius *in Physica* are not listed here.

AETIUS **1.3.18**: 378

ANAXAGORAS **59B1**: 346; **59B11**: 377; **59B12**: 346; **59B17**: 145

ANAXIMENES **13A5**: 381; **13A6**: 381; **13A7**: 381

ARISTOTLE

An. post. **71b17-25**: 297; **71b33-72a5**: 468; **73a34-b3**: 315; **73b25-74a3**: 315; **76b6-10**: 315; **89b29**: 138

An. pr. **70a6**: 220

Cat. **7b15-22**: 456, 507; **7b22-8a12**: 456; **12a27-34**: 314; **14a30**: 155; **14a35-b7**: 154; **15a30**: 142

De an. **427b27-429a9, 434a5-11**: 321

Cael. **270b1-16**: 629; **271a33**: 575; **277a28-9**: 83; **279a27-30**: 576; **288a2-3**: 45; **300b8-10**: 346; **301b22-30**: 509; **303a11-15**: 46; **309a1-14**: 378; **311b8-12**: 512; **312a23-7**: 512; **312a25**: 509

Gen. corr. **315b6-15**: 46, 145, 378, 386; **318a1-5**: 577; **324b35-325a36**: 378; **325a23-6**: 145; **326a9-10**: 378; **336a17-18**: 111

Int. **17b16, 18a33, 21b37-22a1, 24b9**: 194

Part. an. **642a19**: 373; **687a16-17**: 45

Eth. Nic. **1098a18-19**: 349; **1104a18, 1108b11-19**: 198

Metaph. **984a13**: 46; **984a18-19**: 373; **984b15-18**: 579; **984b18-20**: 580; **984b20-2**: 581; **987a32-3**: 346; **989b15**: 582; **1000a9**: 563; **1014a14**: 348; **1018b10-1019a11**: 157; **1019a2-4**: 160, 172; **1026a10-22**: 560; **1072b29**: 566; **1073b26-38**: 55; **1074a14-16**: 131; **1076a4**: 43

Ph. **184b20-1**: 46; **187a26-b7**: 46; **188a17-18**: 47; **188a24-5**: 208; **188b29-30**: 373; **189a15-16**: 47; **190b17-33**: 309; **191b15-16**: 309; **194b13**: 168; **194b29-31**: 574; **198a2-3**: 583; **198a5-13**: 585; **202a13-14**: 6; **202a13-b22**: 427, 480; **204a20-206a8**: 353; **204a34-206a8**: 399; **204b4-206a8**: 207, 558; **205b32-3**: 208; **206a14-b27**: 207, 353; **206a19-21**: 254; **206a21-3**: 256, 564; **206a21-5**: 249; **206a25-7**: 257; **206a26-7**: 426; **206a27-9**: 255, 425; **206a30-3**: 258; **207a7-8**: 353; **207a22**: 207, 353; **207b1-15:** 253; **208a5-22**: 207, 353; **218a18-19**: 234; **220a5, a9-10**: 232; **222a10-20**: 232; **222b30-1**: 316; **223b12-224a2**: 360; **224a21-30**: 9; **224a34-b7**: 53; **224b25-6**: 480; **225a20-b1**: 187; **226b34-227a1**: 513; **226b34-227a6**: 322; **227a7**: 514; **227a10-12**: 323; **227b20-6**: 210; **228a20-2**: 210, 527; **229b23-31**: 195, 197; **230a22-3**: 514; **231a22**: 323; **231a23**: 322; **231a24-b18**: 638; **231b9-10**: 233; **234b10-20**: 28, 279; **235a13-14**: 316; **235b6-27**: 317; **237a10**: 318; **239b9-240a18**: 246; **240b8-241a26**: 30; **241b34-242a49**: 7, 71; **247b1-13**: 279; **250b11-15**: 588; **250b15-251a5**: 589; **251a5-b10**: 590; **251a8-9**: 478; **251a8-b28**: 161; **251a23-b10, b28-252a5**: 57; **251b10-13**: 591; **251b19-26**: 592; **251b28-252a4**: 590; **251b28-252a5**: 59; **252a5**: 593; **252a5-6**: 593; **252a14-16**: 593; **252b9-12**: 594; **252b12-16**: 595; **252b17-28**: 76, 596; **252b28-253a2**: 60, 597; **253a7-20**:

76; **253a14-15**: 77; **253a22-30**: 598; **253a32-b6**: 599; **253b6-254a1**: 600; **254a3-15**: 601; **254a5-8**: 607; **254a15-16**: 602; **254a15-b6**: 602; **254a16-b4**: 603; **254a19-22**: 130; **254a23-33**; 599; **254a33-b4**: 600; **254a35-b4**: 607; **254b5-6**: 604; **254b7**: 605; **255a5-10**: 74; **255a30-b12**: 611; **255b29-31**: 511; **255b29-256a3**: 613; **255b30-1**: 75; **256a4-21**: 614; **256a4-b3**: 529, 608; **256a13-21**: 389; **256a21-b3**: 615; **256b4-13**: 616; **256b4-257a27**: 530; **256b4-258a20**: 392; **256b13-24**: 617; **256b14-15**: 52; **256b14-27**: 607; **256b24-8**: 612; **256b26-258a4**: 609; **256b27-257a26**: 618; **257a26-7**: 619; **257a27-31**: 620; **257a28-31**: 178; **257a31-3**: 621; **257a33-258a20**: 622; **257a33-258a27**: 73; **257a33-258b9**: 3, 66; **257b20-1**: 9; **257b26-258b9**: 607; **257b28-258a27**: 510; **257b32-4**: 9; **256a22-7**: 623; **258a23-4**: 624; **258b4-5**: 609; **258b4-9**: 394, 625; **258b10**: 628; **258b10-259a6**: 394; **258b10-259a20**: 132; **258b11-259a6**: 392, 626; **258b16-20**: 68, 279; **258b31-2**: 54; **258b32-259a6**: 383; **259a6-20**: 12, 627; **259a15-20**: 35; **259a16-17**: 39; **259a18-19**: 528; **259a20-b28**: 630; **259b1-3**: 74; **259b6-15**: 67; **259b32-260a1**: 169, 480; **259b32-260a5**: 607; **259b32-260a10**: 383, 631; **259b32-260a19**: 132; **260a14-17**: 129; **260a20-1**: 136; **260a20-6**: 633; **260a23-6**: 526; **260a26-8**: 133; **260a26-261a25**: 634; **260a29-b5**: 141; **260b16-19**: 356; **260b16-261a26**: 134; **261a23**: 179; **261a27-b26**: 164, 635; **261a28-30**: 61; **261a28-b22**: 137; **261a31-b26**: 204; **261a33**: 186; **261b1-3**: 205; **261b1-15**: 209; **261b3-14**: 186; **261b28-9**: 352; **261b32-6**: 214; **261b36-262a5**: 215; **262a3**: 560; **262a12-b8**: 635; **262a14-b7**: 296, 338; **262a21-5**: 359; **262a21**: 229; **262a32-b1**: 231; **263a5-6**: 247; **263b9-26**: 191, 294, 636; **263b26-264a6**: 637; **264a7-21**: 639; **264a21-33**: 640; **264a23**: 310; **264b1-6**: 641; **264b9-265a2**: 642; **265a7-27**: 643; **265a21-2**: 355; **265a24-7**: 357; **265a29-b1**: 485; **265b8-10**: 645; **265b10-11**: 544, 644; **265b11-16**: 646; **265b16**: 368; **265b17-266a1**: 647; **265b22**: 376; **266a6-9**: 648; **266a10-11**: 649; **266a13**: 401, 403; **266a24-b6**: 650; **266b6-24**: 651; **266b25-267a20**: 652; **267a21-5, b5-7**: 483; **267a21-b9**: 131; **267b6-9**: 556; **267b8**: 554; **267b9-15**: 557; **267b17-26**: 653; **267b19-26**: 398; **267b20-1**: 650

Top. **102a18-30**: 226

DEMOCRITUS and LEUCIPPUS **67A6**: 46; **67A7**: 145, 378; **67A8**: 46; **67A14**: 145; **67A15**: 46; **67A97**: 46, 145; **68A37**: 145, 378; **68A38**: 46; **68A47**: 378; **68A57**: 145; **68A60**: 378; **68A61**: 378; **68A97**: 378, 386; **68B168**: 379

DIOGENES LAERTIUS **7.80**: 293

EMPEDOCLES **31B6**: 47; **31B8, 31B9, 31B11, 31B15**: 145; **31B17**: 47, 374; **31B21, 31B26**: 47; **31B53**: 375

EUDEMUS **fr. 121**: 105; **fr. 122a**: 539; **fr. 122b**: 547; **fr. 123a**: 546; **fr. 123b**: 555

HERACLITUS **22B12**: 348

HIPPOLYTUS *Ref.* **1.7.3**: 381

HOMER *Il.* **2.204**: 43

LEUCIPPUS see DEMOCRITUS and LEUCIPPUS

NICANDER *Ther.* **741**: 252

OLYMPIODORUS *in Meteor.* **278,9**: 252

PHILOPONUS *in Phys.* **179,6**: 252

PLATO
Crat. **397d**: 469; **402a**: 348
Laws **893e1-4**: 148, 180; **893e6-7**: 149; **893e6-894a1**: 181; **893e7-894a6**: 150, 182; **894b10-11**: 150, 182
Phaedr. **245c7-9**: 382; **245d4-6**: 87; **246b6-7**: 124
Tim. **27d6-28a2**: 486; **27d6-28a4**: 561; **28a2-4**: 347; **28a4-8**: 458; **28a5-6**: 562; **28b2-3**: 458; **28b2-c3**: 486; **29d7-e1**: 569; **29e1**: 458; **30b4-6**: 570; **30c3-31a1**: 464; **31c2-4**: 472; **36e5**: 458; **39e3-40b6**: 486; **41a7**: 471, 571; **41a7-b6**: 487; **41a7-b7**: 463; **41b1-6**: 451; **41b4-6**: 439; **41b7-8**: 572; **47c1-4**: 573; **57e2-6**: 519; **57e6-58a1**: 520; **58b6-c4**: 521; **59a1-4**: 522; **80a1**: 500

PLUTARCH *Adv. Col.* **1110F-1111A**: 145

[PLUTARCH] *Strom.* **3**: 381

PROCLUS
Elements of Theology **25-30**: 459; **40, 55:** 461; **112**: 459, 464; **120**: 464; **188***:* 477
in lib. prim. Eucl. lib. comm. **239,15-20**: 333

SEXTUS EMPIRICUS
Math. **8.227**: 176; **9.269: 10.344-50, 10.346, 10.347**: 276
Pyr. **1.41**: 252; **3.111**: 276

SIMPLICIUS
in Cael. **25,23-38,5**: 419; **42,17-49,25**: 419; **56,26-59,23**: 419; **59,15-18**: 433; **66,98-120**: 419, 433; **119,7-144,4**: 419; **156,25-201,10**: 419; **226,19-20**: 559; **242,21-6, 295,11-24**: 145; **569,5-9**: 378

VERGIL *Georgics* **4.281-314, 4.550-6**: 252

www.ingramcontent.com/pod-product-compliance
Ingram Content Group UK Ltd.
Pitfield, Milton Keynes, MK11 3LW, UK
UKHW020138290926
15471UKWH00003B/574

9 781780 938974